SYSTEMS ENGINEERING HANDBOOK

SYSTEMS ENGINEERING HANDBOOK

A GUIDE FOR SYSTEM LIFE CYCLE PROCESSES AND ACTIVITIES

FIFTH EDITION

INCOSE-TP-2003–002-05
2023

Prepared by:

International Council on Systems Engineering (INCOSE)
7670 Opportunity Rd, Suite 220
San Diego, CA, USA 92111-2222

Compiled and Edited by:

DAVID D. WALDEN, ESEP — EDITOR-IN-CHIEF — AMERICAS SECTOR
THOMAS M. SHORTELL, CSEP — DEPUTY EDITOR-IN-CHIEF — AMERICAS SECTOR
GARRY J. ROEDLER, ESEP — EDITOR — AMERICAS SECTOR
BERNARDO A. DELICADO, ESEP — EDITOR — EMEA SECTOR
ODILE MORNAS, ESEP — EDITOR — EMEA SECTOR
YIP YEW-SENG, CSEP — EDITOR — ASIA OCEANIA SECTOR
DAVID ENDLER, ESEP — EDITOR — EMEA SECTOR

Registered Offices
John Wiley & Sons, Inc., 111 River Street, Hoboken, NJ 07030, USA
John Wiley & Sons Ltd, The Atrium, Southern Gate, Chichester, West Sussex, PO19 8SQ, UK

For details of our global editorial offices, customer services, and more information about Wiley products visit us at www.wiley.com.

Wiley also publishes its books in a variety of electronic formats and by print-on-demand. Some content that appears in standard print versions of this book may not be available in other formats.

Library of Congress Cataloging-in-Publication Data

Names: Walden, David D., editor. | International Council on Systems Engineering, editor.
Title: INCOSE systems engineering handbook / edited by INCOSE, David Walden.
Description: Fifth edition. | Hoboken, NJ : John Wiley & Sons Ltd., [2023] | Includes index.
Identifiers: LCCN 2023022915 | ISBN 9781119814290 (paperback) | ISBN 9781119814306 (adobe pdf) | ISBN 9781119814313 (epub)
Subjects: LCSH: Systems engineering--Handbooks, manuals, etc. | Product life cycle--Handbooks, manuals, etc.
Classification: LCC TA168 .I444 2023 | DDC 620/.0042--dc23/eng/20230525
LC record available at https://lccn.loc.gov/2023022915

Cover Design: Wiley
Cover Images: © DANNY HU/Getty Images, © Zero Creatives/Getty Images, © Stebenkov Roman/Shutterstock, © Phonlamai Photo/Shutterstock, © MNBB Studio/Shutterstock, © Titima Ongkantong/Shutterstock

Set in 10/12pt TimesLTStd by Integra Software Services Pvt. Ltd., Pondicherry, India

SKY10071035_032624

CONTENTS

History of Changes   ix

List of Figures   xi

List of Tables   xv

Preface   xvii

How to Use This Handbook   xix

1   Systems Engineering Introduction   1

  1.1   What Is Systems Engineering?   1
  1.2   Why Is Systems Engineering Important?   4
  1.3   Systems Concepts   8
    1.3.1   System Boundary and the System of Interest (SoI)   8
    1.3.2   Emergence   9
    1.3.3   Interfacing Systems, Interoperating Systems, and Enabling Systems   10
    1.3.4   System Innovation Ecosystem   11
    1.3.5   The Hierarchy within a System   12
    1.3.6   Systems States and Modes   14
    1.3.7   Complexity   15
  1.4   Systems Engineering Foundations   15
    1.4.1   Uncertainty   15
    1.4.2   Cognitive Bias   17
    1.4.3   Systems Engineering Principles   17
    1.4.4   Systems Engineering Heuristics   20
  1.5   System Science and Systems Thinking   21

2   System Life Cycle Concepts, Models, and Processes   **25**

2.1   Life Cycle Terms and Concepts   25
    2.1.1   Life Cycle Characteristics   25
    2.1.2   Typical Life Cycle Stages   26
    2.1.3   Decision Gates   29
    2.1.4   Technical Reviews and Audits   31
2.2   Life Cycle Model Approaches   33
    2.2.1   Sequential Methods   35
    2.2.2   Incremental Methods   36
    2.2.3   Evolutionary Methods   38
2.3   System Life Cycle Processes   39
    2.3.1   Introduction to the System Life Cycle Processes   39
        2.3.1.1   Format and Conventions   40
        2.3.1.2   Concurrency, Iteration, and Recursion   42
    2.3.2   Agreement Processes   44
        2.3.2.1   Acquisition Process   45
        2.3.2.2   Supply Process   48
    2.3.3   Organizational Project-Enabling Processes   50
        2.3.3.1   Life Cycle Model Management Process   51
        2.3.3.2   Infrastructure Management Process   54
        2.3.3.3   Portfolio Management Process   57
        2.3.3.4   Human Resource Management Process   60
        2.3.3.5   Quality Management Process   63
        2.3.3.6   Knowledge Management Process   67
    2.3.4   Technical Management Processes   70
        2.3.4.1   Project Planning Process   70
        2.3.4.2   Project Assessment and Control Process   75
        2.3.4.3   Decision Management Process   78
        2.3.4.4   Risk Management Process   81
        2.3.4.5   Configuration Management Process   87
        2.3.4.6   Information Management Process   91
        2.3.4.7   Measurement Process   93
        2.3.4.8   Quality Assurance Process   98
    2.3.5   Technical Processes   101
        2.3.5.1   Business or Mission Analysis Process   103
        2.3.5.2   Stakeholder Needs and Requirements Definition Process   107
        2.3.5.3   System Requirements Definition Process   112
        2.3.5.4   System Architecture Definition Process   118
        2.3.5.5   Design Definition Process   124
        2.3.5.6   System Analysis Process   129
        2.3.5.7   Implementation Process   132
        2.3.5.8   Integration Process   134
        2.3.5.9   Verification Process   138
        2.3.5.10   Transition Process   143
        2.3.5.11   Validation Process   146
        2.3.5.12   Operation Process   152
        2.3.5.13   Maintenance Process   154
        2.3.5.14   Disposal Process   156

3   Life Cycle Analyses and Methods   **159**

   3.1   Quality Characteristics and Approaches   159
         3.1.1   Introduction to Quality Characteristics   159
         3.1.2   Affordability Analysis   160
         3.1.3   Agility Engineering   165
         3.1.4   Human Systems Integration   168
         3.1.5   Interoperability Analysis   171
         3.1.6   Logistics Engineering   172
         3.1.7   Manufacturability/Producibility Analysis   175
         3.1.8   Reliability, Availability, Maintainability Engineering   176
         3.1.9   Resilience Engineering   180
         3.1.10   Sustainability Engineering   184
         3.1.11   System Safety Engineering   185
         3.1.12   System Security Engineering   190
         3.1.13   Loss-Driven Systems Engineering   191
   3.2   Systems Engineering Analyses and Methods   192
         3.2.1   Modeling, Analysis, and Simulation   192
         3.2.2   Prototyping   200
         3.2.3   Traceability   201
         3.2.4   Interface Management   202
         3.2.5   Architecture Frameworks   206
         3.2.6   Patterns   208
         3.2.7   Design Thinking   212
         3.2.8   Biomimicry   213

4   Tailoring and Application Considerations   **215**

   4.1   Tailoring Considerations   215
   4.2   SE Methodology/Approach Considerations   219
         4.2.1   Model-Based SE   219
         4.2.2   Agile Systems Engineering   221
         4.2.3   Lean Systems Engineering   224
         4.2.4   Product Line Engineering (PLE)   226
   4.3   System Types Considerations   229
         4.3.1   Greenfield/Clean Sheet Systems   229
         4.3.2   Brownfield/Legacy Systems   230
         4.3.3   Commercial-off-the-Shelf (COTS)-Based Systems   231
         4.3.4   Software-Intensive Systems   232
         4.3.5   Cyber-Physical Systems (CPS)   233
         4.3.6   Systems of Systems (SoS)   235
         4.3.7   Internet of Things (IoT)/Big Data-Driven Systems   238
         4.3.8   Service Systems   239
         4.3.9   Enterprise Systems   241
   4.4   Application of Systems Engineering for Specific Product Sector or Domain Application   244
         4.4.1   Automotive Systems   245
         4.4.2   Biomedical and Healthcare Systems   248
         4.4.3   Commercial Aerospace Systems   249
         4.4.4   Defense Systems   250

4.4.5   Infrastructure Systems   251
4.4.6   Oil and Gas Systems   253
4.4.7   Power & Energy Systems   254
4.4.8   Space Systems   255
4.4.9   Telecommunication Systems   257
4.4.10   Transportation Systems   258

5   Systems Engineering in Practice   261

5.1   Systems Engineering Competencies   261
       5.1.1   Difference between Hard and Soft Skills   262
       5.1.2   System Engineering Professional Competencies   263
       5.1.3   Technical Leadership   263
       5.1.4   Ethics   264
5.2   Diversity, Equity, and Inclusion   265
5.3   Systems Engineering Relationships to Other Disciplines   266
       5.3.1   SE and Software Engineering (SWE)   266
       5.3.2   SE and Hardware Engineering (HWE)   267
       5.3.3   SE and Project Management (PM)   268
       5.3.4   SE and Industrial Engineering (IE)   270
       5.3.5   SE and Operations Research (OR)   271
5.4   Digital Engineering   273
5.5   Systems Engineering Transformation   274
5.6   Future of SE   275

6   Case Studies   277

6.1   Case 1: Radiation Therapy—the Therac-25   277
6.2   Case 2: Joining Two Countries—the Øresund Bridge   278
6.3   Case 3: Cybersecurity Considerations in Systems Engineering—the Stuxnet Attack
       on a Cyber-Physical System   280
6.4   Case 4: Design for Maintainability—Incubators   282
6.5   Case 5: Artificial Intelligence in Systems Engineering—Autonomous Vehicles   283
6.6   Other Case Studies   285

Appendix A: References   287

Appendix B: Acronyms   305

Appendix C: Terms and Definitions   311

Appendix D: N² Diagram of Systems Engineering Processes   317

Appendix E: Input/Output Descriptions   321

Appendix F: Acknowledgments   335

Appendix G: Comment Form   337

Index   339

HISTORY OF CHANGES

Revision	Revision date	Change description and rationale
Original	Jun 1994	Draft *Systems Engineering Handbook* (SEH) created by INCOSE members from several defense/aerospace companies—including Lockheed, TRW, Northrop Grumman, Ford Aerospace, and the Center for Systems Management—for INCOSE review.
1.0	Jan 1998	Initial SEH release approved to update and broaden coverage of SE process. Included broad participation of INCOSE members as authors. Based on Interim Standards EIA 632 and IEEE 1220.
2.0	Jul 2000	Expanded coverage on several topics, such as functional analysis. This version was the basis for the development of the Certified Systems Engineering Professional (CSEP) exam.
2.0A	Jun 2004	Reduced page count of SEH v2 by 25% and reduced the US DoD-centric material wherever possible. This version was the basis for the first publicly offered CSEP exam.
3.0	Jun 2006	Significant revision based on ISO/IEC 15288:2002. The intent was to create a country- and domain-neutral handbook. Significantly reduced the page count, with elaboration to be provided in appendices posted online in the INCOSE Product Asset Library (IPAL).
3.1	Aug 2007	Added detail that was not included in SEH v3, mainly in new appendices. This version was the basis for the updated CSEP exam.
3.2	Jan 2010	Updated version based on ISO/IEC/IEEE 15288:2008. Significant restructuring of the handbook to consolidate related topics.
3.2.1	Jan 2011	Clarified definition material, architectural frameworks, concept of operations references, risk references, and editorial corrections based on ISO/IEC review.
3.2.2	Oct 2011	Correction of errata introduced by revision 3.2.1.
4.0	Jul 2015	Significant revision based on ISO/IEC/IEEE 15288:2015, inputs from the relevant INCOSE working groups (WGs), and to be consistent with the Guide to the Systems Engineering Body of Knowledge (SEBoK).
5.0	Jul 2023	Significant revision based on ISO/IEC/IEEE 15288:2023 and inputs from the relevant INCOSE working groups (WGs). Significant restructuring of the handbook based inputs from INCOSE stakeholders.

LIST OF FIGURES

1.1   Acceleration of design to market life cycle has prompted development of more automated design methods and tools

1.2   Cost and schedule overruns correlated with SE effort

1.3   Project performance versus SE capability

1.4   Life cycle costs and defect costs against time

1.5   Emergence

1.6   System innovation ecosystem pattern

1.7   Hierarchy within a system

1.8   An architectural framework for the evolving the SE discipline

2.1   System life cycle stages

2.2   Generic life cycle stages compared to other life cycle viewpoints

2.3   Criteria for decision gates

2.4   Relationship between technical reviews and audits and the technical baselines

2.5   Concepts for the three life cycle model approaches

2.6   The SE Vee model

2.7   The Incremental Commitment Spiral Model (ICSM)

2.8   DevSecOps

2.9   Asynchronous iterations and increments across agile mixed discipline engineering

2.10   System life cycle processes per ISO/IEC/IEEE 15288

2.11   Sample IPO diagram for SE processes

2.12   Concurrency, iteration, and recursion

2.13   IPO diagram for the Acquisition process

2.14   IPO diagram for the Supply process

2.15   IPO diagram for Life Cycle Model Management process

2.16   IPO diagram for Infrastructure Management process

2.17   IPO diagram for Portfolio Management process

2.18   Requirements across the portfolio, program, and project domains

2.19   IPO diagram for Human Resource Management process

2.20   IPO diagram for the Quality Management process
2.21   QM Values and Skills Integration
2.22   IPO diagram for Knowledge Management process
2.23   IPO diagram for Project Planning process
2.24   The breakdown structures
2.25   IPO diagram for Project Assessment and Control process
2.26   IPO diagram for the Decision Management process
2.27   IPO diagram for Risk Management process
2.28   Level of risk depends upon both likelihood and consequence
2.29   Intelligent management of risks and opportunities
2.30   Typical relationship among the risk categories
2.31   IPO diagram for Configuration Management process
2.32   IPO diagram for Information Management process
2.33   IPO diagram for Measurement process
2.34   Integration of Measurement, Risk Management, and Decision Management processes
2.35   Relationship of product-oriented measures
2.36   TPM monitoring
2.37   IPO diagram for the Quality Assurance process
2.38   Technical Processes in context
2.39   IPO diagram for Business or Mission Analysis process
2.40   IPO diagram for Stakeholder Needs and Requirements Definition process
2.41   IPO diagram for System Requirements Definition process
2.42   IPO diagram for System Architecture Definition process
2.43   Core architecture processes
2.44   IPO diagram for Design Definition process
2.45   Taxonomy of system analysis dimensions
2.46   IPO diagram for System Analysis process
2.47   IPO diagram for Implementation process
2.48   IPO diagram for Integration process
2.49   IPO diagram for Verification process
2.50   Verification per level
2.51   IPO diagram for Transition process
2.52   IPO diagram for Validation process
2.53   Validation per level
2.54   IPO diagram for Operation process
2.55   IPO diagram for Maintenance process
2.56   IPO diagram for Disposal process
3.1   Quality characteristic approaches across the life cycle
3.2   System operational effectiveness
3.3   Cost versus performance
3.4   Life cycle cost elements
3.5   HSI technology, organization, people within an environment
3.6   Interaction between system, environment, operating conditions, and failure modes and failure mechanisms
3.7   Timewise values of notional resilience scenario parameters
3.8   Schematic view of a generic MA&S process
3.9   System development with early, iterative V&V and integration, via modeling, analysis, and simulation
3.10   Illustrative model taxonomy (non-exhaustive)
3.11   Model-based integration across multiple disciplines using a hub-and-spokes pattern
3.12   Multidisciplinary MA&S coordination along the life cycle
3.13   Sample N-squared diagram

3.14   Sample coupling matrix showing: (a) Initial arrangement of aggregates; (b) final arrangement after reorganization

3.15   Unified Architecture Method

3.16   Enterprise and product frameworks

3.17   S*Pattern class hierarchy

3.18   Examples of natural systems applications and biomimicry

4.1   Tailoring requires balance between risk and process

4.2   IPO diagram for Tailoring process

4.3   SE life cycle spectrum

4.4   Agile SE life cycle model

4.5   Feature-based PLE factory

4.6   Schematic diagram of the operation of a Cyber-Physical System

4.7   The relationship between Cyber-Physical Systems (CPS), Systems of Systems (SoSs), and an Internet of Things (IoT)

4.8   Example of the systems and systems of systems within a transport system of systems

4.9   Service system conceptual framework

4.10   Organizations manage resources to create enterprise value

4.11   Individual competence leads to organizational, system, and operational capability

4.12   Enterprise state changes through work process activities

5.1   The "T-shaped" SE practitioner. From Delicado, et al. (2018). Used with permission. All other rights reserved.   *262*

5.2   Technical leadership is the intersection of technical expertise and leadership skills

5.3   Categorized dimensions of diversity

5.4   The intersection between PM and SE

5.5   IE and SE relationships

6.1   Timeline of vehicle impact

D.1   Input/output relationships between the various SE processes

LIST OF TABLES

1.1 SE standards and guides

1.2 SE return on investment

1.3 Examples for systems interacting with the SoI

1.4 Sources of system uncertainty

1.5 Common cognitive biases

1.6 SE principles and subprinciples

2.1 Representative technical reviews and audits

2.2 Life cycle model approach characteristics

2.3 Eight Attributes of a Quality Management Culture

2.4 Partial list of decision situations (opportunities) throughout the life cycle

2.5 Measurement benefits

2.6 Measurement references for specific measurement focuses

2.7 Requirement statement characteristics

2.8 Requirement set characteristics

2.9 Requirement attributes

3.1 Quality Characteristic approaches

3.2 HSI perspective descriptions

3.3 Resilience considerations

3.4 Implementation process breakout

4.1 Considerations of greenfield and brownfield development efforts

4.2 Considerations for COTS-based development efforts

4.3 SoS types

4.4 Impact of SoS considerations on the SE processes

4.5 Comparison of automotive, aerospace/defense, and consumer electronics domains

4.6 Representative organizations and standards in the automotive industry

4.7 Infrastructure and SE definition correlation

5.1 Differences between the hard skills and soft skills

5.2 Technical leadership model

PREFACE

The objective of the International Council on Systems Engineering (INCOSE) *Systems Engineering Handbook* (SEH) is to describe key Systems Engineering (SE) process activities. The intended audience is the SE practitioner. When the term "SE practitioner" is used in this handbook, it includes the new SE practitioner, a product engineer, an engineer in another discipline who needs to perform SE, or an experienced SE practitioner who needs a convenient reference.

The descriptions in this handbook show what each SE process activity entails, in the context of designing for required performance and life cycle considerations. On some projects, a given activity may be performed very informally; on other projects, it may be performed very formally, with interim products under formal configuration control. This document is not intended to advocate any level of formality as necessary or appropriate in all situations. The appropriate degree of formality in the execution of any SE process activity is determined by the following:

The need for communication of what is being done (across members of a project team, across organizations, or over time to support future activities)

The level of uncertainty

The degree of complexity

The consequences to human welfare

On smaller projects, where the span of required communications is small (few people and short project life cycle) and the cost of rework is low, SE activities can be conducted very informally and thus at low cost. On larger projects, where the span of required communications is large (many teams that may span multiple geographic locations and organizations and long project life cycle) and the cost of failure or rework is high, increased formality can significantly help in achieving project opportunities and in mitigating project risk.

In a project environment, work necessary to accomplish project objectives is considered "in scope"; all other work is considered "out of scope." On every project, "thinking" is always "in scope." Thoughtful tailoring and intelligent application of the SE processes described in this handbook are essential to achieve the proper balance between the risk of missing project technical and business objectives on the one hand and process paralysis on the other hand. Part IV provides tailoring and application guidance to help achieve that balance.

APPROVED FOR THE INCOSE SEH FIFTH EDITION:

Christopher D. Hoffman, CSEP, INCOSE Technical Director, January 2021-January 2023

Olivier Dessoude, INCOSE Technical Director, January 2023-January 2025

Theodore J. Ferrell, INCOSE Assistant Director, Technical Review, January 2021-January 2023

Krystal Porter, INCOSE Assistant Director, Technical Review, January 2023-January 2025

Lori F. Zipes, ESEP, INCOSE Assistant Director, Technical Information, January 2022-January 2024

Tony Williams, ESEP, INCOSE Assistant Director, Product Champion, January 2022-January 2025

HOW TO USE THIS HANDBOOK

PURPOSE

This handbook defines the "state-of-the-good-practice" for the discipline of Systems Engineering (SE) and provides an authoritative reference to understand the SE discipline in terms of content and practice.

APPLICATION

This handbook is consistent with ISO/IEC/IEEE 15288 (2023), *Systems and software engineering—System life cycle processes*, hereafter referred to as ISO/IEC/IEEE 15288, to ensure its usefulness across a wide range of application domains for engineered systems and products, as well as services. ISO/IEC/IEEE 15288 is an international standard that provides system life cycle process outcomes, activities, and tasks, whereas this handbook further elaborates on the activities and practices necessary to execute the processes.

This handbook is also consistent with the *Guide to the Systems Engineering Body of Knowledge*, hereafter referred to as the SEBoK (2023), to the extent practicable. In many places, this handbook points readers to the SEBoK for more detailed coverage of the related topics, including a current and vetted set of references. The SEBoK also includes coverage of "state-of-the-art" in SE.

For organizations that do not follow the principles of ISO/IEC/IEEE 15288 or the SEBoK to specify their life cycle processes, this handbook can serve as a reference to practices and methods that have proven beneficial to the SE community at large and that can add significant value in new domains, if appropriately selected, tailored, and applied. Part IV provides top-level guidance on the application of SE in selected product sectors and domains.

Before applying this handbook in a given organization or on a given project, it is recommended that the tailoring guidelines in Part IV be used to remove conflicts with existing policies, procedures, and standards already in use within an organization. Not every process will apply universally. Careful selection from the material is recommended. Reliance on process over progress will not deliver a system. Processes and activities in this handbook do not supersede any international, national, or local laws or regulations.

USAGE

This handbook was developed to support the users and use cases shown in Table 0.1. Primary users are those who will use the handbook directly. Secondary users are those who will typically use the handbook with assistance from SE practitioners. Other users and use cases are possible.

TABLE 0.1 Handbook users and use cases

User	Type	Use cases
Seasoned SE Practitioner. Those who need to reinforce, refresh, and renew their SE knowledge	Primary	• Adapt or refer to handbook to suit individual applicability • Explore good practices • Identify blind spots or gaps by providing a good checklist to ensure necessary coverage • References to other sources for more in-depth understanding
Novice SE Practitioner: Those who need to start using SE	Primary	• Support structured, coherent, and comprehensive learning • Understand the scope (breadth and depth) of systems thinking and SE practices
INCOSE Certification: Systems Engineering Professional (SEP) certifiers and those being certified	Primary	• Define body of knowledge for SEP certification • Form the basis of the SEP examination
SE Educators: Those who develop and teach SE courses, including universities and trainers	Primary	• Support structured, coherent, and comprehensive learning • Suggest relevant SE topics to trainers for their course content • Serve as a supplemental teaching aid
SE Tool Providers/Vendors: Those who provide tools and methods to support SE practitioners	Primary	• Suggest tools, methods, or other solutions to be developed that help practitioners in their work
Prospective SE Practitioner or Manager: Those who may be interested in pursuing a career in SE or who need to be aware of SE practices	Secondary	• Provide an entry level survey to understand what SE is about to someone who has a basic technical or engineering background
Interactors: Those who perform in disciplines that exchange (consume and/or produce) information with SE practitioners	Secondary	• Understand basic terminologies, scope, structure, and value of SE • Understand the role of the SE practitioner and their relationship to others in a project or an organization

ORGANIZATION AND STRUCTURE

As shown in Figure 0.1, this handbook is organized into six major parts, plus appendices.

Systems Engineering Introduction (Part I) provides foundational SE concepts and principles that underpin all other parts. It includes the what and why of SE and why it is important, key definitions, systems science and systems thinking, and SE principles and concepts.

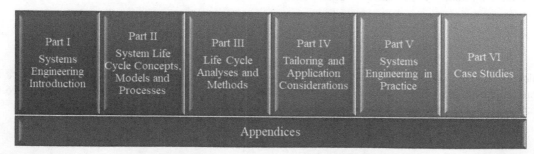

FIGURE 0.1 Handbook structure. INCOSE SEH original figure created by Mornas. Usage per the INCOSE Notices page. All other rights reserved.

System Life Cycle Concepts, Models, and Processes (Part II) describes an informative life cycle model with six stages: concept, development, production, utilization, support, and retirement. It also describes a set of life cycle processes to support SE consistent with the four process groups of ISO/IEC/IEEE 15288: Agreement Processes, Organizational Project Enabling Processes, Technical Management Processes, and Technical Processes.

Life Cycle Analyses and Methods (Part III) describes a set of quality characteristics approaches that need to be considered across the system life cycle. This part also describes methods that can apply across all processes, reflecting various aspects of the concurrent, iterative, and recursive nature of SE.

Tailoring and Application Considerations (Part IV) describes information on how to tailor (adapt and scale) the SE processes. It also introduces various considerations to view and apply SE: SE methodologies and approaches, system types, and project sectors and domains.

Systems Engineering in Practice (Part V) describes SE competencies, diversity, equity, and inclusion, SE relationship to other disciplines, SE transformation, and insight into the future of SE.

Case Studies (Part VI) describes several case studies that are used throughout the handbook to reinforce the SE principles and concepts.

Appendix A contains a list of references used in this handbook. Appendices B and C provide a list of acronyms and a glossary of SE terms and definitions, respectively. Appendix D provides an N^2 diagram of the SE life cycle processes showing an example of the dependencies that exist in the form of shared inputs or outputs. Appendix E provides a list of all the typical inputs/outputs identified for each SE life cycle process. Appendix F acknowledges the various contributors to this handbook. Errors, omissions, and other suggestions for this handbook can be submitted to the INCOSE using instructions found in Appendix G.

SYMBOLOGY

As described in Section 2.3.1.2, SE is a concurrent, iterative, and recursive process. The following symbology is used throughout this handbook to reinforce these concepts

Concurrency is indicated by the parallel lines.
Iteration is indicated by the circular arrows.

Recursion is indicated by the down and up arrows.

TERMINOLOGY

One of the SE practitioner's first and most important responsibilities on a project is to establish nomenclature and terminology that support clear, unambiguous communication and definition of the system and its elements, functions, operations, and associated processes. Further, to promote the advancement of the field of SE throughout the world, it is essential that common definitions and understandings be established regarding general methods and terminology that in turn support common processes. As more SE practitioners accept and use common terminology, SE will experience improvements in communications, understanding, and, ultimately, productivity.

The glossary of terms used throughout this book (see Appendix C) is based on the definitions found in ISO/IEC/IEEE 15288; ISO/IEC/IEEE 24765 (2017); and the SEBoK.

1

SYSTEMS ENGINEERING INTRODUCTION

1.1 WHAT IS SYSTEMS ENGINEERING?

Systems Engineering (SE)

Our world and the systems we engineer continue to become more complex and interrelated. SE is an integrative approach to help teams collaborate to understand and manage systems and their complexity and deliver successful systems. The SE perspective is based on systems thinking—a perspective that sharpens our awareness of wholes and how the parts within those wholes interrelate (incose.org, *About Systems Engineering*). SE aims to ensure the pieces work together to achieve the objectives of the whole. SE practitioners work within a project team and take a holistic, balanced, life cycle approach to support the successful completion of system projects (INCOSE Vision 2035, 2022). SE has the responsibility to realize systems that are *fit for purpose*, namely that systems accomplish their intended purposes and be resilient to effects in real-world operation, while minimizing unintended actions, side effects, and consequences (Griffin, 2010).

Definition of SE

INCOSE Definitions (2019) and ISO/IEC/IEEE 15288 (2023) define:

> **Systems Engineering** is a transdisciplinary and integrative approach to enable the successful realization, use, and retirement of engineered systems, using systems principles and concepts, and scientific, technological, and management methods.

INCOSE Definitions (2019) elaborates:
SE focuses on:

INCOSE Systems Engineering Handbook: A Guide for System Life Cycle Processes and Activities, Fifth Edition.
Edited by David D. Walden, Thomas M. Shortell, Garry J. Roedler, Bernardo A. Delicado, Odile Mornas, Yip Yew-Seng, and David Endler.
© 2023 John Wiley & Sons Ltd. Published 2023 by John Wiley & Sons Ltd.

- establishing, balancing and integrating stakeholders' goals, purpose and success criteria, and defining actual or anticipated stakeholder needs, operational concepts, and required functionality, starting early in the development cycle;
- establishing an appropriate life cycle model, process approach and governance structures, considering the levels of complexity, uncertainty, change, and variety;
- generating and evaluating alternative solution concepts and architectures;
- baselining and modeling requirements and selected solution architecture for each stage of the endeavor;
- performing design synthesis and system verification and validation;
- while considering both the problem and solution domains, taking into account necessary enabling systems and services, identifying the role that the parts and the relationships between the parts play with respect to the overall behavior and performance of the system, and determining how to balance all of these factors to achieve a satisfactory outcome.

SE provides facilitation, guidance, and leadership to integrate the relevant disciplines and specialty groups into a cohesive effort, forming an appropriately structured development process that proceeds from concept to development, production, utilization, support, and eventual retirement.

SE considers both the business and the technical needs of acquirers with the goal of providing a quality solution that meets the needs of users and other stakeholders, is fit for the intended purpose in real-world operation, and avoids or minimizes adverse unintended consequences.

The goal of all SE activities is to manage risk, including the risk of not delivering what the acquirer wants and needs, the risk of late delivery, the risk of excess cost, and the risk of negative unintended consequences. One measure of utility of SE activities is the degree to which such risk is reduced. Conversely, a measure of acceptability of absence of a SE activity is the level of excess risk incurred as a result.

Definitions of System

While the concepts of a *system* can generally be traced back to early Western philosophy and later to science, the concept most familiar to SE practitioners is often traced to Ludwig von Bertalanffy (1950, 1968) in which a system is regarded as a "whole" consisting of interacting "parts."

INCOSE Definitions (2019) and ISO/IEC/IEEE 15288 (2023) define:

> A **system** is an arrangement of parts or elements that together exhibit behavior or meaning that the individual constituents do not.

A system is sometimes considered as a product or as the services it provides.

In practice, the interpretation of its meaning is frequently clarified using an associative noun (e.g., medical system, aircraft system). Alternatively, the word "system" is substituted simply by a context-dependent synonym (e.g., pacemaker, aircraft), though this potentially obscures a system principles perspective.

A complete system includes all of the associated equipment, facilities, material, computer programs, firmware, technical documentation, services, and personnel required for operations and support to the degree necessary for self-sufficient use in its intended environment.

INCOSE Definitions (2019) elaborates:

Systems can be either physical or conceptual, or a combination of both. Systems in the physical universe are composed of matter and energy, may embody information encoded in matter-energy carriers, and exhibit observable behavior. Conceptual systems are abstract systems of pure information, and do not directly exhibit behavior, but exhibit "meaning." In both cases, the system's properties (as a whole) result, or emerge, from:

a) the parts or elements and their individual properties,
b) the relationships and interactions between and among the parts, the system, other external systems (including humans), and the environment.

SE practitioners are especially interested in systems which have or will be "systems engineered" for a purpose. Therefore, INCOSE Definitions (2019) defines:

An **engineered system** is a system designed or adapted to interact with an anticipated operational environment to achieve one or more intended purposes while complying with applicable constraints.

"Engineered systems" may be composed of any or all of the following elements: people, products, services, information, processes, and/or natural elements.

Origins and Evolution of SE

Aspects of SE have been applied to technical endeavors throughout history. However, SE has only been formalized as an engineering discipline beginning in the early to middle of the twentieth century (INCOSE Vision 2035, 2022). The term "systems engineering" dates to Bell Telephone Laboratories in the early 1940s (Fagen, 1978; Hall, 1962; Schlager, 1956). Fagen (1978) traces the concepts of SE within the Bell System back to early 1900s and describes major applications of SE during World War II. The British used multidisciplinary teams to analyze their air defense system in the 1930s (Martin, 1996). The RAND Corporation was founded in 1946 by the United States Air Force and claims to have created "systems analysis." Hall (1962) asserts that the first attempt to teach SE as we know it today came in 1950 at MIT by Mr. Gilman, Director of Systems Engineering at Bell. TRW (now a part of Northrop Grumman) claims to have "invented" SE in the late 1950s to support work with ballistic missiles. Goode and Machol (1957) authored the first book on SE in 1957. In 1990, a professional society for SE, the National Council on Systems Engineering (NCOSE), was founded by representatives from several US corporations and organizations. As a result of growing involvement from SE practitioners outside of the US, the name of the organization was changed to the International Council on Systems Engineering (INCOSE) in 1995 (incose.org, *History of Systems Engineering*; Buede and Miller, 2016).

With the introduction of the international standard ISO/IEC 15288 in 2002, the discipline of SE was formally recognized as a preferred mechanism to establish agreement for the creation of products and services to be traded between two or more organizations—the supplier(s) and the acquirer(s). This handbook builds upon the concepts in the latest edition of ISO/IEC/IEEE 15288 (2023) by providing additional context, definitions, and practical applications. Table 1.1 provides a list of key SE standards and guides related to the content of this handbook.

TABLE 1.1 SE standards and guides

Reference	Title
ISO/IEC/IEEE 15026	Systems and software engineering—Systems and software assurance (Multi-part standard)
ISO/IEC/IEEE 15288	Systems and software engineering—System life cycle processes
IEEE/ISO/IEC 15289	Systems and software engineering—Content of life cycle information items (documentation)
ISO/IEC/IEEE 15939	Systems and software engineering—Measurement process
ISO/IEC/IEEE 16085	Systems and software engineering—Life cycle processes—Risk management
ISO/IEC/IEEE 16326	Systems and software engineering—Life cycle processes—Project management
ISO/IEC/IEEE 21839	Systems and software engineering—System of systems (SoS) considerations in life cycle stages of a system
ISO/IEC/IEEE 21840	Systems and software engineering—Guidelines for the utilization of ISO/IEC/IEEE 15288 in the context of system of systems (SoS)
ISO/IEC/IEEE 21841	Systems and software engineering—Taxonomy of systems of systems
ISO/IEC/IEEE 24641	Systems and software engineering—Methods and tools for model-based systems and software engineering

(Continued)

TABLE 1.1   (Continued)

Reference	Title
ISO/IEC/IEEE 24748–1	Systems and software engineering—Life cycle management—Part 1: Guidelines for life cycle management
ISO/IEC/IEEE 24748–2	Systems and software engineering—Life cycle management—Part 2: Guidelines for the application of ISO/IEC/IEEE 15288
ISO/IEC/IEEE 24748–4	Systems and software engineering—Life cycle management—Part 4: Systems engineering planning
ISO/IEC/IEEE 24748–6	Systems and software engineering—Life cycle management—Part 6: System integration engineering
ISO/IEC/IEEE 24748–7	Systems and software engineering—Life cycle management—Part 7: Application of systems engineering on defense programs
ISO/IEC/IEEE 24748–8 / IEEE 15288.2	Systems and software engineering—Life cycle management—Part 8: Technical reviews and audits on defense programs
ISO/IEC/IEEE 24765	Systems and software engineering—Vocabulary
ISO/IEC/IEEE 26550	Software and systems engineering—Reference model for product line engineering and management
ISO/IEC/IEEE 26580	Software and systems engineering—Methods and tools for the feature-based approach to software and systems product line engineering
ISO/IEC/IEEE 29148	Systems and software engineering—Life cycle processes—Requirements engineering
ISO/IEC/IEEE 42010	Systems and software engineering—Architecture description
ISO/IEC/IEEE 42020	Software, systems and enterprise—Architecture processes
ISO/IEC/IEEE 42030	Software, systems and enterprise—Architecture evaluation framework
ISO/IEC 29110	Systems and Software Engineering Standards and Guides for Very Small Entities (VSEs) (Multi-part set)
ISO/IEC 31000	Risk management
ISO/IEC 31010	Risk management—Risk assessment techniques
ISO/IEC 33060	Process assessment—Process assessment model for system life cycle processes
ISO/PAS 19450	Automation systems and integration—Object-Process Methodology (OPM)
ISO 10007	Quality management—Guidelines for configuration management
ISO 10303-233	Industrial automation systems and integration—Product data representation and exchange—Part 233: Application protocol: Systems engineering
NIST SP 800–160 Vol. 1	Systems Security Engineering: Considerations for a Multidisciplinary Approach in the Engineering of Trustworthy Secure Systems
NIST SP 800–160 Vol. 2	Developing Cyber-Resilient Systems: A Systems Security Engineering Approach
OMG SysML™	OMG Systems Modeling Language
SEBoK	Guide to the Systems Engineering Body of Knowledge (SEBoK)
SAE-EIA 649C	Configuration Management Standard
SAE 1001	Integrated Project Processes for Engineering a System (Note: Replaced ANSI/EIA 632)
ANSI/AIA.A G.043B	Guide to the Preparation of Operational Concept Documents
CMMI	CMMI® V2.0

INCOSE SEH original table created by Mornas, Roedler, and Walden. Usage per the INCOSE Notices page. All other rights reserved.

1.2   WHY IS SYSTEMS ENGINEERING IMPORTANT?

The purpose of SE is to conceive, develop, produce, utilize, support, and retire the right product or service within budget and schedule constraints. Delivering the right product or service requires a common understanding of the current system state and a common vision of the system's future states, as well as a methodology to transform a set of stakeholder needs, expectations, and constraints into a solution. The right product or service is one that accomplishes the required service or mission. A common vision and understanding, shared by acquirers and suppliers, is achieved through application of proven methods that are based on standard approaches across people, processes, and tools. The application of these methods is continuous throughout the system's life cycle.

SE is particularly important in the presence of complexity (see Section 1.3.7). Most current systems are formed by integrating commercially available products or by integrating independently managed and operated systems to provide emergent capabilities which increase the level of complexity (see Sections 4.3.3 and 4.3.6). This increased reliance on off-the-shelf and systems of systems has significantly reduced the time from concept definition to market availability of products. Over the years between 1880 and 2000, average 25% market penetration has been reduced by more than a factor of four as illustrated in Figure 1.1.

In response to complexity and compressed timelines, SE methods and tools have become more adaptable and efficient. Introduction of agile methods (see Section 4.2.2) and SE modeling language standards such the Systems Modeling Language (SysML) have allowed SE practitioners to manage complexity and increase the implementation of a common system vision (see bottom of Figure 1.1). Model Based SE (MBSE) methods adoption continues to grow (see Section 4.2.1), particularly in the early conceptual design and requirements analysis (SEBOK, *Emerging Topics*). MBSE research literature continues to report on the increased productivity and quality of design and promises further progression toward a digital engineering (DE) approach, where data is transparent and cooperation optimized across all engineering disciplines. Standards organizations are updating or developing new approaches that take DE into consideration. SE will have to address this new digital representation of the system as DE becomes the way of doing business (see Section 5.4). The rapid evolution and introduction of Artificial Intelligence (AI) and Machine Learning (ML) into SE further increases complexity of verifiability, safety, and trust of self-learning and evolving systems.

The overall value of SE has been the subject of studies and papers from many organizations since the introduction of SE. A 2013 study was completed at the University of South Australia to quantify the return on investment (ROI) of SE activities on overall project cost and schedule (Honour, 2013). Figure 1.2 compares the total SE effort with cost

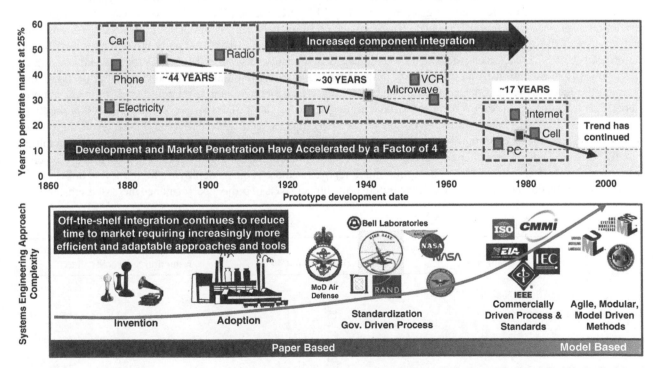

FIGURE 1.1   Acceleration of design to market life cycle has prompted development of more automated design methods and tools. INCOSE SEH original figure created by Amenabar. Usage per the INCOSE Notices page. All other rights reserved.

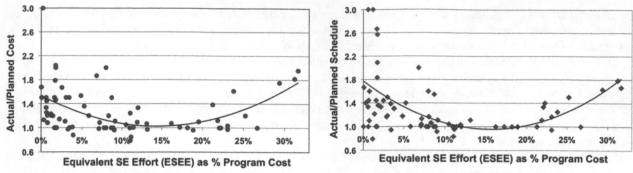

FIGURE 1.2   Cost and schedule overruns correlated with SE effort. From Honour (2013) with permission from University of South Wales. All other rights reserved.

compliance (left figure) and schedule performance (right figure). In both graphs, increasing the percentage of SE within the project results in better success up to an optimum level, above which SE ROI is diminished above those total program expenditure levels due to increased unwarranted processes. Study data shows that SE effort had a significant, quantifiable effect on project success, with correlation factors as high as 80%. Results show that the optimum level of SE effort for a normalized range of 10% to 14% of the total project cost.

The ROI of adding additional SE activities to a project is shown in Table 1.2, and it varies depending on the level of SE activities already in place. If the project is using no SE activities, then adding SE carries a 7:1 ROI; for each cost unit of additional SE, the project total cost will reduce by 7 cost units. At the median level of the projects interviewed, additional SE effort carries a 3.5:1 ROI.

A joint 2012 study by the National Defense Industrial Association (NDIA), the Institute of Electrical and Electronic Engineers (IEEE), and the Software Engineering Institute (SEI) of Carnegie Mellon University (CMU) surveyed 148 development projects and found clear and significant relationships between the application of SE activities and the performance of those projects as seen in Figure 1.3 (Elm and Goldenson, 2012). The study broke the projects by the maturity of their SE processes as measured by the quantity and quality of specific SE work products and considered the complexity of each project and the maturity of the technologies being implemented (n=number of projects). It also assessed the levels of project performance, as measured by satisfaction of budget, schedule, and technical requirements. The left column represents those projects deploying lower levels of SE expertise and capability. Among these projects, only 15% delivered higher levels of project performance and 52% delivered lower levels of project performance. The center column represents those projects deploying moderate levels of SE expertise and capability. Among these projects, the number delivering higher levels of project performance increased to 24% and those delivering lower levels decreased to 29%. The right column represents those projects deploying higher levels of SE expertise and capability. For these projects, the number delivering higher levels of project performance increased substantially

TABLE 1.2   SE return on investment

Current SE effort (% of program cost)	Average cost overrun (%)	ROI for additional SE effort (cost reduction $ per $ SE added)
0	53	7.0
5	24	4.6
7.2 (median of all programs)	15	3.5
10	7	2.1
15	3	−0.3
20	10	−2.8

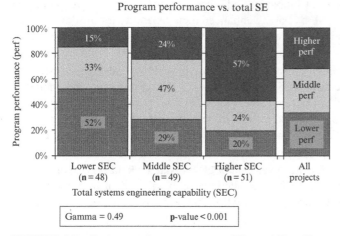

Program performance vs. total SE

FIGURE 1.3 Project performance versus SE capability. From Elm and Goldenson (2012) with permission from Carnegie Mellon University. All other rights reserved.

to 57%, while those delivering lower levels decreased to 20%. As Figure 1.3 shows, well-applied SE increases the probability of successfully developing an engineered system.

A 1993 Defense Acquisition University (DAU) statistical analysis on US Department of Defense (DoD) projects examined spent and committed life cycle cost (LCC) over time (DAU, 1993). As illustrated notionally in Figure 1.4, an important result from this study is that by the time approximately 20% of the actual costs have been accrued, over 80% of the total LCC has already typically been committed. Figure 1.4 also shows that it is less costly to fix or address issues if they are identified early. Good SE practice is the means by which the issues are identified and ensures that the understanding obtained is applied as appropriate during the life cycle, thus reducing technical debt.

INCOSE maintains value proposition statements (INCOSE Value Strategic Initiative Report, 2021) as tailored to different areas and industries. Areas covered include individual INCOSE membership, organizational INCOSE membership, INCOSE SE certification, and the discipline of SE. Industries include commercial, government, and nonprofit organizations. A sample of these findings includes:

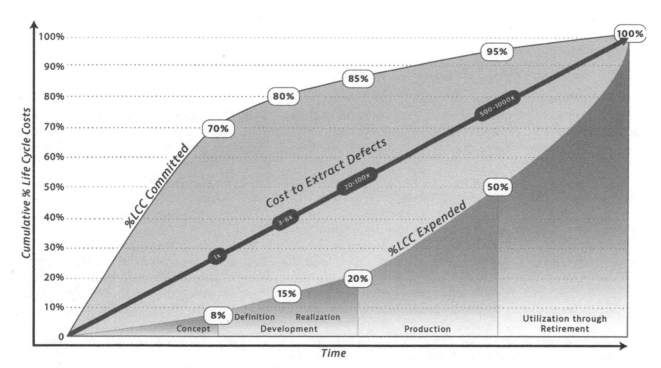

FIGURE 1.4 Life cycle costs and defect costs against time. INCOSE SEH original figure created by Walden derived from DAU (1993). Usage per the INCOSE Notices page. All other rights reserved.

- *Value of SE to the Commercial/Market-Driven Industry*: Companies and other enterprises in commercial industry will benefit from the internal practice of professional SE by having enhanced their capability for the development of innovative products and services for distribution in both mature and immature markets, in a more efficient and competitive manner.
- *Value of SE to Government/Infrastructure/Aerospace/Defense Industry*: SE provides a tailorable, systematic approach to all stages of a project, from concept to retirement. SE can accommodate different approaches including agile and sequential and facilitate commonality and open architectures to ensure lower acquisition, maintenance, and upgrade costs. By confirming correct and complete requirements and requirements allocations, the resulting design has fewer and less significant changes resulting in improved overall cost and schedule performance.
- *Value of SE to Nonprofit/Research Industry*: A nonprofit enterprise will benefit from the internal practice of professional SE by having enhanced their capability for the development of innovative client services in a more efficient and effective manner. An enterprise engaged in basic or applied research will benefit from the internal practice of SE by having enhanced its capabilities for discovery and invention that supports technology development in a more effective manner.

1.3   SYSTEMS CONCEPTS

Important system concepts include the system of interest (SoI), the system environment, and external systems. The boundaries between the system and the surrounding elements are important to understand. These boundaries separate the SoI, enabling systems, interoperating systems, and interfacing systems, supporting the SE practitioner in properly accounting for all the necessary elements which comprise the whole system context. Part of the system concept are the system's modes and states which are fundamental system behavior characteristics important to SE. Systems can be hierarchical in their structural organization, or they can be complex where hierarchy is not always present. The system concepts encompass all types of systems structures and support the SE practitioner with a framework in which to engineer a system.

1.3.1   System Boundary and the System of Interest (SoI)

General System Concepts   An external view of a system must introduce elements that specifically do not belong to the system but do interact with the system. This collection of elements is called the *system environment or context* and can include the users (or operators) of the system. It is important to understand that the system environment or context is not limited to the operating environment, but also includes external systems that interface with or support the system at any time of the life cycle.

The internal and external views of a system give rise to the concept of a *system boundary*. In practice, the system boundary is a "line of demarcation" between the system under consideration, called the system of interest (SoI), and its greater context. It defines what belongs to the system and what does not. The system boundary is not to be confused with the subset of elements that interact with the environment.

The *functionality* of a system is typically expressed in terms of the interactions of the system with its operating environment, especially the users. When a system is considered as an integrated combination of interacting elements, the functionality of the system derives not just from the interactions of individual elements with the environmental elements but also from how these interactions are influenced by the organization (interrelations) of the system elements. This leads to the concept of *system architecture*, which ISO/IEC/IEEE 42020 (2019) defines as:

Fundamental concepts or properties of an entity in its environment and governing principles for the realization and evolution of this entity and its related life cycle processes.

This definition speaks to both the internal and external views of the system and shares the concepts from the definitions of a system (see Section 1.1).

Scientific Terminology Related to System Concepts   In general, *engineering* can be regarded as the practice of creating and sustaining systems, services, devices, machines, structures, processes, and products to improve the quality of life—getting things done effectively and efficiently. The repeatability of experiments demanded by science is critical for delivering practical engineering solutions that have commercial value. Engineering in general, and SE in particular, draw heavily from the terminology and concepts of science.

An *attribute* of a system (or system element) is an observable characteristic or property of the system (or system element). For example, among the various attributes of an aircraft is its air speed. Attributes are represented symbolically by variables. Specifically, a *variable* is a symbol or name that identifies an attribute. Every variable has a domain, which could be but is not necessarily measurable. A *measurement* is the outcome of a process in which the SoI interacts with an observation system under specified conditions. The outcome of a measurement is the assignment of a *value* to a variable. A system is in a *state* when the values assigned to its attributes remain constant or steady for a meaningful period of time (Kaposi and Myers, 2001). In SE and software engineering, the *system elements* (e.g., software objects) have *processes* (e.g., operations) in addition to attributes. These have the binary logical values of being either *idle* or *executing*. A complete description of a system state therefore requires values to be assigned to both attributes and processes. *Dynamic behavior* of a system is the time evolution of the system state. *Emergent behavior* is a behavior of the system that cannot be understood exclusively in terms of the behavior of the individual system elements. See Section 1.3.2 for further information on emergent behavior and Section 1.3.6 for more information on states and modes.

The key concept used for problem solving is the *black box/white box* (also known as *opaque box/transparent box*) system representation. The *black box (opaque box)* representation is based on an external view of the system (attributes). The *white box (transparent box)* representation is based on an internal view of the system (attributes and structure of the elements). Both representations are useful to the SE practitioner and there must be an understanding of the relationship between the two. A system, then, is represented by the external attributes of the system, its internal attributes and structure, and the interrelationships between these that are governed by the laws of science.

1.3.2   Emergence

Emergence describes the phenomenon that whole entities exhibit properties which are meaningful only when attributed to the whole, not to its elements. Every model of human activity system exhibits properties as a whole entity that derive from its element activities and their structure, but cannot be reduced to them (Checkland, 1999). Emergence is a fundamental property of all systems (Sillitto and Dori, 2017). According to Rousseau et al. (2018), emergence derives from the systems science concept of "properties the system has but the elements by themselves do not."

System elements interact between themselves and can create desirable or undesirable phenomena called *emergent properties* such as inhibition, interference, resonance, or reinforcement of any property. Emergent properties can also result from the interaction between the system and its environment. Many engineering disciplines include emergence as a property. For example, system safety (Leveson, 1995) and resilience (Rasoulkahni, 2018) are examples of emergent properties of engineered systems (see Sections 3.1.11 and 3.1.9, respectively).

Definition of the architecture of the system includes an analysis of interactions between system elements in order to reinforce desirable and prevent undesirable emergent properties. According to Rousseau et al. (2019), the systemic virtue of emergent properties are used during systems architecture and design definition to highlight necessary derived functions and internal physical or environmental constraints (see Sections 2.3.5.4 and 2.3.5.5, respectively). Corresponding derived requirements should be added to system requirements baseline when they impact the SoI.

Calvo-Amodio and Rousseau (2019) explain how emergence applies to systems in which complexity is dominant. Complexity dominance, they say, encourages us to consider the significance of the difference between kinds of

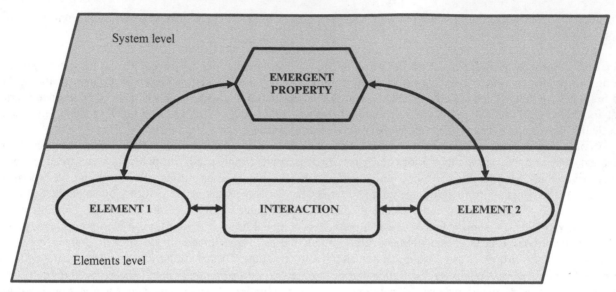

FIGURE 1.5   Emergence. INCOSE SEH original figure created by Jackson. Usage per the INCOSE Notices page. All other rights reserved.

complexity and degrees of complexity systems have. Doing so enables the SE practitioner to use variety engineering to manage complexity accordingly.

Figure 1.5 illustrates how the interaction between elements can result in emergent properties in any kind of system. This figure illustrates the basic rules of emergence. First, individual elements cannot exhibit higher-level system emergence. Second, two or more elements are required for emergence. Finally, emergence occurs at a level above the individual elements.

1.3.3   Interfacing Systems, Interoperating Systems, and Enabling Systems

External systems are systems beyond (or outside of) the SoI boundary. *Interfacing systems* are external systems that share an interface (e.g., physical, material, energy, data/information) with the SoI. Typically, humans also interface with the SoI throughout the SoI's life cycle stages. *Interoperating systems* are interfacing systems that interface with the SoI in its operational environment to perform a common function that supports the SoI's primary purpose. The set of SoI and interoperating systems can be seen as a system of systems (see Section 4.3.6). *Enabling systems* are external systems that facilitate the life cycle activities of the SoI but are not a direct element of the operational environment. The enabling systems provide services that are needed by the SoI during one or more life cycle stages. Some enabling systems share an interface with the SoI and some do not. Examples of enabling systems include collaboration development systems, production systems, and logistics support systems. Table 1.3 gives examples of these types of external systems.

During the life cycle stages for an SoI, it is necessary to concurrently consider interfacing, interoperating, and enabling systems along with the SoI. Otherwise, important requirements may not be identified, which will lead to significant costs in the further course of system development. Typical pitfalls include assuming that a new enabling system will come online in time to support the development of the SoI or that an existing enabling system will be available for the duration of the life cycle of the SoI. A delay in an enabling system coming online or the loss of an existing enabling system can lead to significant issues with the development and deployment of the SoI. In addition, horizontal and vertical integration considerations (see Section 2.3.5.8) may arise from the system context represented by interfacing, interoperating, and enabling systems.

TABLE 1.3   Examples for systems interacting with the SoI

SOI and External Systems	Interfacing System	Interoperating System	Enabling System
Aircraft			
Flight simulator	No	No	Yes
Fuel Truck	Yes	No	Yes
Remote Maintenance	Yes	Yes	Yes
Communication system	Yes	Yes	No
Runway	Yes	No	No
Automobile			
SE Tool	No	No	Yes
Car carrier	Yes	No	Yes
Diagnosis system	Yes	Yes	Yes
Parking assistant	Yes	Yes	No
Windshield snow cover	Yes	No	No

INCOSE SEH original table created by Endler. Usage per the INCOSE Notices page. All other rights reserved.

1.3.4   System Innovation Ecosystem

Sections 1.3.1 and 1.3.3 describe the system boundary and external systems in the overall context of the SoI. This section focuses on learning. Over single, and eventually multiple life cycles, engineered system innovation may be viewed as a form of group learning by "ecosystems" composed of individuals, teams, enterprises, supply chains, markets, and societies. Effective innovation requires effective learning and adaptation at a group level across these ecosystems and brings related challenges. To represent, plan, analyze, and improve such performance, the neutral descriptive System Innovation Ecosystem Pattern has been found to be useful (Schindel and Dove, 2016) (Schindel 2022b). Figure 1.6 provides a high-level view of that multiple-layered descriptive model, further discussed as a formal pattern in Section 3.2.6.

Figure 1.6 identifies three top-level system boundaries:

1. **System 1 – The Engineered System** may be a product developed for a market, a defense system created under contract, a service-providing system, or other system subject to SE life cycle management. It is shown in its larger environment, the Life Cycle Project Management System (System 2). System 1 examples include Medical Devices, Aircraft, Consumer Packaged Goods, and Gas Turbine Engines. This system is typically referred to as the engineered SoI in this handbook.

2. **System 2 – The Life Cycle Project Management System** provides the environment of System 1 over its life cycle, including the life cycle management processes responsible for System 1—described in Part II. System 2, a socio-technical system of people, processes, and facilities, is responsible to learn about System 1 and its environment, and to effectively apply that learning in the life cycle management by System 2. System 2 examples include System Requirements Definition Processes, Verification Processes, Product Manufacturing Processes, Product Distribution Processes, Product Sustainment Systems, Product Life Cycle Management (PLM) Information Systems, and Product Digital Twin Systems.

3. **System 3 – The Enterprise Process and Innovation System** contains System 2 and is responsible for learning about and improving System 2. In that sense, System 3 includes formal life cycle management for the processes of System 2. System 3 contains the "organizational change management" for advancing and adapting System 2 as a recognized formal system in its own right. System 3 examples include Product Life Cycle Management Processes, Program and Project Configuration and Tailoring Processes, Engineering Recruitment, Education, and Advancement Processes, Product Development Methodology Descriptions, Engineering Automation Tooling Acquisition and Development, Development Process Performance Analysis Systems, Regulatory Authorities, Engineering Professional Societies, and Engineering Facilities Construction and Acquisition.

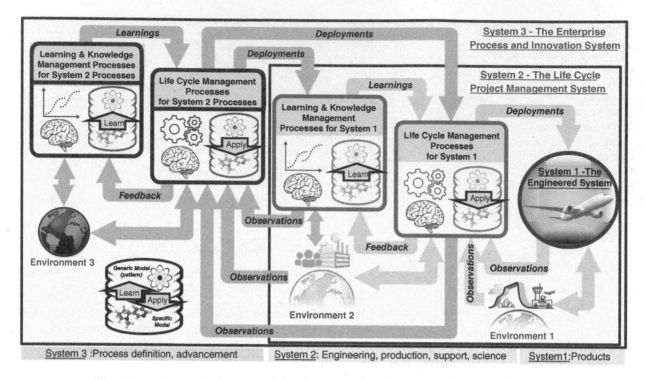

FIGURE 1.6 System innovation ecosystem pattern. From Schindel and Dove (2016) and Schindel (2022b). Used with permission. All other rights reserved.

The System Innovation Ecosystem Pattern emphasizes the learning and execution aspects of the enterprise ecosystem and directly integrates the SE life cycle processes described in Part II of this Handbook. Those processes are applied to two different managed SoIs (System 1 and System 2) and explicate the processes of learning versus application in each of the SE life cycle processes, along with how, and how effectively, execution is coupled with prior learning. The (configurable) System Innovation Ecosystem Pattern intentionally describes any engineering environment, whether effective in its learning and adaptation or not. It is intended as a descriptive, not prescriptive, reference model that can be used to plan and analyze any engineering and life cycle management ecosystem. So, while the "learned models" shown inside System 2 describe knowledge of System 1 (The Engineered System), the models shown inside System 3 describe knowledge of System 2 (The Life Cycle Project Management System).

The formal System Innovation Ecosystem Pattern includes the ability to be configured specific to a local enterprise, project, or supply chain, and for use to plan a series of migration increments representing advancing System 2 capabilities. For more details, refer to Section 3.2.6 and the INCOSE S*Patterns Primer (2022).

1.3.5 The Hierarchy within a System

As explained in Section 1.1, "A system is an arrangement of parts or elements." A *system element* is a member of a set of elements that constitute a system (ISO/IEC/IEEE 15288, 2023). A system element is a discrete part of a system that can be implemented to fulfil specified requirements. Hardware, software, data, humans, processes (e.g., processes for providing service to users), procedures (e.g., operator instructions), facilities, materials, and naturally occurring entities or any combination are examples of system elements.

In the ISO/IEC/IEEE 15288 (2023) usage of terminology, the system elements can be atomic (i.e., not further decomposed), or they can be systems on their own merit (i.e., decomposed into further subordinate system elements).

A system element that needs only a black box (also known as opaque box) representation (i.e., external view) to capture its requirements and confidently specify its real-world solution definition can be regarded as atomic. Decisions to make, buy, or reuse the element can be made with confidence without further specification of the element.

One of the challenges of system definition is to understand what level of detail is necessary to define each system element and the interrelations between elements. The integration of the system elements must establish the relationship between the effects that organizing the elements has on their interactions and how these effects enable the system to achieve its purpose. One approach to defining the elements of a system and their interrelations is to identify a complete set of distinct system elements with regard only to their relation to the whole (system) by suppressing details of their interactions and interrelations. These considerations lead to the concept of *hierarchy* within a system. This is referred to as a *partitioning* of the system and the end result is called a *Product Breakdown Structure* (PBS) (see Section 2.3.4.1). As stated above, each element of the PBS can be either atomic or it can be at a higher level that could be viewed as a system itself. At any given level, the elements are grouped into distinct subsets of elements subordinated to a higher-level system, as illustrated in Figure 1.7. Thus, hierarchy within a system is an organizational representation of system structure using a partitioning relation.

The art of defining a hierarchy within a system relies on the ability of the SE practitioner to strike a balance between clearly and simply defining span of control and resolving the structure of the SoI into a complete set of system elements that can be implemented with confidence. Urwick (1956) suggested a possible heuristic for span of control, recommending that decomposition of any object in a hierarchy be limited to no more than seven subordinate elements, plus or minus two (7 +/–2). Others have also found this heuristic to be useful in other contexts (Miller, 1956). A level of design with too few subordinate elements is unlikely to have a distinct design activity. In this case, both design and verification activities may contain redundancy. In case of too many subordinate elements, it may be difficult to manage all the interfaces between the subordinate elements. In practice, the nomenclature and depth of the hierarchy can and should be adjusted to fit the nature of the system and the community of interest.

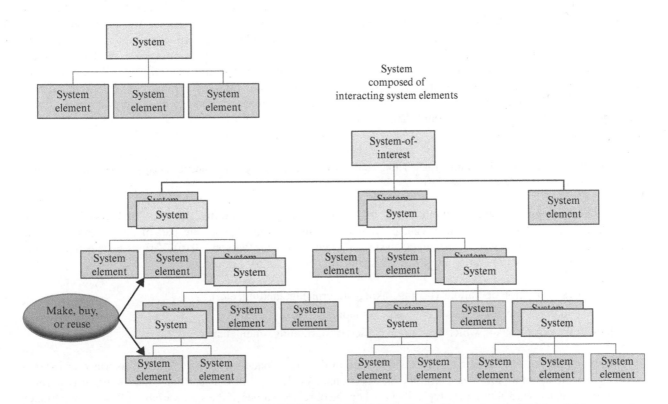

FIGURE 1.7   Hierarchy within a system. From ISO/IEC/IEEE 15288 (2023). Used with permission. All other rights reserved.

The interrelationships of system elements at a given architecture level of decomposition can be referred to as the horizontal view of the system. The horizontal view also includes requirements; integration, verification, or validation activities and results; various other related artifacts; and external elements. How the horizontal elements, activities, results, and artifacts are derived from or lead to higher-level systems and lower-level system element can be referred to as the vertical view of the system.

1.3.6   Systems States and Modes

States and modes are two related concepts that are used for defining and modeling system functional architectures and for modeling and managing system behaviors.

A *state* can be defined as:

> An observable and measurable … attribute used to characterize the current configuration, status, or performance-based condition of a System or Entity. (Wasson, 2016)

States are snapshots of a set of variables or measurements needed to describe fully the system's capabilities to perform the system's functions. *State variables* are the multidimensional list of variables that determine the state of the system. The list of variables does not change over time, but the values that these variables take do change over time (Buede and Miller, 2016). In control theory, the state of a dynamic system is a set of physical quantities, the specification of which (in Newtonian dynamics) completely determines the evolution of the system (Friedland, 2012). From the perspective of MBSE (see Section 4.2.1), "The state of the system is the most concise description of its past history."

The current system state and a sequence of subsequent inputs allow computation of the future states of the system. The state of a system contains all the information needed to calculate future responses without reference to the history of inputs and responses (Chapman, et al., 1992). Bonnet et al. (2017) states, "A state often directly reflects an operating condition or status on structural elements of the system (operational, failed, degraded, absent, etc.). States are also likely to represent the physical condition of a system element (full or empty fuel tank, charged or discharged battery, etc.). States can also be exploited to represent environment constraints (temperature, humidity, etc.)." If the system is transitioning from one state to another as time progresses, then time is one of the key attributes of the system. To monitor the system and manage it, the manager observes a state variable that is comprised of the appropriate collection of the system's attributes (Shafaat and Kenley, 2020).

A *mode* can be defined as:

> A distinct operating capability of the system during which some or all of the system's functions may be performed to a full or limited degree. (Buede and Miller, 2016)

For a personal computer, examples of modes are "off," "on," "waking up," "waiting," "reading from disk," "writing to disk," "computing," "printing," and, of course, "down" (Wymore, 1993). Modes are part of the system functional architecture and can be derived by affinity analysis of system use cases (Wasson, 2016). Various perspectives can be used to define the distinct operating modes of a system (Bonnet, et al., 2017), such as:

- the phases of mission operations (taxiing, taking-off, cruising, landing, etc.),
- the system operating conditions (connected, autonomous, etc.),
- the specific conditions in which the system is used (test, training, maintenance, etc.).

Transitioning from one mode to another is the result of decisions made by the system itself, its users, or external actors in order to adapt to new needs or new contexts (Bonnet, et al., 2017). Decisions that result in the system transitioning from one mode to another are typically based on the observed values of the state variables. When using models to depict system behavior, mode transitions are often based on triggering events that meet specified entry and exit criteria (Wasson, 2016).

1.3.7 Complexity

Systems engineering practitioners encounter a number of systems with simple, complicated, and complex characteristics. Many traditional systems engineering approaches and techniques work well for simple and complicated systems but do not handle complexity in systems (i.e., complex systems) well. Conversly, approaches and techniques that handle complexity well are also used in some complicated system contexts, especially when complex characteristics exist in some aspects of the system. Thus, care must be used to ensure the SE approaches and techniques for the SoI are appropriate and tailored for the type of system, especially with respect to its complexity. Complex systems are defined in the INCOSE publication "A Complexity Primer for Systems Engineers" (INCOSE Complexity Primer, 2021). A complex system has elements, the relationship between the states of which are weaved together so that they are not fully comprehended, leading to insufficient certainty between cause and effect. Complicated systems are less challenging. A complicated system has elements, the relationship between the states of which can be unfolded and comprehended, leading to sufficient certainty between cause and effect. Systems can also be simple. A simple system has elements, the relationship between the states of which, once observed, are readily comprehended. Complex systems can provide beneficial solutions yet also contain challenging characteristics. Complexity can result in positive behavior, such as self-organization and virtuous cycles of activity. However, intricate networks of evolving cause-and-effect relationships can lead to novel, nonlinear, and counterintuitive dynamics over time, resulting in suboptimal system operation, unintended consequences, and system obsolescence. The INCOSE Complexity Primer identifies 14 distinguishing characteristics that define complexity in a system. These characteristics provide insights into complexity, realizing that systems are not wholly complex: they are typically complex in some characteristics and complicated or even simple in others.

Traditional SE process for complicated systems takes a reductionist approach, whereby the problem is procedurally broken down into its parts (i.e., decomposition), solved, and reassembled to form the whole solution. This approach works well for complicated problems, where fixed, deterministic, or predictable patterns of behavior are required. However, these processes often do not perform well in complex environments, such as the challenges involved in designing autonomous vehicles or other socio-technical systems. A fundamentally different approach is required to understand the unexpected emergent interaction between the parts in the context of the whole through iterative exploration and adaptation (Snowden and Boone, 2007).

SE for complex systems requires a balance of linear, procedural methods for sorting through complicated and intricate tasks (e.g., systematic activity) and holistic, nonlinear, iterative methods for harnessing complexity (e.g., systems thinking). Complexity is not antithetical to simplicity, as even relatively simple systems can generate complex behavior. The INCOSE Complexity Primer provides guidance in the methods, approaches, and tools that may benefit complex systems engineering.

1.4 SYSTEMS ENGINEERING FOUNDATIONS

1.4.1 Uncertainty

There is uncertainty associated with much of the systems information and measurement data we use. This section provides a brief summary of the two major types of uncertainty, the sources of systems uncertainty, and decision making under uncertainty.

Types of Uncertainty. There are two types of uncertainties: epistemic and aleatory. In SE, *epistemic uncertainty* is due to our lack of knowledge about the potential demand for a new system and how a technology, system, or process will perform in the future, for example, the knowledge gap about key value attribute or about the acquirer's preferences. *Aleatory uncertainty* is uncertainty due to randomness. If a technology, system, or process can perform a function, there will be always some inherent randomness in every performance measurement. Our system requirements process, and development decisions focus on reducing epistemic uncertainty (overcoming our lack of knowledge), but we can never completely reduce aleatory uncertainty in our development or operational measurement of system performance.

TABLE 1.4   Sources of system uncertainty

Sources of Uncertainty	Major Questions	Potential Uncertainties
Business	Will political, economic, labor, social, technological, environmental, legal or, other factors adversely affect the business environment?	Changes in political viewpoint (e.g., elections) Economic disruptions (e.g., recession). Global disruptions (e.g., supply chain). Changes to laws and regulations. Disruptive technologies. Adverse publicity.
Market	Will there be a market if the product or service works?	User and consumer demand. Threats from competitors (quality and price) and adversaries (e.g., hackers and terrorists). Continuing stakeholder support.
Management	Does the organization have the people, processes, and culture to manage a major system?	Organization culture. SE and management experience and expertise. Mature baselining processes (technical, cost, schedule). Reliable cost estimating processes.
Performance (Technical)	Will the product or service meet the required desired performance?	Defining future requirements in dynamic environments. Understanding of the technical baseline. Technology maturity to meet performance. Adequate modeling, simulation, test, and evaluation capabilities to predict and evaluate performance. Availability of enabling systems needed to support use.
Schedule	Can the system that provides the product or service be delivered on time?	Concurrency in development. Impact of uncertain events on schedule. Time and budget to resolve technical and cost risks.
Development and Production Cost	Can the system be delivered within the budget? Will the cost be affordable?	Changes in missions. Technology maturity. Hardware and software development processes. Industrial/supply chain capabilities. Production facilities capabilities and processes.
Operations and Support Cost	Can the owner afford to operate and support the system? Will the cost be affordable?	Increasing operations and support (e.g., resource or environmental) costs. Resiliency of the design to new missions and tasks. Changes in maintenance or logistics strategy/needs.
Sustainability	Will the system provide sustainable future value?	Availability of future resources and impact on the natural environment.

Sources of Uncertainty and Risk. There are many sources of epistemic uncertainty that impact SE in the system life cycle. Table 1.4 provides a partial list of some of the major uncertainties that confront project managers and SE practitioners and describes some of the implications for SE.

Decisions Under Uncertainty

As can be seen from Table 1.4, uncertainties impact every SE decision process. Taking decisions before having enough knowledge is potentially very risky. Key decisions that have a strong impact on the solution require reducing uncertainty by closing the knowledge gap to an appropriate level. However, SE practitioners must be able to make decisions under uncertainty and should record a corresponding risk with those decisions (see Sections 2.3.4.3 and 2.3.4.4).

1.4.2 Cognitive Bias

SE practitioners need to obtain information from stakeholders throughout the system life cycle. SE practitioners and stakeholders (individual or groups) are subject to cognitive biases when interpreting uncertain information. The best defense from cognitive biases is understanding what they are and how they can be avoided and setting up organizational projects to obtain unbiased assessments. Cognitive biases are mental errors in judgment under uncertainty caused by our simplified information processing strategies (sometimes called heuristics) and are consistent and predictable (Tversky and Kahneman, 1974). There are many lists of cognitive biases, including one that lists 50 sources (Hallman, 2022). Cognitive biases can affect both individual and teams of SE practitioners (McDermott, et al., 2020). Cognitive biases can contribute to incidents, failures, or disasters as a result of distorted decision making and can lead to undesirable outcomes. Cognitive biases are included in a field called Behavioral Decision-Making. Table 1.5 lists some of the most common cognitive biases.

For major systems decisions, more formal methods are required to avoid cognitive biases. Both Tversky and Kahneman (1974) and Thaler and Sunstein (2008) describe mitigation methods suitable to different environments. The most effective methods are external group methods. For example, NASA (2003) recommends the Independent Technical Authority (ITA) to warn decision makers of the potential for failure. The ITA must be both financially and organizationally independent of the project manager. Another method, adopted by the aviation industry, is called the Crew Resource Management (CRM) method. With the CRM method, all crew members, including the co-pilot, are responsible for warning the pilot of imminent danger.

1.4.3 Systems Engineering Principles

SE is a relatively young discipline. The emergence of a set of SE principles has occurred over the past 30 years within the discipline. In reviewing various published SE principles, a set of criteria emerged for SE principles. SE principles cover broad application within the practice; they are not constrained to a particular system type, to the system development or operational context, or to a particular life cycle stage. SE principles transcend these system characteristics and inform a worldview of the discipline. Thus, a SE principle:

- transcends a particular life cycle model or stage,
- transcends system types,
- transcends a system context,
- informs a world view on SE,
- is not a "how to" statement,
- is supported by literature or widely accepted by the community (i.e., has proven successful in practice across multiple organizations and multiple system types),
- is focused, concise, and clearly worded.

SE principles are a form of guidance proposition which provide guidance in application of the SE processes and a basis for the advancement of SE. SE has many kinds of guidance propositions that can be classified by their sources, e.g., heuristics (derived from practical experience as discussed in Section 1.4.4), conventions (derived from social agreements), values (derived from cultural perspectives), and models (based on theoretical mechanisms). Although these all support purposeful judgment or action in a context, they can vary greatly in scope, authority, and conferred capability. They can all be refined, and as they mature, they gain in their scope, authority, and capability, while the set becomes more compact. A key moment in their evolution occurs with gaining insight into why they work, at which point they become principles. Principles can have their origins associated in referring to them as "heuristic principles," "social principles," "cultural principles," and "scientific principles," although in practice it is usually sufficient to just refer to them as SE principles. SE principles are derived from principles of these various origins providing a diverse set of transcendent principles based on both practice and theory.

TABLE 1.5 Common cognitive biases

Cognitive Bias	Description	Implication for the SE Practitioner.
Framing	How we ask the question or describe the decision matters.	Carefully word questions and problem description to avoid influencing the response.
Representativeness	People draw conclusions based on representative characteristics and often ignore relevant facts or the base rates.	Discuss the relevant facts and data before requesting a judgment about an uncertainty or risk. Use Bayes Law to update our beliefs after we receive new data. Teams that reflect Diversity, Equity, and Inclusion principles can help reduce the bias for the team (see Section 5.2).
Availability	We place too much weight on vivid, striking, and recent events.	Ask about the relevant facts and data before requesting a judgment about an uncertainty or risk. Design systems to provide the relevant data.
Anchoring	The initial estimate affects the final estimates.	Never begin by asking about the expected outcome. Instead obtain information about the worst or best outcomes first to understand the range of outcomes.
Motivational	When making probability judgments, people have incentives to provide estimates that will benefit themselves	Understand the potential bias of an individual providing an assessment. For example, a technology developer has an incentive to overestimate technology readiness if a more conservative estimate could result in loss of funding.
Optimism	We overestimate the likelihood of good outcomes and underestimate the likelihood bad outcomes.	Seek data on similar bad outcomes. Obtain assessments from experts not involved in the decision.
Confirmation	We seek or put more weight on data that confirms our beliefs.	Actively seek data that would disprove our current belief in all tests and evaluations.
Group Think	A group of people make irrational or unsound decisions to suppress dissent and maintain group harmony.	Seek dissenting opinions inside the group and seek outside assessments.
Authority	We trust and are more often influenced by the opinions of people in positions of authority	Assess the opinion independent of the source.
Rankism	Assumption that person of higher rank is always correct in decisions	Seek to determine correct decision

In addition, SE principles differ from systems principles in important ways (Watson, et al., 2019). System principles address the behavior and properties of all kinds of systems, looking at the scientific basis for a system and characterizing this basis in a system context via specialized instances of a general set of system principles. SE principles build on systems principles that are general for all kinds of systems (Rousseau, 2018) (Watson, 2020) and for all kinds of human activity systems (Senge, 1990) (Calvo-Amodio and Rousseau, 2019).

INCOSE compiled an early list of principles consisting of 8 principles and 61 subprinciples in 1993 (Defoe, 1993). These early principles were important considerations recognized in practice for the success of system developments and ultimately became the basis for the SE processes. These early principles were focused on particular aspects of the

SE process and particular life cycle stages. The INCOSE work on SE principles considered these earlier sources and compiled a set of SE principles that are transcendent. The INCOSE SE Principles (2022) documents each SE principle with a description, evidence that supports the principle (e.g., observable evidence of the application, proof from scientific evidence), and implications in SE practice for application of the principle. There are presently 15 SE principles and 20 subprinciples as shown in Table 1.6.

TABLE 1.6   SE principles and subprinciples

1 SE in application is specific to stakeholder needs, solution space, resulting system solution(s), and context throughout the system life cycle.
2 SE has a holistic system view that includes the system elements and the interactions amongst themselves, the enabling systems, and the system environment.
3 SE influences and is influenced by internal and external resources, and political, economic, social, technological, environmental, and legal factors.
4 Both policy and law must be properly understood to not over-constrain or under-constrain the system implementation.
5 The real system is the perfect representation of the system.
6 A focus of SE is a progressively deeper understanding of the interactions, sensitivities, and behaviors of the system, stakeholder needs, and its operational environment.
   Sub-Principle 6(a): Mission context is defined based on the understanding of the stakeholder needs and constraints
   Sub-Principle 6(b): Requirements and models reflect the understanding of the system
   Sub-Principle 6(c): Requirements are specific, agreed to preferences within the developing organization
   Sub-Principle 6(d): Requirements and system design are progressively elaborated as the development progresses
   Sub-Principle 6(e): Modeling of systems must account for system interactions and couplings
   Sub-Principle 6(f): SE achieves an understanding of all the system functions and interactions in the operational environment
   Sub-Principle 6(g): SE achieves an understanding of the system's value to the system stakeholders
   Sub-Principle 6(h): Understanding of the system degrades during operations if system understanding is not maintained.
7 Stakeholder needs can change and must be accounted for over the system life cycle.
8 SE addresses stakeholder needs, taking into consideration budget, schedule, and technical needs, along with other expectations and constraints.
   Sub-Principle 8(a): SE seeks a best balance of functions and interactions within the system budget, schedule, technical, and other expectations and constraints.
9 SE decisions are made under uncertainty accounting for risk.
10 Decision quality depends on knowledge of the system, enabling system(s), and interoperating system(s) present in the decision making process.
11 SE spans the entire system life cycle.
   Sub-Principle 11(a): SE obtains an understanding of the system
   Sub-Principle 11(b): SE defines the mission context (system application)
   Sub-Principle 11(c): SE models the system
   Sub-Principle 11(d): SE designs and analyzes the system
   Sub-Principle 11(e): SE tests the system
   Sub-Principle 11(f): SE supports the production of the system
   Sub-Principle 11(g): SE supports operations, maintenance, and retirement
12 Complex systems are engineered by complex organizations.
13 SE integrates engineering and scientific disciplines in an effective manner.
14 SE is responsible for managing the discipline interactions within the organization.
15 SE is based on a middle range set of theories.
   Sub-Principle 15 (a): SE has a systems theory basis
   Sub-Principle 15 (b): SE has a physical logical basis specific to the system
   Sub-Principle 15 (c): SE has a mathematical basis
   Sub-Principle 15 (d): SE has a sociological basis specific to the organization

These principles provide a start in defining a transcendent disciplinary basis for SE. Application of the principles aids in determining a system life cycle model, implementing SE processes, and defining organizational constructs to help the SE practitioner successfully develop and sustain the SoI.

1.4.4   Systems Engineering Heuristics

Summary   Heuristics provide a way for an established profession to pass on its accumulated wisdom. This allows practitioners to gain insights from what has been found to work well in the past, and apply the lessons learned. Heuristics usually take the form of short expressions in natural language. These can be memorable phrases encapsulating shortcuts, "rules of thumb," or "words of the wise," giving general guidelines on professional conduct or rules, advice, or guidelines on how to act under specific circumstances. Heuristics usually do not express all there is to know, yet they can act as a useful entry point for learning more. At their best, heuristics can act as aids to decision making, value judgments, and assessments.

Interest in SE heuristics currently centers on their use in two contexts: (1) encapsulating engineering knowledge in an accessible form, where the underlying practice is widely accepted and the underlying science understood, and (2) overcoming the limitations of more analytical approaches, where the science is still of limited use. This is especially applicable as we extend the practice of SE to providing solutions to inherently complex, unbounded, ill-structured, or very difficult problems.

Background   Engineering first emerged as a series of skills acquired while transforming the ancient world, principally through buildings, cities, infrastructure, and machines of war. Since then, mankind has sought to capture the knowledge of "how to" to allow each generation to learn from its predecessors, enabling more complex structures to be built with increasing confidence while avoiding repeated real-world failures. For example, early cathedral builders encapsulated their knowledge in a small number of "rules of thumb," such as "maintain a low center of gravity" and "put 80% of the mass in the pillars." Designs were conservative, with large margins. When the design margins were exceeded (e.g., out of a desire to build higher and more impressive structures), a high price was sometimes paid, with the collapse of a roof, a tower, or even a whole building. From such failures, new empirical rules emerged. Much of this took place before the science behind the strength of materials or building secure foundations was understood. Only in recent times have computer simulations revealed the contribution toward certain failures played by such dynamic effects as wind shear on tall structures.

Since then, engineering and applied sciences have co-evolved: with science providing the ability to predict and explain performance of engineered artefacts with greater assurance and engineering developing new and more complex systems, requiring new scientific explanations and driving research agendas. In the modern era, complex and adaptive systems are being built which challenge conventional engineering sciences, and we are turning to social and behavioral sciences, management sciences, and increasingly systems science to deal with some of the new forms of complexity involved and guide the profession accordingly.

Current Use   Renewed interest in the application of heuristics to the field of SE stems from the seminal work of Maier and Rechtin (2009), and their book remains the best single published source of such knowledge. Their motivation was to provide guidance for the emerging role of system architect as the person (or team) responsible for coordinating engineering effort toward devising solutions to complex problems and overseeing their implementations. They observed that it was in many cases better to apply heuristics than attempt detailed analysis. The reason for this is the number of variables involved and the complexity of the interactions between stakeholders, internal dynamics of system solutions, and the organizations responsible for their realization. Some examples of SE heuristics are:

- ***Don't assume that the original statement of the problem is necessarily the best, or even the right one.*** This has to be handled with tact and respect for the user, but experience shows that failure to reach mutual understanding early on is a fundamental cause of failure, and strong relationships forged in the course of doing such coordination with stakeholders can pay off when solving more difficult issues which might arise later on.

- *In the early stages of a project, unknowns are a bigger issue than known problems.* Sometimes developing a clear understanding of the environment, all of the stakeholders, and the ramifications of possible solutions uncovers many unanticipated issues.
- *Model before build, wherever possible.* System Science postulates "The only complete model of the system in its environment is the system in its environment," which leads into using evolutionary life cycles, rapid deployment of prototypes, agile life cycles, and so on. This heuristic opens a door into twenty-first-century systems.

A repository of heuristics can act as a knowledge base, especially if media (such as video clips or training materials) or even interactive media (to encourage discussion and feedback) are included. A heuristics repository should link to other established knowledge sources and be tagged with other metadata to allow flexible retrieval. It should be organized to reflect accepted areas of SE competency and allow users to assemble a personal set of heuristics most meaningful to them, being relevant to their professional or personal sphere of activity.

1.5   SYSTEM SCIENCE AND SYSTEMS THINKING

This section considers the nature and relationship between systems science and systems thinking and describes how they relate to SE.

Relationship between Systems Science, Systems Thinking, and SE

The association of concepts such as system, boundary, relationships, environment/context, hierarchy, emergence, communication, and control, among others, when interrelated with purpose, gives rise to a *systems worldview* (Rousseau, et al., 2018). Interrelating concepts with purpose changes how we investigate and reason about things, producing *systems thinking*. Systems thinking enables us to recognize systems patterns across different phenomena, problem contexts, and disciplines. Studying these patterns has produced the systems sciences of General System Theory, Cybernetics, and Complexity Theory and their related systems methodologies, models, and methods. The application of systems thinking and systems science concepts, principles, methodologies, models, and methods in engineering is one of the bases for the practice of SE. Applying SE, and reflecting on the results, help us improve systems science and systems thinking, further enhancing our ability to design and intervene in complex systems—a virtuous cycle. Through this virtuous cycle, we develop principles to better our SE applications (Rousseau, et al., 2022).

Figure 1.8 depicts this virtuous cycle as a multifaceted and purposeful activity to deliver elegant solutions to complex problems, supported by principles that guide why, what, and how we do SE. To connect our purpose to our actions, we adopt a systemic approach, because complexity and elegance are both systems phenomena. Our systemic approach is of course guided by our systems principles. The kinds and relationships of principles, as well as how they inform and are informed by SE practice, is depicted in Figure 1.8. We select and organize these based on our intentions as expressed by our motivational principles. We use our transdisciplinary principles to select and organize our technique principles. In this way, the systemic relationships between our principles support how our principles guide the systemic relationships between our purpose, approach, and practice. The systemic roles our principles play in our discipline thus support the systematic evolution of our value in society.

The success of SE applications reinforces the credibility of the systems worldview which in turn enhances the SE practitioner's ability to conceptualize why a solution is needed, how a solution can be conceptualized, and what tools and/or methods to use to solve complex problems and achieve elegant solutions.

Systems Science

Questions about the nature of systems, organization, and complexity are not specific to the modern age. As Warfield (2006) put it:

> Virtually every important concept that backs up the key ideas emergent in systems literature is found in ancient literature and in the centuries that follow.

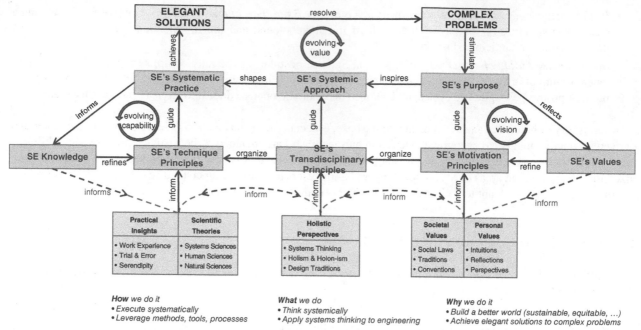

FIGURE 1.8   An architectural framework for the evolving the SE discipline. From Rousseau, et al. (2022). Used with permission. All other rights reserved.

Systems science can be defined as a transdisciplinary approach interested in understanding all aspects of systems with the goals of (1) identifying, exploring, and understanding patterns of behavior crossing disciplinary fields and areas of application, and (2) establishing a general theory applicable to all types of systems whether physical, natural, engineered, or social. Attempts to establish a systems science have taken on both reductionist and holistic forms, and both are valuable. For instance, a clock is a system, but its workings can be explained through reductionism. On the other hand, a holistic approach helps us understand why we need clocks, how clocks exist (operate/sustain/degrade) in their environment throughout their life cycle, and how the esthetics of their design evolve over time. Both the reductionist and the holistic approaches to explanation involve systemic arguments but each starts from different directions—bottom-up and outside-in. Complexity Theory has had some success developing a science of systems using a reductionist approach. Agent-based modeling, pioneered at the Santa Fe Institute, works from the "bottom-up" and seeks to explain the behavior of whole systems in terms of the rules of interaction of the "agents" that constitute the system.

Where reductionist (traditional) methods prove unsuccessful, systems science relies on the holistic approach. A holistic approach is adept at connecting and contextualizing systems, system elements, and their environments to understand difficult to explain patterns of organized complexity. This was the approach taken by Ludwig von Bertalanffy in developing General System Theory and Norbert Wiener in developing Cybernetics. The biologist von Bertalanffy was one of the first to argue for a general science of systems. He explained the scientific need for systems-based research as an alternative to traditional analytical procedures in science. This alternative method would overcome the limitations that result from explaining a system by breaking it down to its constituent parts and then being reconstituted from its parts, either materially or conceptually:

> This is the basic principle of *classical* science, which can be circumscribed in different ways: resolution into isolable causal trains or seeking for *atomic* units in the various fields of science, etc. (von Bertalanffy, 1969)

This makes it impossible to account for the emergent properties that systems display as a result of the interrelationships between their parts (see Section 1.3.2). Instead, von Bertalanffy promoted an alternative worldview concerned with the laws that apply to systems behavior in general. Such a General System Theory was possible, von Bertalanffy thought, and would be particularly valuable, because of the large number of parallelisms that appear across systems independent of the types and quantities of system elements in the systems:

> Thus, there exist models, principles, and laws that apply to generalized systems or their subclasses, irrespective of their particular kind, the nature of their component elements, and the relations or 'forces' between them. It seems legitimate to ask for a theory, not of systems of a more or less special kind, but of universal principles applying to systems in general. In this way we postulate a new discipline called General System Theory. (von Bertalanffy, 1971)

The study of general systems was to focus on such principles as:

> growth, regulation, hierarchical order, equifinality, progressive differentiation, progressive mechanization, progressive centralization, closed and open systems, competition, evolution toward higher organization, teleology, and goal-directedness. (Hammond, 2003)

The systems sciences, including General Systems Theory, Cybernetics, and Complexity Theory, seek to provide key foundational concepts to build a common language and intellectual foundations to make rigorous systems theories and tools accessible to practitioners. Where they succeed, they can serve as the foundation for a meta-discipline such as SE, transdisciplinary in nature, that unifies scientific and engineering practices. SE, informed by systems science, would be in a powerful position to enhance its theory and practice in ways that would make it applicable to the most complex of systems.

Finally, identifying SE's principles and heuristics can offer a useful approach to categorize systems-related knowledge and to focus research efforts. Systems principles and heuristics are special cases of guiding propositions. A guiding proposition provides guidance for purposeful judgment or action in a context and offers a wider perspective to that of a principle or a heuristic. Guiding propositions vary in (1) scope—the range of SE contexts they work, (2) authority—how compelling they are, and (3) capability—how predictable the outcomes of applying them are (Rousseau, et al., 2022). Readers can consult details on SE Principles in Section 1.4.3 and details on SE Heuristics in Section 1.4.4.

Systems Thinking Divergences

Systems thinking is a key enabler of SE. It is one of the core competencies defined in INCOSE SE Competency Framework (2018). Systems thinking applies the properties, concepts, and principles of systems to the given situation as a framework for curiosity—to get insight and understanding about the situation.

There needs to be a balance between the being systematic with the application of SE processes (as described in Part II) and being systemic, applying systems thinking to drive these processes. As SE practitioners, it is vital to possess the knowledge and skills necessary to perform holistic analysis and guide systemic intervention. Systems thinking lacks a unified definition; however, the following captures the nature of systems thinking and some key ideas:

> Systems thinking is a field characterized by a baffling array of methods and approaches. We posit that underlying all, however, are four universal rules called DSRP (distinctions, systems, relationships, and perspectives). We make distinctions between and among things and ideas, each implying the existence of another. We identify systems, which are composed of parts and wholes. We recognize relationships composed of actions and reactions. We take perspectives consisting of a point (from which we see) and a view (that which is seen). (Cabrera, et al., 2015)

This definition incorporates aspects of complex problem situations, such as "distinctions" and "perspectives," which it is essential to take account of, but which systems science may never be able to incorporate into its scientific models.

Based on the pioneering work of Ludwig von Bertalanffy in General System Theory, Norbert Wiener in Cybernetics, Jay Forrester in System Dynamics, Peter Checkland in Soft Systems Thinking, and others, a variety of systems methodologies, models, and methods have been formalized to perform systemic analyses and interventions. The SE practitioner can take advantage of this diversity providing they are aware of what the different methodologies, models, and methods do well, and what they are less good at. To assist systems thinking practitioners in selecting the most appropriate systems approaches, Jackson and Keys (1984) offered an initial classification of systems methodologies, the Systems of Systems Methodologies (SOSM), according to their strengths in addressing the complexity of systems and in reconciling divergences among stakeholder viewpoints. Jackson (2019) has since updated the SOSM, to reflect developments in Complexity Theory, by incorporating lessons from the Cynefin framework (Kurtz and Snowden, 2003). This use of different systems approaches in informed combinations, according to their strengths and weaknesses and the nature of the problem situation, is called Critical Systems Thinking (CST) (Jackson, 2003, 2019). CST is a multi-perspectival, multi-methodological, and multi-method approach.

While most of the prominent systems thinking approaches are rooted and/or contextualized within the management sciences, these approaches apply equally to SE practice. This is because the problems faced by SE practitioners, such as the need to incorporate cultural, social, political, and project management perspectives into systems models and other SE tasks, are common to the management sciences.

According to Jackson (2019), systems methodologies translate hypotheses about the nature of problem situations, and how they can be improved, into practical action. There are a number of systems methodologies available, for example, system dynamics, the viable system model, soft systems methodology, and critical systems heuristics. Each is based upon different assumptions about the world and how best to intervene in it. Together, these methodologies can recognize and respond to the range of issues encountered during the exploration of complex problem situations. These systems approaches can then be used, individually or in combination, in the problem situation. When the systems approaches are used in combination, the weighting of each system approach in the hybrid solution will be tailored based on the technical, organizational, cultural, and political factors within the organization and the relative dominance of those factors. According to systems thinkers, if SE can embrace the full range of systems methodologies, models, and methods, it will be in a much better position to tackle the hyper-complexity plaguing projects, organizations, and society in the contemporary world.

2

SYSTEM LIFE CYCLE CONCEPTS, MODELS, AND PROCESSES

2.1 LIFE CYCLE TERMS AND CONCEPTS

The overall purpose of Systems Engineering (SE) is to enable successful realization of the system while optimizing among competing stakeholder objectives. One way in which realization is managed is by breaking the overall effort into transformational steps, or stages, then checking for satisfactory fulfillment of system characteristics at the end of each stage, as well as checking whether risk is acceptable and the system is ready to enter other stages. Stages do not need to be executed sequentially or singularly. They can be executed multiple times as needed, and often in parallel. The critical feature of this approach is that progress is gated by specific decision points, generally called decision gates. By analogy with the stages that living things go through, called a life cycle, the set of stages for a system is termed a system life cycle. In summary, engineered systems progress in some manner through a set of stages, conceptually forming a system life cycle, with decision gates determining the completion of one stage and start of another. This part of the SE Handbook gives details for each of these parts of the system life cycle concept, as well as pointing out the role of the SE practitioner throughout a system's life cycle. Further details can be found in ISO/IEC/IEEE 24748–1 (2018).

2.1.1 Life Cycle Characteristics

As the introduction states, life cycles are defined in terms of the stages that mark progress in achieving the system characteristics. A commonly encountered set of life cycle stages is shown in Figure 2.1. These stages are also shown in ISO/IEC/IEEE 15288 (2023) and in ISO/IEC/IEEE 24748–1 (2018).

System life cycle stages can be entered based on the needs of the SoI or any system element. Stages can be entered into as many times as needed. Stages often are not sequential and can occur concurrently or as needed. Stages can overlap and stages can be entered at any point in the life cycle. The retirement stage does not require the entire SoI to be retired, it can be any system element, and retirement does not need to be in the order the systems are delivered.

INCOSE Systems Engineering Handbook: A Guide for System Life Cycle Processes and Activities, Fifth Edition.
Edited by David D. Walden, Thomas M. Shortell, Garry J. Roedler, Bernardo A. Delicado, Odile Mornas, Yip Yew-Seng, and David Endler.
© 2023 John Wiley & Sons Ltd. Published 2023 by John Wiley & Sons Ltd.

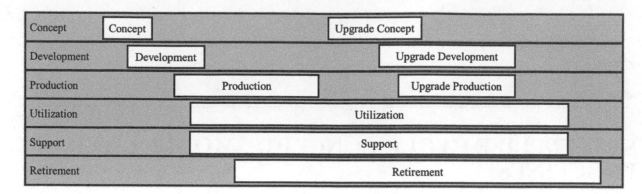

FIGURE 2.1   System life cycle stages. INCOSE SEH original figure created by Yokell. Usage per the INCOSE Notices page. All other rights reserved.

Typically, life cycle stages have both entry and exit decision gates. The entry decision gate is intended to help ensure that the entry criteria are met and the resources needed for the stage are available. The exit decision gate is intended to help ensure that the objectives of the stage have been achieved and the risk of going forward is acceptable. Decision gates are discussed in more detail in Section 2.1.3.

Figure 2.2 compares the generic life cycle stages to other life cycle viewpoints. Typical decision gates are represented along the bottom.

Major system elements may have their own life cycles. These life cycles have to be managed so that an integrated SoI is achieved and used over a span of time. When the SoI is, or is part of, an SoS (see Section 4.3.6), the influences from the evolution of the SoS need to be considered in the life cycle of the SoI. Each constituent system of the SoS has its own life cycle. Further, enabling systems (see Section 1.3.3) also have their own life cycles, which must be integrated with that of the SoI.

Requirements must be flowed down to the elements to be integrated and the decision gates should support progressive integration into the final SoI in a timely manner to help ensure that the elements can be progressively integrated. The decision gates associated with the various life cycle models should be synchronized, whatever the types of system element or parts of the life cycle are involved to support progressive integration into the final SoI.

Note that the above figures are notional and do not attempt to scale the relative time spans of the stages. For example, a system could move from initial concept to a fielded system in a few years, then remain in utilization, being supported and possibly upgraded, for decades (e.g., jet aircraft, nuclear power facility, day care nursery). A different system could have a series of development efforts, each resulting in relatively short periods of utilization and retirement (e.g., mobile phone, consumer electronics). While that is of interest from a programmatic viewpoint, it is secondary to the rationale of breaking the life cycle into stages to allow decisions to be made at key points.

2.1.2   Typical Life Cycle Stages

As shown in Figure 2.1, a system progresses through various life cycle stages that span the conception, development, production, utilization, support, and retirement of the SoI. This section highlights specific characteristics of each life cycle stage. Note that other life cycle models use different names for the stages and the associated characteristics of the stage. For other types of stages, such as those illustrated in Figure 2.2, the discussion here needs to be adapted as appropriate. Additional discussion of life cycle stage characteristics is in ISO/IEC/IEEE 24748–1 (2018).

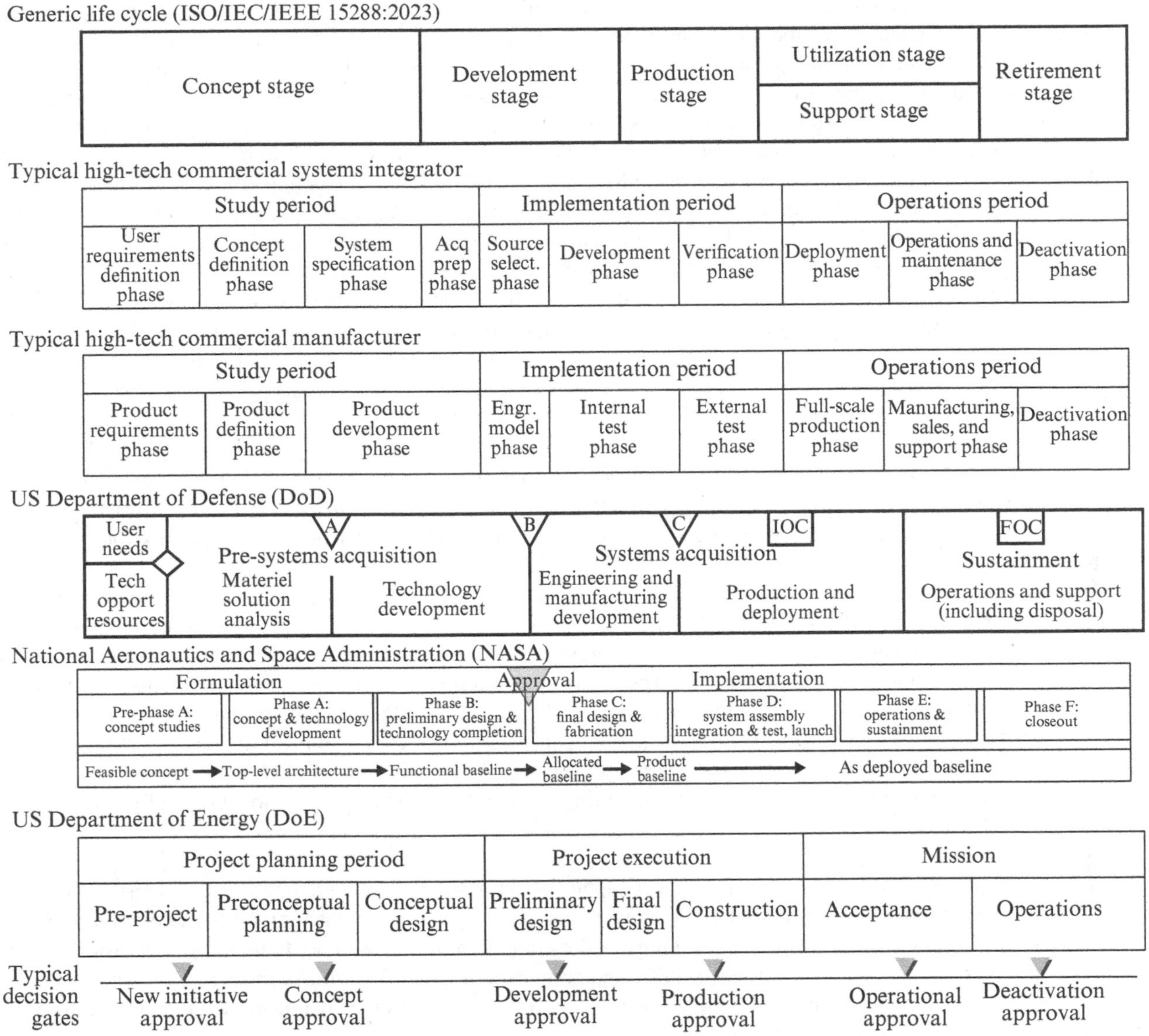

Generic life cycle (ISO/IEC/IEEE 15288:2023)
Concept stage
Development stage
Production stage
Utilization stage
Support stage
Retirement stage
Typical high-tech commercial systems integrator
Study period
Implementation period
Operations period
User requirements definition phase
Concept definition phase
System specification phase
Acq prep phase
Source select. phase
Development phase
Verification phase
Deployment phase
Operations and maintenance phase
Deactivation phase
Typical high-tech commercial manufacturer
Study period
Implementation period
Operations period
Product requirements phase
Product definition phase
Product development phase
Engr. model phase
Internal test phase
External test phase
Full-scale production phase
Manufacturing, sales, and support phase
Deactivation phase
US Department of Defense (DoD)
User needs
Tech opport resources
A
B
C
IOC
FOC
Pre-systems acquisition
Materiel solution analysis
Technology development
Systems acquisition
Engineering and manufacturing development
Production and deployment
Sustainment
Operations and support (including disposal)
National Aeronautics and Space Administration (NASA)
Formulation
Approval
Implementation
Pre-phase A: concept studies
Phase A: concept & technology development
Phase B: preliminary design & technology completion
Phase C: final design & fabrication
Phase D: system assembly integration & test, launch
Phase E: operations & sustainment
Phase F: closeout
Feasible concept
Top-level architecture
Functional baseline
Allocated baseline
Product baseline
As deployed baseline
US Department of Energy (DoE)
Project planning period
Project execution
Mission
Pre-project
Preconceptual planning
Conceptual design
Preliminary design
Final design
Construction
Acceptance
Operations
Typical decision gates
New initiative approval
Concept approval
Development approval
Production approval
Operational approval
Deactivation approval

FIGURE 2.2 Generic life cycle stages compared to other life cycle viewpoints. Derived from Forsberg, et al. (2005) with permission from John Wiley & Sons. All other rights reserved.

Concept Stage The concept stage can include exploratory research and begins with recognition of a need for a new or modified mission or business capability. Unless the solution is immediately at hand, which is the first thing to analyze, new potential solutions will need to be sought. Exploration needs to address both short- and long-range factors, including technical, economic, market, and resource considerations, including human resources. Surveys, trade-off studies, business or mission analyses, and other means of exploring the solution space are used. It is key that the problem space is clearly defined (existing issue or new opportunity), the solution space is characterized, business or mission requirements, and stakeholder needs and requirements are identified. From this, an estimate of the cost, schedule, and performance across the life cycle can be derived. Throughout the concept stage, it is critical to perform ongoing and robust assessment and management of risks. Getting feedback from current and potential stakeholders (e.g., customers, users, suppliers) significantly aids in developing solution concepts. The maturity and availability of enabling systems over the system life cycle must also be considered.

Typical outputs from the concept stage include preliminary concept artifacts (e.g., Operational Concept (OpsCon), Support Concept), SE methodology approach considerations, feasibility assessments (e.g., models, simulations, prototypes), preliminary architecture solutions, and stakeholder requirements. The concept stage could create key preliminary system requirements and could outline design solutions and acquisition strategies. Enabling systems are also addressed, as are first estimates of cost and schedule over the whole life cycle.

The concept stage is a particularly critical part of the system life cycle because the decisions made during the stage will shape, with increasing difficulty to change, the possibilities for all the remaining stages. It is difficult to project the possibilities for as-yet untried solutions, though these may provide the greatest long-term benefit. At the same time, it is easy to fall into the trap of projecting incremental changes to what has worked in the past and is used now, which can significantly limit the future possibilities.

Development Stage   The development stage defines an SoI that meets its agreed-to stakeholder needs and requirements and can be produced, utilized, supported, and retired. System analyses, including trade-off analysis, as well as further modeling, simulation, and prototyping are performed to achieve system balance and to optimize the design for key parameters.

The main aspect of the development stage is to mature the system concepts and stakeholder needs and requirements into an engineering baseline that can be produced, utilized, and supported over the desired span of its useful life, and finally retired in a responsible manner. The goal is not perfection, but rather to adequately meet the stakeholder needs and requirements in a manner that is supportable. The engineering baseline includes system requirements, architecture, design, documentation, and plans for subsequent stages. Outputs can include an SoI prototype, enabling system requirements (or the enabling systems themselves), plans for integration, verification, validation, transition, acquisition, logistics support, risk management, staffing and training, and detailed cost estimates and schedules for future stages. These outputs can occur incrementally, supporting a phased realization of the SoI, especially for complex systems.

Production Stage   The production stage begins with approval to translate the baselines of the development stage into an actual system, or those parts of the SoI where approval is given (which is not uncommon for a complex system). The approval includes the enabling systems and must address all areas of the baseline. In this stage, the SoI becomes reality, is qualified for use, and is ready for installation and transition under the utilization stage. The outputs of this stage are the realized portions of the SoI (with its enabling systems) as well as the documentation that will go forward for use in the utilization, support, and retirement stages.

Utilization Stage   The utilization stage begins with the transition to use of a system, or the parts of a system approved for use. This includes any enabling systems that will support use of the system being used in its intended environment to provide its intended capabilities. Product modifications are often introduced throughout the utilization stage, which generally is much longer than the other stages. Such changes can remedy deficiencies, enhance the capabilities, or extend the life of the system. Throughout, it is critical to maintain documentation from prior stages, as well as to ensure that Technical Management Processes, such as Configuration Management and Risk Management, and SE support remain in place and are robustly applied. The utilization stage proceeds in parallel with the support stage and ends, possibly by steps for different parts of the SoI, with the retirement stage.

Support Stage   The support stage begins with provisioning of support for the SoI's utilization. Planning and acquisition actions for the system support are often taken before utilization is allowed to start. In this stage, deficiencies and failures are noted and used as the basis for either remediation of the problems, or to build a case for evolutionary modification. Modifications may be proposed to resolve supportability problems, to reduce operational costs, or to extend the life of a system. These changes require SE assessments to avoid loss of system capabilities while under operation, or violation of non-performance related requirements. The support stage ends when a decision is made that the system is at the end of its useful life or that it should no longer be supported.

Retirement Stage   The retirement stage is where the system or a system element and its related services are removed from operation. SE activities in this stage are primarily focused on ensuring that disposal requirements, which can be extensive, are satisfied. However, it is often of value to ensure that documentation generated during at least the utilization and support stages is archived. That information can be invaluable when belated recognition arises that there is a need for new system.

Planning for retirement is part of the system definition during the concept and development stages. Experience has repeatedly demonstrated the consequences when system retirement is not considered from the outset. Early in the twenty-first century, many countries have changed their laws to hold the developer of a SoI accountable for proper end-of-life disposal of the system.

2.1.3 Decision Gates

It is good practice to have risk-managing decision points that occur at the beginning and end of each stage. This approach ensures that progress is gated by specific decision points that are clearly visible. These decision points help ensure the readiness to proceed with a stage and that the stage accomplishes is objective as it finishes. They often take place within the context of "project milestones," "project reviews," or "milestone reviews." Key is to help ensure that decisions are clearly made and documented and that they relate directly to the criteria established to begin or end a particular stage of a system's life cycle. Note that some approaches, such as agile (see Section 4.2.2), accomplish their decision points in a different cadence and tend to avoid the terms "milestones" and "decision gates." In agile development, frequent interaction with stakeholders can change the frequency (more frequent) and scope (smaller scope), and formality (less formal) of decision gates.

Typical goals of decision gates are to confirm that:

- increase in system maturity is within the defined threshold;
- the project deliverables satisfy the business case;
- the resources are sufficient to for the stage and subsequent stages;
- unresolved issues that need to be addressed in that stage are addressed; and
- overall risk for proceeding forward in the system life cycle is acceptable.

As shown in Figure 2.3, decision criteria can also include stage entry/exit criteria, entry/exit criteria from other stages, and risk assessment. Figure 2.3 shows the following cases:

- the entry criteria are met, but the start of the stage is delayed;
- when the entry criteria are met, the decision to start the stage is made;
- although the entry criteria are not met, the stage is started;
- although the exit criteria are met, the decision to end the stage is delayed;
- when the exit criteria are met, the decision to end the stage is made;
- the decision to end the stage is made before the exit criteria are met.

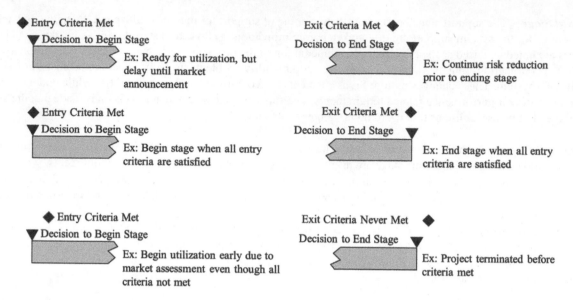

FIGURE 2.3   Criteria for decision gates. INCOSE SEH original figure created by Yokell. Usage per the INCOSE Notices page. All other rights reserved.

At each decision gate, the options can include:

- Begin subsequent stage or stages;
- Continue this stage, possibly after some reformulation;
- Go to or restart a preceding stage;
- Hold the project activity;
- Terminate the project.

The option selected depends on the quality of the results of the effort so far (based on the answers to the exit questions) plus the risk of moving forward (based on the answers to the entrance questions). Stages do not need to be sequential. Transitions often occur, but it is common to have stages occurring concurrently. In complex systems, the decision can also be more differentiated. For example: move part of the effort forward; hold on some; and terminate or reform others.

Decision gate approval follows review by qualified experts, involved stakeholders, and management. Approval should be based on evidence of compliance to the criteria of the review. Balancing the formality and frequency of decision gates is seen as a critical success factor for all SE process areas. The consequences of conducting a superficial review, omitting a critical discipline, or skipping a decision gate are usually long-term and costly.

It is important to note that there may be significant changes in the project's environment. This may impact the project's business case, system scope, or resources needed. Consequently, the related decision criteria should be updated and evaluated at every decision gate. Inadequate consideration can set up subsequent failures—usually a major factor in cost overruns and delays.

Upon successful completion of a decision gate, some artifacts (e.g., documents, analysis results, models, or other products of a system life cycle stage) will have been approved as the basis upon which future work must build. These artifacts are placed under configuration management along with the decisions made and the associated rationale and assumptions (see Section 2.3.4.5).

2.1.4   Technical Reviews and Audits

Technical reviews and audits are used to assess technical progress, coordinate activities, and determine the technical status of a system of interest (SoI). According to ISO/IEC 24748–8 / IEEE 15288.2 (2014):

> A technical review is "a series of systems engineering activities conducted at logical transition points in a system life cycle, by which the progress of a [project] is assessed relative to its technical requirements using a mutually agreed-upon set of criteria" and

> An audit is "a detailed review of processes, product definition information, documented verification of compliance with requirements, and an inspection of products to confirm that products have achieved their required attributes or conform to released product configuration definition information."

The technical reviews and audits to be performed occur throughout the system life cycle and should be captured in the project's Systems Engineering Management Plan (SEMP) and reflected in the project's schedule (see Section 2.3.4.1). They may be part of a decision gate review (see Section 2.1.3). A representative set of technical reviews and audits are listed in Table 2.1. They should be tailored for the needs of the project and the methodologies being used. ISO/IEC/ IEEE 24748–1 (2018), Annex F and ISO/IEC 24748–8 / IEEE 15288.2 (2014) provide useful guidance for the planning and tailoring of reviews to the needs of the project and its stakeholders.

Figure 2.4 depicts the relationship between these reviews and audits identified in ISO/IEC 24748–8 / IEEE 15288–2 (2014) and the typical technical baselines across the system life cycle applicable for a sequential life cycle model. This depiction will vary significantly for incremental life cycle models.

TABLE 2.1   Representative technical reviews and audits

Defense Projects per ISO/IEC/IEEE 24748-8/IEEE 15288.2 (2014)	Space Projects per NASA (2007b)	Incremental Commitment Spiral Model per Boehm, et al. (2014)
Alternative Systems Review (ASR)	Mission Concept Review (MCR)	Exploration Commitment Review (ECR)
System Requirements Review (SRR)	System Requirements Review (SRR)	Valuation Commitment Review (VCR)
System Functional Review (SFR)	Mission Definition Review (MDR)	Foundation Commitment Review (FCR)
Preliminary Design Review (PDR)	System Definition Review (SDR)	Development Commitment Review$_n$ (DCR$_n$)
Critical Design Review (CDR)	Preliminary Design Review (PDR)	Operations Commitment Review$_n$ (OCR$_n$)
Test Readiness Review (TRR)	Critical Design Review (CDR)	
Functional Configuration Audit (FCA)	System Integration Review (SIR)	
System Verification Review (SVR)	Operational Readiness Review (ORR)	
Production Readiness Review (PRR)	Flight Readiness Review (FRR)	
Physical Configuration Audit (PCA)	Mission Readiness Review (MRR)	
	Post-Launch Assessment Review (PLAR)	
	Critical Events Readiness Review (CERR)	
	Post-Flight Assessment Review (PFAR)	
	Decommissioning Review (DR)	
	Disposal Readiness Review (DRR)	

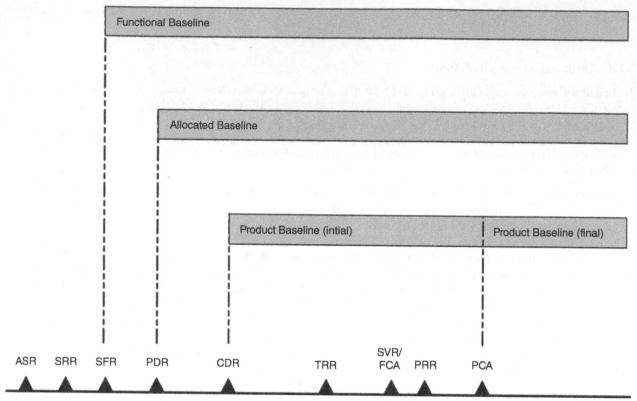

FIGURE 2.4   Relationship between technical reviews and audits and the technical baselines. From ISO/IEC 24748–8 / IEEE 15288.2 (2014). Used with permission. All other rights reserved.

Technical reviews and audits provide an opportunity to assess the following:

- The SoI is meeting its requirements
- The SoI is meeting stakeholder expectations, internal and external
- The SoI will have acceptable quality characteristics (QCs)
- The SoI is at an appropriate level of maturity
- The SoI is at an acceptable level of risk
- There is a clear path toward verifying and validating the SoI and its elements

Good practices for technical reviews and audits include:

- Plan the review or audit, including getting concurrence on a mutually agreeable location and date
- Application of multiple instances of the reviews and audits, both at multiple levels of the systems hierarchy and during each increment or iteration
- Elimination of unnecessary reviews or audits
- Establish clear entry and exit criteria for each review and audit
- Establish roles and responsibilities for the preparation, conduct, and acceptance of each review

- Make the reviews be risk-driven (risk is at an acceptable level) or event-driven (the entry criteria has been satisfied), not schedule-driven (must happen on a certain date)
- Consider "dry-runs" to make the review as efficient as possible
- Include subject matter experts (SMEs) and independent reviewers
- Include all members of the team, including acquirers and suppliers
- Capture clear actions, with ownership and due dates, for all issues that arise
- Follow up on actions that were raised

Each technical review or audit should include knowledgeable participants as well as participants with sufficient objectivity to assess satisfaction of the pre-established review criteria. Based on the purpose and level of the review, the participants may include representatives from the acquirer or supplier organizations, or both. A list of possible participants is provided below:

- Project Manager
- Lead SE Practitioner / Chief Engineer / Lead Engineer
- Review or Audit Chair
- Recorder (person charged with capturing the results of the review or audit)
- Acquirer Representative(s)
- Supplier Representative(s)
- Project Verification and Validation Lead
- Other Technical Leads

2.2 LIFE CYCLE MODEL APPROACHES

Section 2.1 introduces the concept of life cycle stages. The life cycle models are thus the framework within which the individual life cycle stages and transitions between them are planned and implemented. There are many different life cycle models, each suitable for different situations. A common way to differentiate them is to divide the life cycle model approaches into three groups: sequential, incremental, and evolutionary. Figure 2.5 provides the general concept for each of these approaches, and Table 2.2 summarizes their distinguishing characteristics.

ISO/IEC/IEEE 24748–1 (2018) provides further information on sequential (identified as "once-through"), incremental, and evolutionary life cycle model approaches. Sections 2.2.1 to 2.2.3 elaborate on how these approaches can be applied to manage the work within each life cycle stage.

There are many factors that help determine which life cycle models are suitable for a specific system or project. Clause 6 of ISO/IEC/IEEE 24748–1 (2018) provides informational considerations that may influence the selection and adaptation of life cycle model, including:

a) stability of, and variety in, operational environments;
b) risks, commercial or performance, to the concern of stakeholders;
c) novelty, size, and complexity;
d) starting date and duration of utilization;
e) integrity issues such as safety, security, privacy, usability, availability;
f) emerging technology opportunities;
g) profile of budget and organizational resources available;
h) availability of the services of enabling systems;

i) roles, responsibilities, accountabilities, and authorities in the overall life cycle of the system;

j) the need to conform to other standards.

Other sources define characteristics and tailoring factors that can be used to guide tailoring. As an example, Project Management Institute (PMI) has published their Situation Context Framework (SCF) (2022) that "defines how to select and tailor a situation-dependent strategy for software development. The SCF is used to provide context for organizing your people, process, and tools for a software-based solution delivery team." Seven dimensions (team size,

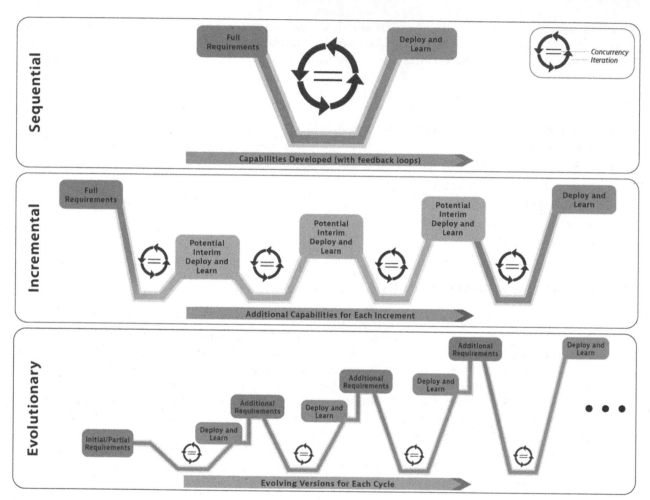

FIGURE 2.5   Concepts for the three life cycle model approaches. INCOSE SEH original figure created by Endler. Usage per the INCOSE NOTICES page. All other rights reserved.

TABLE 2.2   Life cycle model approach characteristics

Life cycle approach	Requirement set at start	Planned iterations	Multiple deployments
Sequential	Full	Single	No
Incremental	Full	Multiple	Potential
Evolutionary	Partial	Multiple	Typically

INCOSE SEH original table created by Endler derived from ISO/IEC/IEEE 24748-1 (2018). Usage per the INCOSE Notices page. All other rights reserved.

geographic distribution, organizational distribution, skill availability, compliance, domain complexity, and solution complexity) with scaling factors in each dimension are defined within this framework.

As there is no universal approach, it is recommended that each organization continuously questions itself as to which approach, or combination of approaches, is most suitable. Part IV of this handbook addresses tailoring and application considerations in more detail.

2.2.1 Sequential Methods

The sequential approach is focused on the general flow of the processes with feedback loops, but a single delivery. Sequential life cycle models break down SE activities into linear sequential stages, where each stage depends on the deliverables of the previous stages, along with feedback from subsequent stages.

On projects where it is necessary to coordinate large teams of people working in multiple companies, sequential approaches provide an underlying framework to provide discipline to the life cycle processes. Sequential life cycle models are characterized by a systematic approach that adheres to specified processes as the system moves through a series of representations from requirements through design to finished product. Specific attention is given to the completeness of documentation, traceability from requirements, and verification of each representation after the fact.

The strengths of sequential life cycle models are predictability, stability, repeatability, and high assurance. Process improvement focuses on increasing process capability through standardization, measurement, and control. These models rely on "master plans" to anchor their processes and provide project-wide communication. Historical data is usually carefully collected and maintained as inputs to future planning to make projections more accurate (Boehm and Turner, 2004).

The waterfall model, introduced by Royce (1970), was used to characterize the advantages and disadvantages of sequential approaches. The waterfall model has been used successfully in the manufacturing and construction industries, where the highly structured physical environments meant that design changes became prohibitively expensive much sooner in the development process. In addition, safety-critical products, such as the Therac-25 medical equipment (see the case study in Section 6.1), can only meet modern certification standards by following a thorough, documented set of plans and specifications. Such standards mandate strict adherence to process and specified documentation to achieve safety or security.

The SE Vee model (named due to its shape representing the letter "V" in the English language), introduced in Forsberg and Mooz (1991), described in Forsberg, et al. (2005), and shown in Figure 2.6, is another example of a sequential approach used to visualize key areas for SE focus, associating each development stage with a corresponding testing stage. The Vee highlights the need for continuous validation with the stakeholders, the need to define verification plans during requirements development, and the importance of continuous risk and opportunity assessment.

There are several variations of the Vee model. Typically, the "left" side of the Vee is called system definition and the "bottom" and "right" side of the Vee are called system realization. In the Vee model, time and system maturity conceptually proceed from left to right (down the left side of the Vee and up the right side of the Vee). However, all the system life cycle processes are performed concurrently and iteratively at each level of the system hierarchy and all the system life cycle processes are recursively applied at each level of the system hierarchy (see Section 2.3.1.2). One of the strengths of the Vee model is its depiction of the relationships between the left and right sides of the Vee.

The left side of the Vee depicts the evolving baseline from stakeholder requirements, to system requirements, to the identification of a system architecture, to definition of elements that will comprise the final system. The development team then can move from the highest level of the system requirements down to the lowest level of detail. Risk and opportunity management investigations address development options to provide assurance that the baseline performance being considered can indeed be achieved and to initiate alternate concept studies at the lower levels of detail to determine the best approach. Stakeholder discussions (in-process validation) occur to ensure that the proposed baselines are acceptable to the organization, customer, user, and other stakeholders. Changes to enhance system performance or to reduce risk or cost are welcome for consideration, but after baselining these must go through formal change control, since others may be building on previously defined and released design decisions. The bottom of the Vee depicts either the recursive application of the systems life cycle processes at the next level of the system hierarchy or the implementation of atomic system elements (see Section 1.3.5). The broadening at the base of the figure shows the growth in the

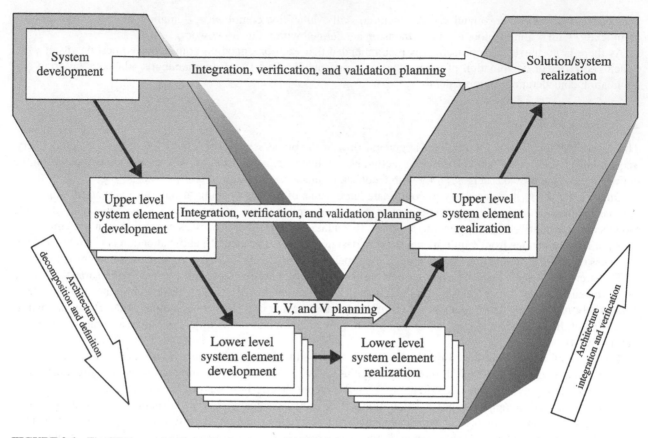

FIGURE 2.6   The SE Vee model. From Forsberg, et al. (2005) with permission from John Wiley & Sons. All other rights reserved.

number of system elements. Note that system elements can also be bought or reused. The right side of the Vee depicts the evolving baseline of system elements that are implemented, integrated, verified, and validated. In each stage of the system life cycle, the SE processes iterate to ensure that a concept or design is feasible and that the stakeholders remain supportive of the solution as it evolves.

ISO/IEC/IEEE 24748–2 (2018), Clause 6.4.3.1 provides further details on sequential life cycle models, including typical applicable systems as well as risks and opportunities associated with these models.

2.2.2   Incremental Methods

Incremental approaches have been in use since the 1960s (Larman and Basili, 2003). They represent a practical and useful approach that allows a project to provide an initial capability (or a limited set of capabilities) followed by successive deliveries to reach the desired SoI. The goal of an incremental approach is to provide rapid value and responsiveness. Generally, each increment adds capabilities intended to converge on a stakeholder satisfying result for the increment. Based on a set of requirements, a candidate set of increments is defined and the initial increment is initiated. Subsequent increments are initiated, and the process is repeated, until a complete system is deployed or until the organization decides to terminate the effort. Intermediate increments can potentially be deployed to support learning.

An incremental approach works well when an organization intends to market new versions of a product at planned intervals. Typically, the capabilities of the final delivery are known at the beginning. However, as there is significant technical risk, the development of the capabilities is performed incrementally to allow for the latest technology insertion or potential changes in needs or requirements. A core part of the planning process for an incremental approach establishes the cycle times for increments. Increments are beneficially timed in development projects to accommodate coordinated events such as integration testing and evaluation, capability deployment, experimental deployment, or release to production. Iteration cycles are beneficially timed to minimize rework cost as a project learns experimentally and empirically. Project planning and management often benefit from a constant cadence among increments.

One example of an incremental approach is the Incremental Commitment Spiral Model (ICSM) (Boehm, et al., 2014), which extends the classic Spiral Model for software introduced in Boehm (1987) for SE. A view of the ICSM is shown in Figure 2.7. In the ICSM, each increment addresses requirements and solutions concurrently, rather than sequentially. ICSM also considers products and processes; hardware, software, and human aspects; and business case analyses of alternative product configurations or product line investments. The stakeholders consider the risks and risk mitigation plans and decide on a course of action. If the risks are acceptable and covered by risk mitigation plans, the project proceeds into the next spiral (increment).

ISO/IEC/IEEE 24748–2 (2018), Clause 6.4.3.2 provides further details on incremental life cycle models, including typical applicable systems as well as risks and opportunities associated with these models.

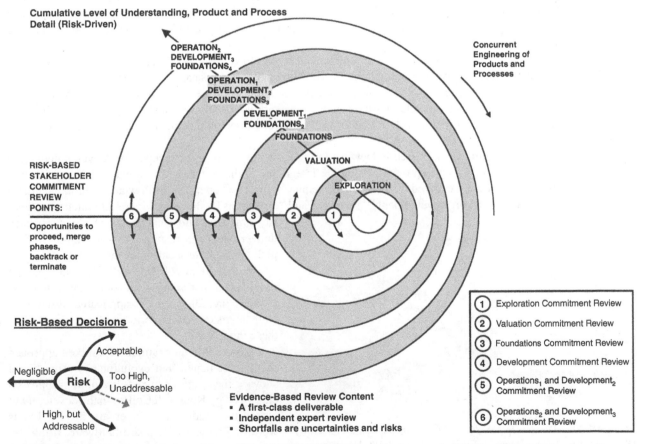

FIGURE 2.7   The Incremental Commitment Spiral Model (ICSM). From Boehm, et al. (2014) with permission from Pearson Education. All other rights reserved.

2.2.3   Evolutionary Methods

In the sequential and incremental approaches described previously, the full set of required capabilities of the final system is assumed to be mostly known at the start of the effort. In some cases, especially in novel systems, the final system requirements may be unknown or only partially known. An evolutionary approach provides the adaptability and flexibility needed for the development in these situations. For example, the high-temperature tiles of the NASA space shuttle were developed using an evolutionary approach (Forsberg, 1995). An evolutionary approach is often used in research and development (R&D) projects and SoS developments. Software development efforts are increasingly using agile methods, which are a type of evolutionary development.

In evolutionary approaches, cycles are typically planned on a regular periodic basis, each resulting in a deployable version. The requirements for the SoI are typically only partially known and are increasingly refined with each cycle. At the beginning, the goal of each cycle may be more or less unknown. Therefore, it is particularly important that the experience gained with the earlier cycles is taken into account for the subsequent cycles. Similar to the incremental approach, versions may be developed sequentially or in parallel. This is a particular challenge for those involved in the project, since new capabilities are typically assigned to exactly one version. If this assignment is lost or becomes unclear, this leads to confusion and negatively impacts the schedule and cost targets. Thus, a well-functioning configuration control is essential, also since multiple versions can be operated and supported simultaneously (see Section 2.3.4.5). Aspects to be considered include operating manuals, maintenance instructions, spare parts, disposal instructions, etc.

The evolutionary approach offers significant advantages if it is possible to obtain steady and high-quality feedback from relevant stakeholders. For example, the first versions can be used to demonstrate basic feasibility, such as a minimal viable product (MVP), and facilitate market entry. Likewise, emerging technical innovations can be planned for later versions.

When developing subsequent versions, it is recommended to carefully examine whether the previous versions should be completely replaced by newer ones. Alternatively, subsequent versions can be developed such that a partial or even complete upgrade of the previous versions to the new version is possible. For this, it is necessary that these things are considered during the early cycles. Criteria such as adaptability, flexibility, and modularity should be carefully considered to enable the long-term evolution of the system. Decisions should be made in the context of life cycle cost (see Section 3.1.2).

An example of an evolutionary approach is DevOps (a blend of the terms and concepts for "development" and "operations"). The goal of DevOps is to provide continuous integration of the system and continuous delivery of capabilities. DevOps is typically characterized by three key principles: shared ownership, workflow automation, and rapid feedback. DevSecOps (a blend of "development," "security," and "operations"), shown in Figure 2.8, integrates security practices into DevOps. In DevSecOps, each delivery team is responsible and empowered to pick appropriate security means.

ISO/IEC/IEEE 24748–2 (2018), Clause 6.4.3.4 provides further details on evolutionary life cycle models, including typical applicable systems as well as risks and opportunities associated with these models.

Figure 2.9 is an example of a mixed approach (both incremental and evolutionary). This figure shows the agile mixed-discipline approach employed by Rockwell Collins in the development of military radios (Dove, et al., 2017). Teams working on electronic-board hardware, firmware, and software have different timings for hardware increments and firmware and software epics

FIGURE 2.8   DevSecOps INCOSE SEH original figure created by D'Souza derived from Banach (2019) and Anx (2021). Usage per the INCOSE Notices page. All other rights reserved.

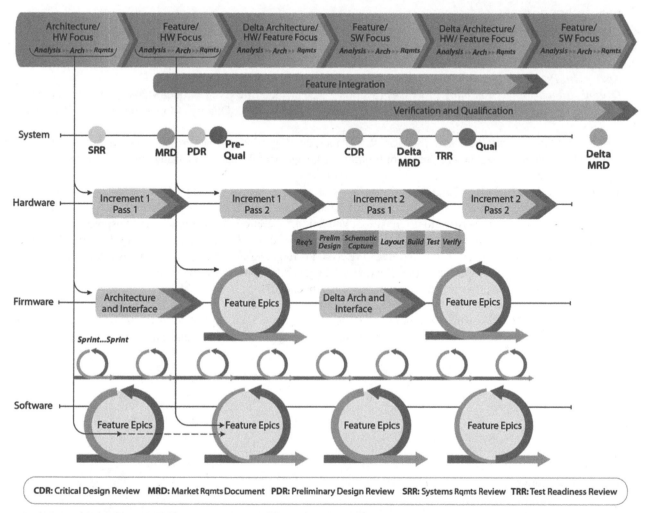

FIGURE 2.9 Asynchronous iterations and increments across agile mixed discipline engineering. From Dove, et al. (2017). Used with permission. All other rights reserved.

(versions). The teams accomplish integrated work-in-process testing with the latest increments and versions from each of the disciplines.

All these life cycle approaches are supported by the processes defined in ISO/IEC/IEEE 15288 (2023) and this handbook. The life cycle model should be chosen so that it is sufficiently adaptable and flexible. Section 4.1 provides more information on tailoring life cycle models.

2.3 SYSTEM LIFE CYCLE PROCESSES

2.3.1 Introduction to the System Life Cycle Processes

A process is a series of activities and tasks performed to achieve one or more outcomes for a stated purpose. In SE, the system life cycle processes are one of the enablers to help manage a system solution across the life cycle stages. The processes are intended to be applied concurrently, iteratively, and recursively with other enablers (e.g., tools, technology) throughout the stages of the life cycle (see Section 2.3.1.2).

ISO/IEC/IEEE 15288 (2023) identifies four process groups for the system life cycle, providing "a common process framework for describing the life cycle of engineered systems, adopting a Systems Engineering approach." Each of these process groups is the subject of a section within Part 2. A graphical overview of these processes is given in Figure 2.10:

Agreement Processes (Section 2.3.2) include Acquisition and Supply.

Organizational Project-Enabling Processes (Section 2.3.3) include Life Cycle Model Management, Infrastructure Management, Portfolio Management, Human Resource Management, Quality Management, and Knowledge Management.

Technical Management Processes (Section 2.3.4) include Project Planning, Project Assessment and Control, Decision Management, Risk Management, Configuration Management, Information Management, Measurement, and Quality Assurance.

Technical Processes (Section 2.3.5) include Business or Mission Analysis, Stakeholder Needs and Requirements Definition, System Requirements Definition, System Architecture Definition, Design Definition, System Analysis, Implementation, Integration, Verification, Transition, Validation, Operation, Maintenance, and Disposal.

The application of these system life cycle processes is supported by SE practitioners having the relevant competencies. The competencies are defined in the INCOSE Systems Engineering Competency Framework (SECF) (2018). Note that the professional competencies (see Section 5.1.2) generally apply to all the processes.

Note: Acronyms for the process names are provided in Appendix D.

2.3.1.1 *Format and Conventions*

A common section structure has been applied to describe the system life cycle processes in this handbook. The following structure provides consistency in the discussion of these processes:

Process overview
  Purpose
  Description
  Inputs/outputs
  Process activities
  Common approaches and tips
Process elaboration

To ensure consistency with ISO/IEC/IEEE 15288, the purpose statements from the standard are included verbatim for each process described herein. The titles of the process activities listed in each section are also consistent with ISO/IEC/IEEE 15288. The process activities describe "what" should be done, not "how" to do it. In some cases, additional items have been included to provide summary-level information regarding industry good practices and evolutions in the application of SE processes. Process elaborations provide additional details on topics that are unique to the specific life cycle process. See Part III for topics that cross-cut multiple life cycle processes.

In addition, each system life cycle process is illustrated by an input–process–output (IPO) diagram showing typical inputs, process activities, and typical outputs. A sample is shown in Figure 2.11. To understand a given process, readers are encouraged to study the complete information provided in the combination of figures and text and not rely solely on the IPO diagrams.

Typical inputs and outputs are listed by name within the respective IPO diagrams with which they are associated. The typical inputs and outputs are consistent with ISO/IEC 33060 (2020) when possible. Note that the IPO diagrams throughout this handbook represent "a" way that the SE processes can be performed, but not necessarily "the" way that they must be performed. The system life cycle processes produce "results" that are often captured in "documents" or "artifacts" or "models," rather than producing "documents" simply because they are identified as outputs. A complete list of all inputs and outputs with their respective descriptions appears in Appendix E.

The controls and enablers shown in Figure 2.11 govern all processes described herein and, as such, are not repeated on the subsequent IPO diagrams. Typically, IPO diagrams do not include controls and enablers, but since they are not

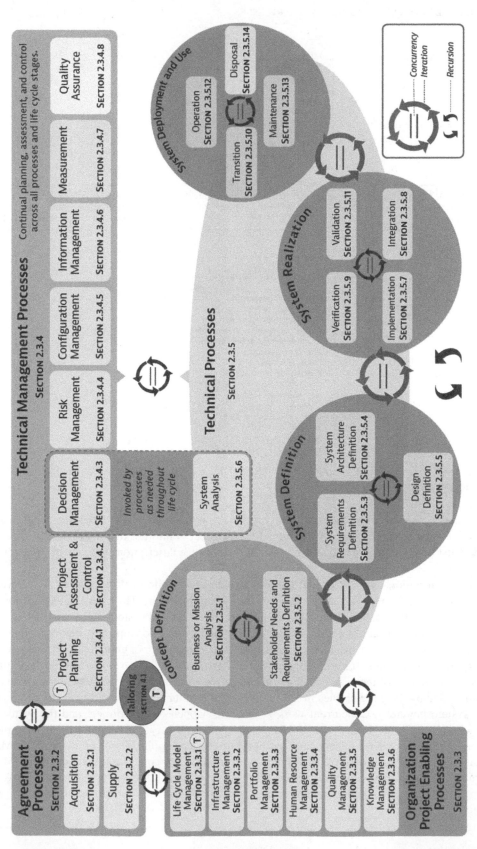

FIGURE 2.10 System life cycle processes per ISO/IEC/IEEE 15288. INCOSE SEH original figure created by Roedler and Walden. Usage per the INCOSE Notices page. All other rights reserved.

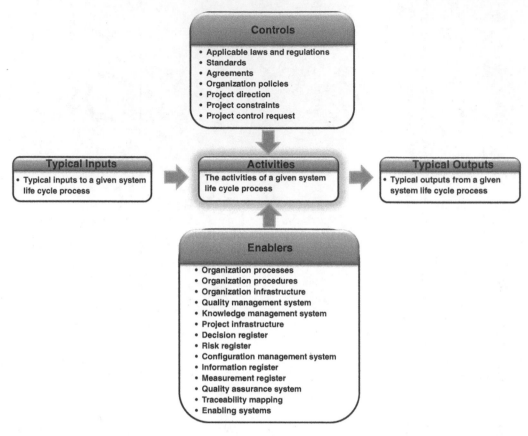

FIGURE 2.11 Sample IPO diagram for SE processes. INCOSE SEH original figure created by Walden, Shortell, and Yip. Usage per the INCOSE Notices page. All other rights reserved.

repeated in the IPO diagrams throughout the rest of the handbook, we have chosen to label them IPO diagrams. The enablers work together with the inputs to be transformed by the process into the outputs under the direction of the controls. A complete list of all controls and enablers with their respective descriptions appears in Appendix E.

2.3.1.2 Concurrency, Iteration, and Recursion Too often, the system life cycle processes are viewed as being applied in a sequential, linear manner at a single level of the system hierarchy. However, valuable information and insight need to be exchanged between the processes in order to ensure a good system definition that effectively and efficiently meets the stakeholder needs and requirements. The application of concurrency, iteration, and recursion to the system life cycle processes helps to ensure communication that accounts for ongoing learning and decisions. This facilitates the incorporation of learning from further analysis and process application as the technical solution evolves. Figure 2.12 shows an illustration of the concurrent, iterative, and recursive nature of the system life cycle processes.

Concurrency (indicated by the parallel lines in the figure) is the parallel application of two or more processes at a given level in the system hierarchy. Concurrent work is likely to happen on any project and the system life cycle processes can likewise be performed in a concurrent manner. It is not necessary for processes to be performed serially, especially when one process is not dependent on another for information or results. For example, the Risk Management process and Measurement process usually are performed in a continual and concurrent manner. This is illustrated in Figure 2.34, in which both of these processes occur concurrently, yet provide information to one another. Additionally, the system architecture should enable concurrency through modularization, encapsulation, commonality/reuse, and other design methods.

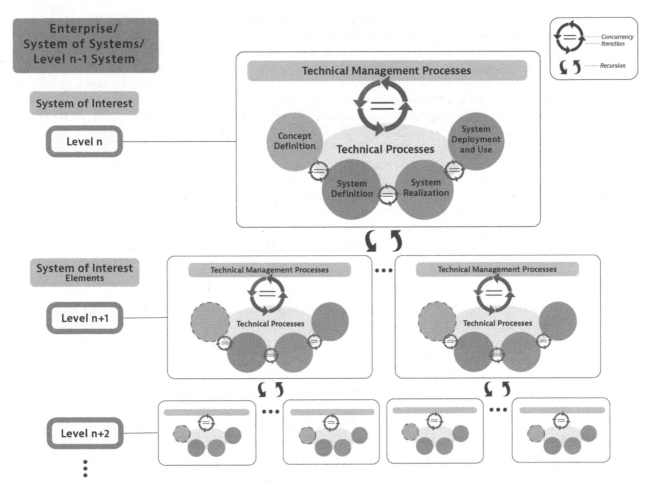

FIGURE 2.12   Concurrency, iteration, and recursion. INCOSE SEH original figure created by Roedler and Walden. Usage per the INCOSE Notices page. All other rights reserved.

Iteration (indicated by the circular arrows in the figure) is the repeated application of and interaction between two or more processes at a given level in the system hierarchy. Iteration is needed to accommodate stakeholder decisions and evolving understanding, account for architectural decisions or constraints, and resolve trade-offs for affordability, adaptability, feasibility, resilience, etc. There can be iteration between any of the processes. For example, there is often iteration between System Requirements Definition and System Architecture Definition processes. The system architecture will reflect these design iterations through identification of functions, their allocation to system elements, assignment to logical and physical interfaces, and verification as intended in the design. In this case, there is a concurrent application of the processes with iteration between them, where the evolving system requirements help to shape the architecture through identified constraints and functional and quality requirements. For example, the system architecture may need to be changed due to detailed electrical modeling indicating that a particular system element's load exceeds its allocated power budget and forces a design change or reallocation of the network power assignments in the overall system. The tradeoffs between candidate architectures or elements of the architecture, in turn, may identify requirements that are not feasible, driving further requirements analysis with trade-offs that change some requirements. Likewise, the Design Definition process could identify the need to reconsider decisions and trade-offs in the

System Requirements Definition or System Architecture Definition processes. Any of these can invoke additional application of the System Analysis and Decision Management processes.

Recursion (indicated by the down and up arrows in the figure) is the repeated application of the set of life cycle processes, tailored as appropriate, at successive levels in the system hierarchy. The Technical Management and Technical Processes are expected to be recursively applied for each successive level of the system hierarchy until the level is reached where the decision is made to make, buy, or reuse a system element (see Section 1.3.5). During the recursive application of the processes, the outputs at one level become inputs for the next successive level (below for system definition, above for system realization).

Horizontal integration ensures completeness before diving deeper and *vertical integration* ensures consistency between levels in this concurrent, iterative, and recursive environment (see Section 2.3.5.8). For example, one may have to define functions, their inputs and outputs, their associated performance and conditions of operations *before* writing the associated requirement. Then one can define the related verification (method, conditions, criteria), from which one can postulate the next-lower-level architecture to assess feasibility and perform the related function, performance, or requirements decomposition and allocation. However, teams may need to go down multiples levels to validate that the functions or elements at the lower levels are going to be suitable solutions for the SoI and its stakeholders.

2.3.2   Agreement Processes

The initiation of a project begins with the identification of a problem or opportunity to be addressed, which results in the development of needs to be satisfied. Once a need is identified and resources are committed to establish a project, it is possible to define the terms and conditions of an acquisition and supply relationship through the Agreement Processes, which are defined in ISO/IEC/IEEE 15288 as follows:

> [5.7.2] [Agreement] Processes define the activities necessary to establish an agreement between two organizations.

The Agreement Processes in this handbook and in ISO/IEC/IEEE 15288 are focused on the acquisition and supply of systems, system elements, products, or services, although agreements could be established for other objectives. With respect to the acquisition and supply relationship, the acquirer and supplier could be two independent organizations (i.e., no common parent organization or enterprise) or two organizations from the same parent organization or enterprise.

The Agreement Processes are utilized under many conditions, including when:

- an organization cannot satisfy a defined need itself,
- a supplier can satisfy a defined need in a more economical or timely manner,
- a higher authority has directed the use of a specific supplier, and
- an organization needs materials or specialized services.

An overall objective of Agreement Processes is to identify the interfaces between the acquirer and supplier(s) and establish the terms and conditions of these relationships, including identifying the inputs required and the outputs that will be provided.

Agreement negotiations are handled in various ways depending on the specific organizations and the formality of the agreement. In a formal agreement, there is usually a contract negotiation activity to refine the contract terms and conditions. Note also that the Agreement Processes can be used for coordinating within an organization between different business units or functions. In this case, the agreement will usually be more informal, not requiring a formal or specific contract.

An important contribution of ISO/IEC/IEEE 15288 is the recognition that SE practitioners are relevant contributors to the Agreement Processes (Arnold and Lawson, 2003). The SE practitioner is usually in a supporting role to the project management practitioner during negotiations and is responsible for impact assessments for changes, trade studies on alternatives, risk assessments, and other technical input needed for decisions.

Acceptance criteria are critical elements to each party because they protect both sides of the business relationship—the acquirer from being coerced into accepting a product with poor quality and the supplier from the unpredictable actions of an indecisive acquirer. It is important to note that the acceptance criteria are negotiated during the Agreement Processes. During negotiations, it is also critical that both parties are able to track progress toward an agreement. Identifying where further work toward achieving consensus in the documents and clauses is vital.

Two Agreement Processes are identified by ISO/IEC/IEEE 15288: the Acquisition process and the Supply process. These processes, subject of Sections 2.3.2.1 and 2.3.2.2, respectively, are two sides of the same coin. They conduct the essential business of the organization related to the SoI. They establish the relationships between organizations relevant to the acquisition and supply of products and services, regardless of whether the agreement is formal (as in a contract) or informal. Each process establishes the context and constraints of the agreement under which the other system life cycle processes belonging to the project scope are performed. Note that an organization can be both a supplier and acquirer for a given system. For example, an organization may be contracted to supply a system to an end customer. However, that organization may choose to acquire some of the system elements, materials, or services for developing or producing the system. So, that organization is the supplier to the end customer of the system and is the acquirer to those organizations providing it system elements, materials, or services.

Changes may happen during the execution of an agreement including acquirer change requests, deviations and waivers from the supply chain, or changes in the context of the project that were foreseen in risk analysis or not. Upon decision of the parties, this may lead to modifications to the initial state of the agreement. For that purpose, a statement of compliance may be initiated and updated all along the project describing the agreed changes and can include requirements impacted by a modification, the reference of modification, compliance verification by the supplier, and compliance validation by the acquirer.

2.3.2.1   *Acquisition Process*

Overview

Purpose As stated in ISO/IEC/IEEE 15288,

> [6.1.1.1] The purpose of the Acquisition process is to obtain a product or service in accordance with the acquirer's requirements.

The Acquisition process is invoked to establish an agreement between two organizations under which one party acquires products and/or services from the other. The acquirer experiences a need for an operational system, for services in support of an operational system, for elements of a system being developed by a project, or for services in support of project activities.

Description This section is written from the perspective of the acquirer organization. An acquiring organization applies due diligence in the selection of a supplier to avoid costly failures and impacts to the organization's budgets and schedules and other issues. Therefore, the role of the acquirer demands familiarity with the Technical, Technical Management, and Organizational Project-Enabling Processes, as it is through them that the supplier will execute the agreement.

Inputs/Outputs Inputs and outputs for the Acquisition process are listed in Figure 2.13. Descriptions of each input and output are provided in Appendix E.

Process Activities The Acquisition process includes the following activities:

- *Prepare for the acquisition.*
  - Develop and maintain acquisition policies, plans, and procedures to meet the organization strategies, goals, and objectives as well as the needs of the project management and SE organizations.

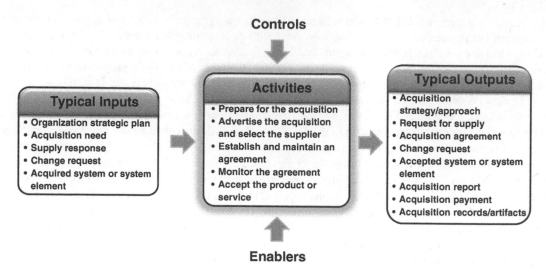

FIGURE 2.13   IPO diagram for the Acquisition process. INCOSE SEH original figure created by Shortell, Walden, and Yip. Usage per the INCOSE Notices page. All other rights reserved.

- Collect needs in a request for supply—such as a Request for Proposal (RFP) or Request for Quotation (RFQ) or some other mechanism—to obtain the supply of the service and/or product. Through the use of the Technical Processes, the acquiring organization produces a set of requirements and models that will form the basis for the technical information of the agreement.
- Identify a list of potential suppliers—suppliers may be internal or external to the acquirer organization.

• *Advertise the acquisition and select the supplier.*

- Distribute the request for supply and select appropriate suppliers—using selection criteria, rank suppliers by their suitability to meet the overall need and establish supplier preferences and corresponding justifications. Viable suppliers should be willing to conduct ethical negotiations, able to meet obligations, and willing to maintain open communications throughout the Acquisition process. Note that the approach may be less formal when a function within the organization is a candidate for the supply need.
- Evaluate supplier responses to the request for supply—ensure the offered product and/or service can meet acquirer needs and complies with industry and other standards. Assessments from the Project Portfolio Management (see Section 2.3.3.3) and Quality Management (see Section 2.3.3.5) processes and review results from the requesting organization are necessary to determine the suitability of each response and the ability of the supplier to meet the stated commitments. Record results from the evaluation of responses to the request for supply. This can range from formal documentation to less formal interorganizational interactions (e.g., between design engineering and marketing).
- Select the preferred supplier(s) based on acquisition criteria.

• *Establish and maintain an agreement.*

- Establish an agreement. Ensure an understanding of expectations, including acceptance criteria.
- This agreement ranges in formality from a written contract to a verbal agreement. Appropriate to the level of formality, the agreement establishes requirements, development and delivery milestones, verification, validation and acceptance conditions, process requirements (e.g., configuration management, risk management, measurements), exception-handling procedures, agreement change management procedures, payment schedules, and handling of data rights and intellectual property so that both parties understand the basis for executing the agreement. For a written contract, this occurs when the contract is signed.

- Identify the necessary changes to the agreement and evaluate the related impacts on the agreement.
- Update the agreement with the supplier as necessary.
- *Monitor the agreement.*
  - Manage Acquisition process activities, including decision making for agreements, relationship building and maintenance, interaction with organization management, responsibility for the development of plans and schedules, and final approval authority for deliveries accepted from the supplier.
  - Maintain communications with supplier, stakeholders, and other organizations regarding the project.
  - Status progress against the agreed-to schedule to identify risks and issues, to measure progress toward mitigation of risks and adequacy of progress toward delivery and cost and schedule performance, and to determine potential undesirable outcomes for the organization. The Project Assessment and Control process (see Section 2.3.4.2) provides necessary evaluation information regarding cost, schedule, and performance.
  - Amend agreements when impacts on schedule, budget, or performance are identified.
- *Accept the product or service.*
  - Accept delivery of products and services—in accordance with all agreements and relevant laws and regulations.
  - Render payment—or other agreed consideration in accordance with agreed payment schedules.
  - Accept responsibility in accordance with all agreements and relevant laws and regulations.
  - When an Acquisition process cycle concludes, a final review of performance is conducted to extract lessons learned for continued process performance.
  - Retire the agreement.

Note: The project is closed through the Portfolio Management process (see Section 2.3.3.3), which manages the full set of projects of the organization.

Common approaches and tips:

- Establish acquisition guidance and procedures that inform acquisition planning, including recommended milestones, standards, assessment criteria, and decision gates. Include approaches for identifying, evaluating, choosing, negotiating, managing, and terminating suppliers.
- Establish a technical point of responsibility within the organization for monitoring and controlling individual agreements. This person maintains communication with the supplier and is part of the decision-making team to assess technical development and progress in the execution of the agreement.

Note: There can be multiple points of responsibility for an agreement that focus on technical, programmatic, marketing, etc.

- Define and track measures that indicate progress on agreements. Avoid measures that are not focused on the true information needs. Leading indicators are preferable (see Section 2.3.4.7).
- Include technical representation in the selection of the suppliers to critically assess the capability of the supplier to perform the required task.
- Past performance of the supplier is highly important, but changes to key supplier personnel should be identified and evaluated to understand any impact with respect to the current request for supply.
- Communicate clearly with the supplier about priorities and avoid conflicting statements or making frequent changes in the statement of need that introduce risk into the process.
- Maintain traceability between the supplier's responses to the acquirer's solicitation. This can reduce the risk of contract modifications, cancellations, or follow-on contracts to fix the product or service.

- Smart contracts can be used to establish and maintain an agreement. A smart contract is a transaction protocol intended to execute automatically and control or document legally relevant events and actions according to the terms of a contract or an agreement (Tapscott and Tapscott, 2018). The objectives of smart contracts are the reduction of need in trusted intermediaries, arbitrations and enforcement costs, fraud losses, and the reduction of malicious and accidental exceptions (Fries and Paal, 2019).

Elaboration:

The Project Manager's role is to define, execute, and manage the acquisition. This is focused on the project needs to deliver the system, system elements, products, or services that meet the end user requirements and achieve the acquisition milestones. This is done in collaboration with the SE practitioners and the selected contractor to ensure the technical expectations and key performance parameters are achieved. The team needs to define plans and methods collectively, and refine them as more is learned about the nature and challenges inherent in the system or capability being built and its intended operating environment. For more information on PM-SE integration, see 5.3.3.

When the acquisition involves systems or system elements where technology or a system capability is not mature enough, it is necessary to account for uncertainty and the need for additional risk management actions in the planning. This includes allowing additional margin in the development/production timeframe, such as ample lead time in anticipation of inherent challenges, especially when technology maturation is required. These challenges may also include limited availability of adequate resources for the supplier (skilled labor and/or technologies), a need for customization of supplier products or equipment, poor or early understanding of interface requirements, integration challenges, and required verification and/or validation of the development. If there is no flexibility in the delivery date, then trade-offs may be needed to provide the system capabilities in an incremental manner.

Technical supplier management is about ensuring the supplier meets the allocated project requirements and that the supplier is effectively managed. This is usually achieved through the Statement of Work (SOW) and a set of requirements. The SOW is a mechanism to ensure progress is being made and describes the necessary work, quality, standards, designs, models, evidence, reviews, timescales, and meetings, etc. that the supplier is expected to provide contingent on the contract. To prove the system/system element functional, performance, and operational requirements are met, the supplier will also need to provide compliance matrices and verification and validation evidence.

2.3.2.2   Supply Process

Overview

Purpose As stated in ISO/IEC/IEEE 15288,

[6.1.2.1] The purpose of the Supply process is to provide an acquirer with a product or service that meets agreed requirements.

The Supply process is invoked to establish an agreement between two organizations under which one party supplies products or services to the other. Within the supplier organization, a project is conducted according to the recommendations of this handbook with the objective of providing a product or service to the acquirer that meets the agreed requirements. In the case of a mass-produced commercial product or service, a marketing, or similar, function may represent the acquirer and establish stakeholder expectations.

Description This section is written from the perspective of the supplier organization. The Supply process is highly dependent upon the Technical, Technical Management, and Organizational Project-Enabling Processes as it is through them that the work of executing the agreement is accomplished. This means that the Supply process is the larger context in which the other processes are applied under the agreement.

Inputs/Outputs Inputs and outputs for the Supply process are listed in Figure 2.14. Descriptions of each input and output are provided in Appendix E.

Process Activities The Supply process includes the following activities:

Controls

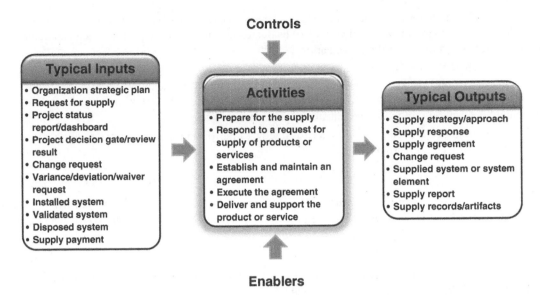

Typical Inputs	Activities	Typical Outputs
• Organization strategic plan • Request for supply • Project status report/dashboard • Project decision gate/review result • Change request • Variance/deviation/waiver request • Installed system • Validated system • Disposed system • Supply payment	• Prepare for the supply • Respond to a request for supply of products or services • Establish and maintain an agreement • Execute the agreement • Deliver and support the product or service	• Supply strategy/approach • Supply response • Supply agreement • Change request • Supplied system or system element • Supply report • Supply records/artifacts

Enablers

FIGURE 2.14 IPO diagram for the Supply process. INCOSE SEH original figure created by Shortell, Walden, and Yip. Usage per the INCOSE Notices page. All other rights reserved.

- *Prepare for the supply.*
  - Develop and maintain strategic plans, policies, and procedures to meet the needs of potential acquirer organizations, as well as internal organization goals and objectives including the needs of the project management and technical SE organizations.
  - Identify opportunities.
- *Respond to a request for supply of products or services.*
  - Select appropriate acquirers willing to conduct ethical negotiations, able to meet financial obligations, and willing to maintain open communications throughout the Supply process.
  - Evaluate the acquirer requests and propose a product or service that meets acquirer needs and complies with industry and other standards. Assessments from the Portfolio Management, Human Resource Management, Quality Management, and Business or Mission Analysis processes are necessary to determine the suitability of this response and the ability of the organization to meet these commitments.
- *Establish and maintain an agreement.*
  - Establish an agreement. Ensure an understanding of expectations, including acceptance criteria.
  - Identify the necessary changes to the agreement and evaluate the related impacts on the agreement.
  - Update the agreement with the acquirer as necessary.
- *Execute the agreement.*
  - Start the project and invoke the other processes defined in this handbook.
  - Manage the Supply process and related activities including the development of plans and schedules, decision making for agreements, relationship building and maintenance, interaction with organization management, and final approval authority for deliveries made to acquirer.
  - Maintain communications with acquirers, suppliers, stakeholders, and other organizations regarding the agreement.

- Carefully evaluate the terms of the agreement to identify risks and issues, progress toward mitigation of risks, and adequacy of progress toward delivery. Also evaluate cost and schedule performance and determine potential undesirable outcomes for the organization.
- *Deliver and support the product or service.*
  - After acceptance and transfer of the final products and/or services, the acquirer will provide payment or other consideration in accordance with all agreements, schedules, and relevant laws and regulations. A support agreement is often ongoing after the transfer of products and/or services.
  - When a Supply process cycle concludes, a final review of performance is conducted to extract lessons learned for continued process performance.
  - Retire the agreement.

Note: The agreement is closed through the Portfolio Management process (see Section 2.3.3.3), which manages the full set of systems and projects of the organization. When the project is closed, action is taken to close the agreement.

Common approaches and tips:

- Relationship building and trust between the parties is a nonquantifiable quality that, while not a substitute for good processes, makes human interactions agreeable.
- Develop technology white papers or similar artifacts to demonstrate and describe to the (potential) acquirer the range of capabilities in areas of interest. Use traditional marketing approaches to encourage acquisition of mass-produced products.
- When expertise is not available within the organization (legal and other governmental regulations, laws, etc.), retain subject matter experts to provide information and specify requirements related to agreements.
- Invest sufficient time and effort into understanding acquirer needs before the agreement. This can improve the estimations for cost and schedule and positively affect agreement execution. Evaluate any technical specifications for the product or service for clarity, completeness, and consistency.
- Involve personnel who will be responsible for agreement execution to participate in the evaluation of and response to the acquirer's request. This reduces the start-up time once the project is initiated, which in turn is one way to recapture the cost of writing the response.
- Make a critical assessment of the ability of the organization to execute the agreement; otherwise, the high risk of failure and its associated costs, delivery delays, and increased resource commitment needs will reflect negatively on the reputation of the entire organization.

Elaboration:
Agreements fall into a large range, from formal to very informal based on verbal understanding (e.g., from a written contract to a verbal agreement). Agreements may call for a fixed price, cost plus fixed fee, incentives for early delivery, penalties for late deliveries, and other financial motivators. Appropriate to the level of formality, the agreement establishes requirements, development and delivery milestones, verification, validation and acceptance conditions, process requirements (e.g., configuration management, risk management, measurements), exception-handling procedures, agreement change management procedures, payment schedules, and handling of data rights and intellectual property so that both parties understand the basis for executing the agreement. For a written contract, this occurs when the contract is signed.

2.3.3 Organizational Project-Enabling Processes

The Organizational Project-Enabling Processes are defined in ISO/IEC/IEEE 15288 as follows:

[5.7.3] The Organizational Project-Enabling Processes are concerned with providing the resources needed to enable the project to meet the needs and expectations of the organization's stakeholders. The Organizational Project-Enabling Processes are typically concerned at a strategic level with the management and improvement of the organization's undertaking, with the provision and deployment of resources and assets, and with its management of risks in competitive or uncertain situations.… The Organizational Project-Enabling Processes establish the environment in which projects are conducted.

This section focuses on the capabilities of an organization relevant to enabling the system life cycle; they are not intended to address general business management objectives, although sometimes the two overlap. Six Organizational Project-Enabling Processes are identified by ISO/IEC/IEEE 15288. They are Life Cycle Model Management, Infrastructure Management, Portfolio Management, Human Resource Management, Quality Management, and Knowledge Management. As defined in ISO/IEC/IEEE 15288 and this handbook, these processes provide the resources and organizational support to enable the projects that are focused on the system life cycle. The organization will tailor these processes and their interfaces to meet specific strategic and tactical objectives in support of the organization's projects (see Section 4.1).

2.3.3.1 *Life Cycle Model Management Process*

Overview

Purpose As stated in ISO/IEC/IEEE 15288,

[6.2.1.1] The purpose of the Life Cycle Model Management process is to define, maintain, and help ensure availability of policies, life cycle processes, life cycle models, and procedures for use by the organization with respect to the scope of ISO/IEC/IEEE 15288.

Description This process (i) establishes and maintains a set of policies and procedures at the organization level that support the organization's ability to acquire and supply products and services and (ii) provides integrated system life cycle models necessary to meet the organization's strategic plans, policies, goals, and objectives for all projects and all system life cycle stages. The processes are defined, adapted, and maintained to support the requirements of the organization, SE organizational units, individual projects, and personnel. The Life Cycle Model Management process is supplemented by recommended methods and tools. The resulting guidelines in the form of organization policies and procedures are still subject to tailoring by projects (see Section 4.1).

Inputs/Outputs Inputs and outputs for the Life Cycle Model Management process are listed in Figure 2.15. Descriptions of each input and output are provided in Appendix E.

Process Activities The Life Cycle Model Management process includes the following activities:

- *Establish the life cycle process.*
  - Establish policies and procedures for managing and deploying life cycle processes.
  - Establish the life cycle processes with process performance metrics to assess effectiveness and efficiency.
  - Define roles, responsibilities, accountabilities, and authorities to enable the implementation of the life cycle processes.
  - Establish entrance and exit criteria for decision gates.
  - Define an appropriate set of life cycle models that are comprised of stages.
  - Establish tailoring guidance for projects

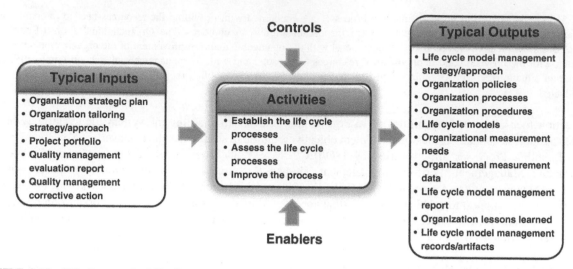

FIGURE 2.15   IPO diagram for Life Cycle Model Management process. INCOSE SEH original figure created by Shortell, Walden, and Yip. Usage per the INCOSE Notices page. All other rights reserved.

- *Assess the life cycle process.*
  - Use assessments and reviews of the life cycle models' performance to confirm the adequacy and effectiveness of the Life Cycle Model Management process.
  - Identify opportunities to improve the organization life cycle model management guidelines on a continuing basis based on individual project assessments, individual feedback, metrics, and changes in the organization strategic plan.
- *Improve the process.*
  - Prioritize and implement the identified improvement opportunities.
  - Communicate with all relevant organizations regarding the creation of and changes in the life cycle model management guideline.

NOTE: ISO/IEC/IEEE 15288 provides more details for the Life Cycle Model Management process that are aligned with the activities listed above.

Common approaches and tips:

- Base the policies and procedures on an organization-level strategic and business area plan that provides a comprehensive understanding of the organization's goals, objectives, stakeholders, competitors, future business, and technology trends.
- Ensure that policy and procedure compliance review is included as part of the business decision gate criteria.
- Develop a Life Cycle Model Management process information database with essential information that provides an effective mechanism for disseminating consistent guidelines and providing announcements about organization-related topics, as well as industry trends, research findings, and other relevant information. This provides a single point of contact for continuous communication regarding the life cycle model management guidelines and encourages the collection of valuable feedback, metrics, and the identification of organization trends.
- Establish an organization center of excellence for the Life Cycle Model Management process. This organization can become the focal point for the collection of relevant information, dissemination of guidelines, and analysis

of assessments, performance, and feedback. They can also develop checklists and other templates to support project assessments to ensure that the predefined measures and criteria are used for evaluation.

- Manage the network of external relationships by assigning personnel to identify standards, industry and academia research, and other sources of organization management information and concepts needed by the organization. The network of relationships includes government, industry, and academia. Each of these external interfaces provides unique and essential information for the organization to succeed in business and meet the continued need and demand for improved and effective systems and products for its stakeholders. It is up to the Life Cycle Model Management process to fully define and utilize these external entities and interfaces (i.e., their value, importance, and capabilities that are required by the organization):
  - Legislative, regulatory, and other government requirements
  - Industry SE and management-related standards, training, and capability maturity models
  - Academic education, research results, future concepts and perspectives, and requests for financial support
- Establish an organization communication plan for the policies and procedures. Most of the processes in this handbook include dissemination activities. An effective set of communication methods is needed to ensure that all stakeholders are well informed.
- Include stakeholders, such as engineering and project management organizations, as participants in developing the life cycle model management guidelines. This increases their commitment to the recommendations and incorporates a valuable source of organizational experience.
- Develop alternative life cycle models based on the type, scope, complexity, and risk of a project. This decreases the need for tailoring by engineering and project organizations.

Elaboration

Value Proposition for Organizational Processes. The value propositions to be achieved by instituting organization-wide processes for use by projects are as follows:

- Provide repeatable/predictable performance across the projects in the organization (this helps the organization in planning and estimating future projects and in demonstrating reliability to stakeholders)
- Leverage practices that have been proven successful by certain projects and instill those in other projects across the organization (where applicable)
- Enable process improvement across the organization
- Improve ability to efficiently transfer staff across projects as roles are defined and performed consistently
- Enable leveraging lessons that are learned from one project for future projects to improve performance and avoid issues
- Improve startup of new projects (less reinventing the wheel)

In addition, the standardization across projects may enable cost savings through economies of scale for support activities (tool support, process documentation, etc.).

Benchmarking. SE benchmarking involves comparing an organization's system life cycle processes and practices to those of other entities that are considered as good performers, internally or externally, or comparing to industry standards or good practices. SE benchmarking results and comparisons can be used to generate ideas for driving process improvement to maximize efficiency and effectiveness.

Standard SE Processes. An organization engaged in SE provides the requirements for establishing, maintaining, and improving the standard SE processes and the policies, practices, and supporting functional processes necessary to meet the needs throughout the organization. Further, it defines the process for tailoring the standard SE processes for use on projects addressing the specific needs of the project and for making improvements to the project-tailored SE processes.

Analysis of Process Performance. A high-performing organization also reviews the process (as well as work products), conducts assessments and audits (e.g., assessments based on CMMI (2018), ISO/IEC 33060 (2020), and ISO 9000 (2015) audits), retains corporate memory through the understanding of lessons learned, and establishes how benchmarked processes and practices of related organizations can affect the organization. Successful organizations should analyze their process performance, its effectiveness and compliance to organizational and higher directed standards, and the associated benefits and costs and then develop targeted improvements.

The basic requirements for standard and project-tailored SE process control, based on CMMI (2018), ISO/IEC 33060 (2020), or other resources, are as follows:

- Process responsibilities for projects:
  - Identify SE processes.
  - Document the implementation and maintenance of SE processes.
  - Use a defined set of standard methods and techniques to support the SE processes.
  - Apply accepted tailoring guidelines to the standard SE processes to meet project-specific needs.
- Good process definition includes:
  - Inputs and outputs
  - Entrance and exit criteria
- Process responsibilities for organizations and projects:
  - Assess strengths and weaknesses in the SE processes.
  - Compare the SE processes to benchmark processes used by other organizations.
  - Institute SE process reviews and audits of the SE processes.
  - Institute a means to capture and act on lessons learned from SE process implementation on projects.
  - Institute a means to analyze potential changes for improvement to the SE processes.
  - Institute measures that provide insight into the performance and effectiveness of the SE processes.
  - Analyze the process measures and other information to determine the effectiveness of the SE processes.

Although it should be encouraged to identify and capture lessons learned throughout the performance of every project, the SE organization must plan and follow through to collect lessons learned at predefined milestones in the system life cycle. The SE organization should periodically review lessons learned together with the measures and other information to analyze and improve SE processes and practices. The results need to be communicated and incorporated into training. It should also establish good practices and capture them in an easy-to-retrieve form.

For more information on process definition, assessment, and improvement, see the resources in the bibliography, including the CMMI and ISO/IEC TS 33060.

2.3.3.2 Infrastructure Management Process

Overview

Purpose As stated in ISO/IEC/IEEE 15288,

[6.2.2.1] The purpose of the Infrastructure Management process is to provide the infrastructure and services to projects to support organization and project objectives throughout the life cycle.

Description The work of the organization is accomplished through projects, which are conducted within the context of the infrastructure environment. This infrastructure needs to be defined and understood within the organization and the project to ensure alignment of the working units and achievement of overall organization strategic objectives. This process exists to establish, communicate, and continuously improve the system life cycle process environment.

Infrastructure Management is an organizational project-enabling process and foundational to all SE process management and improvement. Effective infrastructure management is imperative to an organization's ability to change and for that change to be positive, durable, and impactful. Each element of infrastructure is a SoI and both Technical Management and Technical Processes, as stated in ISO/IEC/IEEE 15288 apply to the establishment and maintenance of the infrastructure. Additionally, the Infrastructure Management process includes the physical, political, and process improvement infrastructures.

Inputs/Outputs Inputs and outputs for the Infrastructure Management process are listed in Figure 2.16. Descriptions of each input and output are provided in Appendix E.

Process Activities The Infrastructure Management process includes the following activities:

- *Establish the infrastructure.*
  - Define infrastructure requirements.
  - Define infrastructure elements (e.g., facilities, tools, hardware, software, services, and standards)
  - Define, gather, and negotiate infrastructure resource needs with the organization and projects.
  - Identify, obtain, and provide the infrastructure resources and services to ensure organization goals and objectives are met.
  - Control the infrastructure elements, resources, and services.
  - Conduct inventory management to include enumeration, lists, storage to establish ownership, accessibility, and expectations.
  - Manage resource and service conflicts and shortfalls with steps for resolution.
  - Conduct infrastructure management inventories including identification, status, type, location, access, and condition.
- *Maintain the infrastructure.*
  - Continue to assess whether the project infrastructure needs are met.

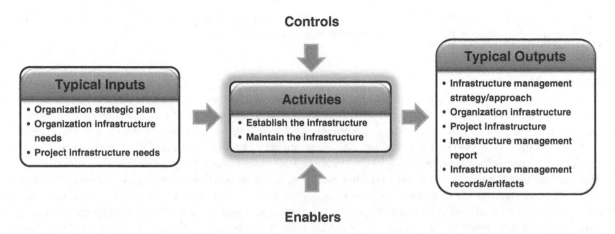

FIGURE 2.16   IPO diagram for Infrastructure Management process. INCOSE SEH original figure created by Shortell, Walden, and Yip. Usage per the INCOSE Notices page. All other rights reserved.

- Identify and provide improvements or changes to the infrastructure resources as the project requirements change.
- Manage infrastructure resource availability to ensure organization goals and objectives are met. Conflicts and resource shortfalls are managed with steps for resolution.
- Allocate infrastructure resources and services to support all projects.
- Evaluate the condition of the infrastructure.
- Perform cost analysis toward the cost of infrastructure management.
- Control multi-project infrastructure resource management communications to effectively allocate resources throughout the organization; and identify potential future or existing conflict issues and problems with recommendations for resolution.
- Provide change control for the infrastructure management.
- Conduct risk analysis regarding infrastructure management.
- Evaluate infrastructure management alternatives through analysis of alternatives. This evaluation and analysis compliments risk management and cost reduction activities.

Common approaches and tips:

- Qualified resources may be leased (insourced or outsourced) or licensed in accordance with the investment strategy.
- Establish an organization infrastructure architecture. Integrating the infrastructure of the organization can make the execution of routine business activities more efficient.
- Establish a resource management information system with enabling support systems and services to maintain, track, allocate, and improve the resources for present and future organization needs. Computer-based equipment tracking, facilities allocation, and other systems are recommended for organizations with over 50 people.
- Attend to physical factors, including facilities and human factors, such as ambient noise level and computer access to specific tools and applications.
- Begin planning in early life cycle stages of all system development efforts to address utilization and support resource requirements for system transition, facilities, infrastructure, information/data storage, and management. Enabling resources should also be identified and integrated into the organization's infrastructure.
- Engage project management, risk management, and business management processes to fully integrate Infrastructure Management processes to ensure organizational adoption.

Elaboration

Infrastructure Management Concepts. Projects all need resources to meet their objectives. Project planners determine the resources needed by the project and attempt to anticipate both current and future needs. The Infrastructure Management process provides the mechanisms whereby the organization infrastructure is made aware of project needs and the resources are scheduled to be in place when requested. While this can be simply stated, it is less simply executed. Conflicts must be negotiated and resolved, equipment must be obtained and sometimes repaired, buildings need to be refurbished, and information technology services are in a state of constant change. The infrastructure management organization collects the needs, negotiates to remove conflicts, and is responsible for providing the enabling organization infrastructure without which nothing else can be accomplished. Since resources are not free, their costs are also factored into investment decisions. Financial resources are addressed under the Portfolio Management process (see Section 2.3.3.3), but all other resources, except for human resources which are addressed under the Human Resource Management process (see Section 2.3.3.4), are addressed under this process.

Infrastructure management is complicated by the number of sources for requests, the need to balance the skills of the labor pool against the other infrastructure elements (e.g., computer-based tools), the need to maintain a balance between the budgets of individual projects and the cost of resources, the need to keep apprised of new or modified policies and procedures that might influence the skills inventory, and myriad unknowns.

Resources are allocated based on requests. Infrastructure management collects the needs of all the projects in the active portfolio and schedules or acquires nonhuman assets, as needed. Additionally, the infrastructure management process maintains and manages the facilities, hardware, and support tools required by the portfolio of organization projects. Infrastructure management is the efficient and effective deployment of an organization's resources when and where they are needed. Such resources may include inventory, production resources, or information technology. The goal is to provide materials and services to a project when they are needed to keep the project on target and on budget. A balance should be found between efficiency and robustness. Infrastructure management relies heavily on forecasts into the future of the demand and supply of various resources.

The organization environment and subsequent investment decisions are built on the existing organization infrastructure, including facilities, equipment, personnel, and knowledge. Efficient use of these resources is achieved by exploiting opportunities to share enabling systems or to use a common system element on more than one project. These opportunities are enabled by good communications within the organization. Integration and interoperability of supporting systems, such as financial, human resources (see Section 2.3.3.4), and training, is critically important to executing organization strategic objectives. Feedback from active projects is used to refine and continuously improve the infrastructure.

Further, trends in the market may suggest changes in the supporting environment. Assessment of the availability and suitability of the organization infrastructure and associated resources provides feedback for improvement and reward mechanisms. All organization processes require mandatory compliance with government and corporate laws and regulations. Decision making is governed by the organization strategic plan.

Infrastructure Management Process Maturity. The Infrastructure Management process primarily focuses on the establishment and deployment of infrastructure rather than the construction or actual use of the infrastructure. Since the quality of a product is related to the structure and use of the infrastructure employed, the maturity and quality of the process employed toward management of the infrastructure can help provide higher quality process inputs, outputs, and outcomes.

2.3.3.3 *Portfolio Management Process*

Overview

Purpose As stated in ISO/IEC/IEEE 15288,

> [6.2.3.1] The purpose of the Portfolio Management process is to initiate and sustain necessary, sufficient, and suitable projects to meet the strategic objectives of the organization.

Portfolio management also provides organizational output regarding the set of projects, systems, and technical investments of the organization to external stakeholders, such as parent organizations, investors/funding sources, and governance bodies.

Description Projects create the products or services that meet the objectives and generate revenue for an organization. Thus, the conduct of successful projects requires an adequate allocation of funding and resources and the authority to deploy them to meet project objectives. Most business entities manage the commitment of financial resources using well-defined and closely monitored processes.

The Portfolio Management process also performs ongoing evaluation of the projects and systems in its portfolio. Based on periodic assessments, projects are determined to justify continued investment if they have the following characteristics:

- Contribute to the organization strategy
- Progress toward achieving established goals

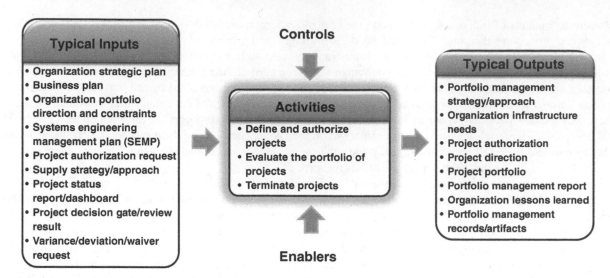

FIGURE 2.17   IPO diagram for Portfolio Management process. INCOSE SEH original figure created by Shortell, Walden, and Yip. Usage per the INCOSE Notices page. All other rights reserved.

- Comply with project directives from the organization
- Are conducted according to an approved plan
- Provide a service or product that is still needed and providing acceptable investment returns

Otherwise, projects may be redirected or, in extreme instances, terminated.

Inputs/Outputs Inputs and outputs for the Portfolio Management process are listed in Figure 2.17. Descriptions of each input and output are provided in Appendix E.

Process Activities The Portfolio Management process includes the following activities:

- *Define and authorize projects.*
    - Obtain business area plans and organization strategic plans—use the strategic objectives to identify candidate projects to fulfill them.
    - Identify, assess, prioritize, and select investment opportunities consistent with the organization strategic plan.
    - Establish project scope, define project management accountabilities and authorities, and identify expected project outcomes.
    - Establish the domain area of the product line defined by its main features and their suitable variability.
    - Allocate adequate funding and other resources to selected projects.
    - Identify interfaces and opportunities for multi-project synergies.
    - Specify the project governance process including organizational status reporting and reviews.
    - Authorize project execution.
- *Evaluate the portfolio of projects.*
    - Evaluate ongoing projects to provide rationale for continuation, redirection, or termination.
    - Provide direction and supporting actions for continuation or redirection, as applicable for successful completion.

- *Terminate projects*.
  - Close, cancel, or suspend projects that are completed or designated for termination.

Common approaches and tips:

- Logic modeling techniques that capture how an organization works can aid development or evaluation of business area plans at multiple levels of interest, ranging from the project- to portfolio-level plans (see, for example, PMBOK® (2021) *S*ection 4.2 for a list of commonly used models). The logic models typically describe the fundamental theory/assumptions, planned work (resources, inputs and activities) linked with intended results (outputs, outcomes, and impact).
- When investment opportunities present themselves, prioritize them based on measurable criteria such that projects can be objectively evaluated against a threshold of acceptable performance. This assessment is done in the context of the business area planning to focus resources to best meet present and future objectives.
- Expected project outcomes should be based on clearly defined, measurable criteria to ensure that an objective assessment of progress can be determined. Specify the investment information that will be assessed for each milestone. Initiation should be a formal milestone that does not occur until all resources are in place as identified in the project plan.
- Establish organizational coordination mechanisms to manage the synergies between active projects in the organization portfolio. Complex and large organization architectures require the management and coordination of multiple interfaces and make additional demands on investment decisions. These interactions occur within and between the projects.
- Use a product line engineering approach (see Section 4.2.4) when stakeholders need the same or similar systems (e.g., common features), with some customizations (e.g., variants). The goal is to manage a product line as one product definition with planned variants as opposed to multiple separate products managed individually, thereby streamlining and simplifying the management effort.
- Include risk assessments (see Section 2.3.4.4) in the evaluation of ongoing projects. Projects that contain risks that may pose a challenge in the future might require redirection. Cancel or suspend projects whose disadvantages or risks to the organization outweigh the investment.
- Include opportunity assessments (see Section 2.3.4.4) in the evaluation of ongoing projects. Addressing project challenges may represent a positive investment opportunity for the organization. Avoid pursuing opportunities that are inconsistent with the capabilities of the organization and its strategic goals and objectives or contain unacceptably high technical risk, resource demands, or uncertainty.
- Allocate resources based on the requirements of the projects; otherwise, the risk of cost and schedule overruns may have a negative impact on quality and performance of the project.
- Establish effective governance processes that directly support investment decision making and communications with project management.

Elaboration
Define the Business Cases and Assess Against Business Area Plans. Portfolio management tries to maximize the benefit obtained by the organization from the use of financial assets and other resources within the organization. Thus, business cases for potential projects are evaluated for cost-benefit and the business need before a project is approved for the proposed SoI. Each decision gate reviews the business case as the project matures. The result is reverification or perhaps restatement of the business case.

The business case may be validated in a variety of ways. For large projects, sophisticated engineering models, or even prototypes of key system elements, help prove that the objectives of the business case can be met, and that the system will work as envisioned prior to committing large amounts of resources to full-scale engineering and

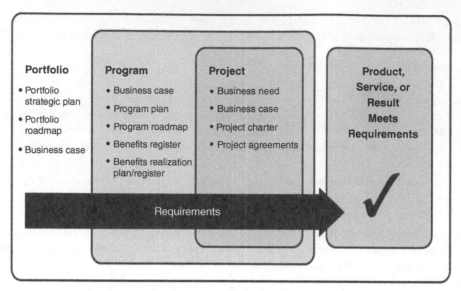

FIGURE 2.18   Requirements across the portfolio, program, and project domains. From PMI (2016). Used with permission. All other rights reserved.

manufacturing development. For smaller projects, when the total investment is modest, proof-of-concept models may be constructed during the concept stage to prove the validity of business case assumptions.

Investment opportunities are not all equal, and organizations are limited in the number of projects that can be conducted concurrently. Further, some investments are not well aligned with the overall strategic plan of the organization. For these reasons, opportunities are evaluated against the portfolio of existing agreements and ongoing projects, taking into consideration the attainability of the stakeholders' requirements.

Project Management and SE considerations. Portfolios may have multiple projects. As previously stated, projects are added to the portfolio after the candidate project can show that it is both feasible and meets organizational business needs. In many organizations projects with defined scope are organized in programs focused on a set of objectives that are part of the organization's strategic plan. As stated in the PMI (2017), the focus of portfolio management is "doing the right work" as opposed to program or project management which is more concerned with "doing work right."

The disciplines of project management and SE have overlapping responsibilities regarding portfolio management. To save time, share knowledge, facilitate the accomplishment of shared objectives, and achieve success, a strong partnership should exist between each of these disciplines (see Section 5.3.3).

At the portfolio level, the scope is extensive with consideration external to the organization and internal across the organization's enterprise. At the other end of the spectrum, the focus is internal to the project with consideration for the context of the product/service/result. An example of this is to look at the range in scope in requirements development, as shown in Figure 2.18.

At the portfolio level, the portfolio's strategic plan and roadmap address business and mission needs and provides direction and organizational focus, and plans/actions to realize the direction. Requirements often start at the concept or portfolio level as a high-level view associated with investment or business opportunities.

2.3.3.4   Human Resource Management Process

Overview

Purpose As stated in ISO/IEC/IEEE 15288,

[6.2.4.1] The purpose of the Human Resource Management process is to provide the organization with necessary human resources and to maintain their competencies, consistent with strategic needs.

Description Projects all need resources to meet their objectives. This process deals with human resources. Nonhuman resources, including tools, databases, communication systems, financial systems, and information technology, are addressed using the Infrastructure Management process (see Section 2.3.3.2).

Project planners determine the resources needed for the project by anticipating both current and future needs. The Human Resource Management process provides the mechanisms whereby the organization management is made aware of project needs and personnel are scheduled to be in place when requested. While this can be simply stated, it is less simply executed. Conflicts must be resolved, personnel must be trained, and employees are entitled to vacations and time away from the job.

The human resource management organization collects the needs, negotiates to remove conflicts, and is responsible for providing the personnel, without which nothing else can be accomplished. Since qualified personnel are not free, their costs are also factored into investment decisions.

Inputs/Outputs Inputs and outputs for the Human Resource Management process are listed in Figure 2.19. Descriptions of each input and output are provided in Appendix E.

Processes Activities The Human Resource Management process includes the following activities:

- *Identify skills.*
    - Identify and record skills of existing personnel to establish a "skills inventory."
    - Review current and anticipated projects to determine and record the skill needs across the portfolio of projects. The INCOSE Systems Engineering Competency Framework (SECF) (2018) and Systems Engineering Competency Assessment Guide (SECAG) (2023) can be used as resources to identify SE skills.
    - Evaluate skill needs against available personnel with the prerequisite skills to determine if training, hiring, or other skill acquisition activities are indicated.
- *Develop skills.*
    - Establish a strategy/approach for skills development.
    - Plan for the skill development per the strategy.
    - Obtain (or develop) and deliver training, education, and mentoring to close identified gaps of project personnel.

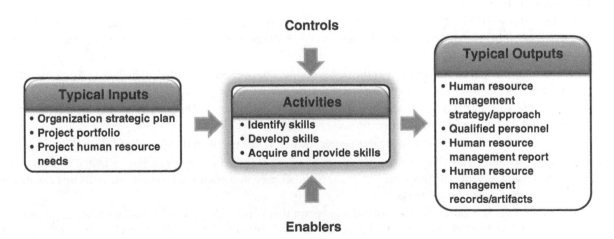

FIGURE 2.19 IPO diagram for Human Resource Management process. INCOSE SEH original figure created by Shortell, Walden, and Yip. Usage per the INCOSE Notices page. All other rights reserved.

- Identify skills, abilities, and behaviors needed for competencies. The INCOSE Systems Engineering Assessment Guide is a recommended resource for this.Identify training and development resources to match desired skills, abilities, and behaviors development. The INCOSE Professional Development Portal can help identify potential resources.
- Identify assignments that lead toward career progression.
- Create succession plans to ensure that the desired skill set and flow of skill development through the organization is sustained into the future.
- Create and maintain skill development records.

- *Acquire and provide skills.*
  - Provide human resources to support all projects.
  - Train or hire qualified personnel when gaps indicate that skill needs cannot be met with existing personnel.
  - Maintain and manage a skilled personnel pool to staff ongoing projects.
  - Assign personnel to projects based on personnel development and project needs.
  - Create and maintain staff assignment records.
  - Motivate personnel by providing career development and reward programs.
  - Resolve personnel conflicts between or within projects.
  - Maintain communication across projects to effectively allocate human resources throughout the organization and identify potential future or existing conflicts and problems with recommendations for resolution.
  - Schedule other related assets or, if necessary, acquire them.
- *Develop and Manage Competencies.*
  - Create and maintain job role definitions related to competencies required.
  - Identify organization competency gaps.
  - Align organization competencies with strategic objectives.
  - Maintain organization-level competency definitions and frameworks.

Common approaches and tips:

- The availability and suitability of personnel is one of the critical project assessments and provides feedback for improvement and reward mechanisms.
- Consider using an IPDT environment as a means to reduce the frequency of project rotation, recognize progress and accomplishments and reward success, and establish apprentice and mentoring programs for newly hired employees and students.
- Maintain both a listing of skill needs and the paths to obtain the necessary expertise, including a pipeline of candidates, training provisions, consultants, temporary outsourcing, reassignments, etc.
- Personnel are allocated based on requests and conflicts are negotiated. The goal is to provide personnel to a project when they are needed to keep the project on target and on budget.
- Try to avoid the overcommitment of project personnel, especially people with specialized skills.
- Skills inventory and career development plans are important documentation that can be validated by engineering and project management. The INCOSE SECF and SECAG are comprehensive resources of skills that can be used to develop career development plans.
- Maintain an organization career development program that is not sidetracked by project demands. Develop a policy that all personnel receive training or educational benefits on a regular cycle. This includes both undergraduate and graduate studies, in-house training courses, certifications, tutorials, workshops, and conferences.

- Remember to provide training on organization policies and procedures and system life cycle processes.
- Establish a resource management information infrastructure with enabling support systems and services to maintain, track, allocate, and improve the resources for present and future organization needs.
- Use the slack time in the beginning of a project to provide training to ensure necessary skills.
- Career development plans should be managed and aligned to the objectives of both the employee and the organization. Career development plans should be reviewed, tracked, and refined to provide a mechanism to help manage the employee's career within the organization.

Elaboration

Human Resource Management Concepts. The Human Resource Management process maintains and manages the people required by the portfolio of organization projects. Human resource management is the efficient and effective deployment of qualified personnel when and where they are needed. A balance should be found between efficiency and robustness. Human resource management relies heavily on forecasts into the future of the demand and supply of various resources.

The primary objective of this process is to provide a pool of qualified personnel to the organization. This is complicated by the number of sources for requests, the need to balance the skills of the labor pool against the other infrastructure elements (e.g., computer-based tools), the need to maintain a balance between the budgets of individual projects and the cost of resources, the need to keep apprised of new or modified policies and procedures that might influence the skills inventory, and myriad unknowns.

Project managers face their resource challenges competing for scarce talent in the larger organization pool. They must balance access to the experts they need for special studies with stability in the project team with its tacit knowledge and project memory. Today's projects depend on teamwork and optimally multidisciplinary teams. Such teams are able to resolve project issues quickly through direct communication between team members. Such intrateam communication shortens the decision-making cycle and is more likely to result in improved decisions because the multidisciplinary perspectives are captured early in the process.

2.3.3.5 Quality Management Process

Overview

Purpose As stated in ISO/IEC/IEEE 15288,

> [6.2.5.1] The purpose of the Quality Management process is to assure that products, services, and implementations of the Quality Management process meet organizational and project quality objectives and achieve customer satisfaction.

Description The overarching process for achieving quality goals is the Quality Management (QM) process and its supporting methods, values, and subprocesses. Properly communicated, through policy and procedure, it makes visible the goals of the organization to achieve customer satisfaction. These goals, when supported by measurable activities, provide feedback for maintaining consistency in work processes and delivering quality outcomes. Since primary drivers in any project are time, cost, and quality, inclusion of a comprehensive QM process and its subprocesses is essential to every organization and must be sustained by a work culture that is disciplined in the proper execution of QM foundational principles and values. System life cycle processes are concerned with quality issues, and this is sufficient justification for spending the time, money, and energy into establishing QM fundamentals in an organization, its processes, and its people.

The QM process for SE ensures that all SE processes are deployed consistently by capable staff that can then produce systems designs that fulfill the stakeholder's requirements and lead to development and build processes that are aligned to produce high levels of performance throughout the organization.

Inputs/Outputs Inputs and outputs for the Quality Management process are listed in Figure 2.20. Descriptions of each input and output are provided in Appendix E.

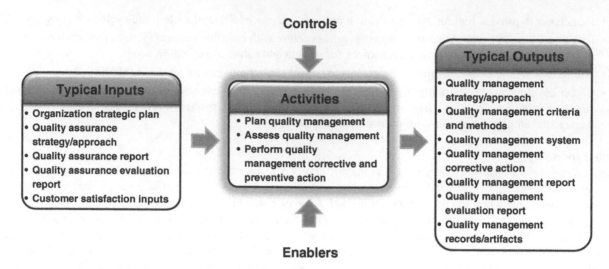

FIGURE 2.20   IPO diagram for the Quality Management process. INCOSE SEH original figure created by Shortell, Walden, and Yip. Usage per the INCOSE Notices page. All other rights reserved.

Process Activities The Quality Management process includes the following activities:

- *Plan quality management.*
    - Identify, assess, and prioritize quality guidelines consistent with the organization strategic plan. Establish QM guidelines-policies, standards, and procedures.
    - Establish organization and project QM goals and objectives, including QM Culture emphasis.
    - Establish organization and project QM responsibilities and authorities.
- *Assess quality management.*
    - Evaluate project assessments.
    - Assess customer satisfaction against compliance with requirements and objectives.
    - Continuously improve the QM guidelines.
- *Perform quality management corrective action and preventive action.*
    - Recommend appropriate action, when indicated.
    - Maintain open communications within the organization and with stakeholders.

Common Approaches and Tips

- Management's commitment to quality is reflected in the integration of QM principles in the strategic planning and budgeting of the organization, and the allocation of educational resources to achieve and sustain a reliable QM culture.
- A quality policy, mission, strategies, goals, and objectives provide essential inputs along with a description of an organization's fundamental values for supporting a growing QM culture.

Elaboration

QM Generally accepted theory and practice. The four generally accepted foundational values of quality are its definition, its system, the standard for quality, and the method for measuring quality. Philip Crosby called them the Four Absolutes of Quality (Crosby, 1979).

1. *The definition of quality is meeting the stakeholder's requirements, needs and expectations.* Organizations (and individuals) are both producers and users of systems. One organization or person (acting as an acquirer) can task another (acting as a supplier) for products or services. This transaction is achieved using agreements that promise to fulfil the stakeholder's requirements in exchange for something of value, usually money. Quality pioneer W. Edwards Deming stressed that meeting stakeholder needs represents the defining criterion for quality and that all members of an organization need to participate actively in "constant and continuous" quality improvement (Deming, 1986).

2. *The system of quality is prevention.* One of the two QM prevention methods is Quality Assurance (QA). QA can be described as "putting good things into our processes" so that they perform as designed and conform to our stakeholder's requirements. QA was born in the aerospace industry and was originally referred to as "reliability engineering." It is generally associated with activities such as failure testing and pre-inspecting batches of materials and system elements that are then certified for use, thus preventing errors and defects from occurring by building-in quality. The QA methodology also includes infusing processes with reliable human resources and appropriate policies, procedures, and training (SEH Section 2.3.4.8). Quality Control (QC) is the QM method for "taking bad things out of our processes after they occur" to prevent the defects that are discovered from reaching our stakeholders. QC includes checking, monitoring, and inspecting for defects and the removal, replacement, or rework of defective outcomes. One method of monitoring and statistically evaluating the stability and potential defect rates of processes is Statistical Process Control (SPC). Many manufacturing and high-volume service organizations use SPC to help achieve quality. Traditional SPC techniques include real-time, random sampling to test a fraction of the output for variances within critical tolerances (Juran, 1974).

3. *The standard for Quality is Zero Defects (ZD).* It is important to make a distinction between the tracking of defects from feedback loops to improve our processes and progress toward a ZD count, and the more fundamental human term which is the Zero Defects Attitude (ZDA) (Kennedy, 2005). A ZDA is not about achieving perfection; it is a commitment to make each stakeholder's experience as close to what was promised as possible. No one can achieve perfection, nor attain and sustain ZD, so we cannot expect perfection from any of our staff or processes. Like the "pride of workmanship," people with a ZDA have a "heart attitude" that desires to prevent all defects and to reach the highest level of personal performance and customer satisfaction. People with a ZDA want to keep their promises to everyone and make things right when we fail. A ZDA, coupled with appropriate metrics and plans to progress toward ZD, will result in continuous and incremental improvement.

4. *The method for measuring quality is the price of non-conformance* (Crosby, 1979). It is a calculation of the expenses incurred by defects and their related rework, replacement, warranties, customer service, etc. The American Society for Quality calls it the "cost of poor quality." It is an essential factor in calculating the actual "cost of quality" which is determined by comparing the "price of non-conformance (or doing things wrong) that includes expense caused by re-work, defects, and warranties, with the "price of conformance (or doing things right)" which is a calculation of the expenses related to improving processes and applying preventive methods. The cost of quality includes a calculation of quantitative and qualitative parameters that are measured in both financial and human values. When the cost of doing things right is equal to or less than the price of non-conformance then, as Crosby said, "Quality is Free."

QM Culture. SE practitioners need to have sufficient process knowledge and a QM knowledge base to be able to evaluate prevention options and make continuous, incremental improvements. When engineering disciplines are supported by planning and budgeting skills that resonate with the organization, we can achieve Process Quality with effective, efficient, and profitable outcomes, low defect rates, and delighted stakeholders. Deming, in his "14 Points" emphasized the need to "create constancy of purpose for improving products and services" and that it should be supported by "a vigorous program of education and self-improvement for everyone" (Deming, 1986). A high-performing work culture is measured by identifiable attributes or values within an organization's leadership style and workforce that directly influence the reliability of outputs. Kennedy (2005) leverages Deming's mandate and the work of Crosby by defining the Eight Attributes of a Quality Management Culture that are described in Table 2.3. Figure 2.21 shows a QM culture resulting from QM values and skills integration.

TABLE 2.3   Eight Attributes of a Quality Management Culture

1. Zero Defects Attitude: A measure of our commitment to keep our promises and to initiate systems with the goal of preventing defects from reaching our customers.
2. Vocational Certainty: A measure of our faithfulness to our career agenda. A QM culture is disciplined about developing their skills and talents and acquiring earned confidence.
3. Process Quality: A measure of our mastery of planning and budgeting disciplines and how effectively we apply them to create viable work processes.
4. Administrative Consistency: A measure of our attention to details. QM cultures carefully listen to their customer to identify and conform to their requirements and assure customer satisfaction.
5. Executive Credibility: A measure of our sincerity and skill with people. Sincerity comes naturally from the heart, but skills can be sharpened and improved to gain reliable influence.
6. Personal Authenticity: A measure of our resolve to be consistent with our customers and coworkers. Authentic QM cultures work diligently to make exceptional service feel normal.
7. Ethical Dependability: A measure of our trustworthiness in practical matters. QM cultures are what we turn to when we want things to work right, run on time, and be there when needed.
8. Create a Keeping the Promise Culture: A measure of the mutual respect, accountability, and professionalism in a work culture. These are the practiced values of effective QM cultures.

From Kennedy (2005). Used with permission. All other rights reserved.

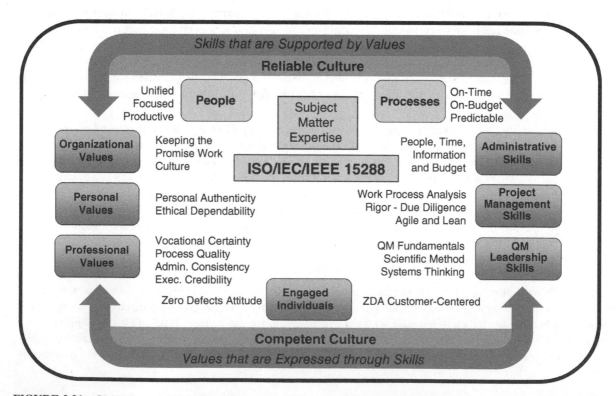

FIGURE 2.21   QM Values and Skills Integration. From Kennedy (2005). Used with permission. All other rights reserved.

2.3.3.6 Knowledge Management Process

Overview

Purpose As stated in ISO/IEC/IEEE 15288,

[6.2.6.1] The purpose of the Knowledge Management process is to create the capability and assets that enable the organization to exploit opportunities to re-apply existing knowledge.

Description Knowledge Management (KM) includes the identification, capture, creation, representation, dissemination, and exchange of knowledge across targeted groups of stakeholders. It draws from the insights and experiences of individuals and/or organizational groups or projects. The knowledge includes both explicit knowledge (conscious realization of the knowledge, often captured in artifacts and able to be communicated) and tacit knowledge (internalized in an individual or team without conscious realization) and can come from either individuals (through experience) or organizations (through processes, practices, and lessons learned) (Alavi and Leidner, 1999) (Roedler, 2010).

Within an organization, explicit knowledge is usually captured in its training, processes, practices, methods, policies, and procedures. In contrast, tacit knowledge is embodied in the individuals or teams of the organization and requires specialized techniques to identify and capture the knowledge, if it is to be passed along within the organization.

KM efforts typically focus on organizational objectives such as improved performance, competitive advantage, innovation, the sharing of good practices or lessons learned, avoidance of relearning practices, integration, and continuous improvement of the organization (Gupta and Sharma, 2004). KM captures knowledge that would otherwise be lost. So, it is generally advantageous for an organization to adopt a KM approach that includes building the framework, assets, and infrastructure to support the KM.

In this handbook, KM is viewed from an organizational project-enabling perspective, that is, how the organization supports the project (or program) environment with the resources in its KM system. The support provided to the project can come in several ways, including:

- Knowledge captured from technical experts.
- Lessons learned captured from previous similar projects.
- Domain engineering information that is applicable for reuse on the project, such as part of a product line or system family (see Section 4.2.4).
- Architecture or design patterns that are commonly encountered.
- Other reusable assets that may be applicable to the SoI.

Inputs/Outputs Inputs and outputs for the Knowledge Management process are listed in Figure 2.22. Descriptions of each input and output are provided in Appendix E.

Process Activities The knowledge management process includes the following activities:

- *Plan knowledge management.*
  - Establish a KM strategy that defines the approach and priorities for how the organization and projects within the organization will interact to ensure the right level of knowledge is captured to provide useful knowledge assets.
  - Establish the scope of the KM strategy—the organization and projects need to identify the specific knowledge information to capture and manage. Considerations include the importance and cost effectiveness of capturing the knowledge. If there is no identified project that will benefit from the knowledge asset, then it probably should not be considered.
- *Share knowledge and skills throughout the organization.*

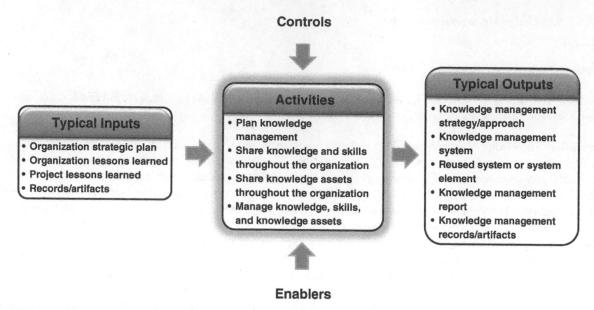

FIGURE 2.22 IPO diagram for Knowledge Management process. INCOSE original figure created by Shortell, Walden, and Yip. Usage per the INCOSE Notices page. All other rights reserved.

– Capture, maintain, and share knowledge and skills per the strategy. The infrastructure should be established to include mechanisms to easily identify, access, and determine the applicability of the knowledge and skills.

- *Share knowledge assets throughout the organization.*
  – Establish a taxonomy for the reapplication of knowledge.
  – Establish a representation for domain models and domain architectures to help ensure an understanding of the domain and identify and manage opportunities for common system elements and their representations, such as architecture or design patterns, reference architectures, and common requirements.
  – Define or acquire the knowledge assets applicable to the domain, including system and software elements, and share them across the organization. As the system and system elements are defined in the Technical Processes, the information items that represent those definitions should be captured and included as knowledge assets for the domain. The infrastructure should be established to include mechanisms to easily identify, access, and determine the applicability of the assets.

- *Manage knowledge, skills, and knowledge assets.*
  – As the domain, family of systems, or product line changes, ensure the associated knowledge assets are revised or replaced to reflect the latest information. In addition, the associated domain models and architectures also may need to be revised.
  – Assess and track where the knowledge assets are being used. This can help understand the utility of specific assets, as well as determine whether they are being applied where they are applicable.
  – Determine whether the knowledge assets reflect current technology and continue to evolve.

Common approaches and tips:

- The planning for KM may include:
  – Plans for obtaining and maintaining knowledge assets for their useful life.
  – Characterization of the types of assets to be collected and maintained along with a scheme to classify them.

- Criteria for accepting, qualifying, and retiring knowledge assets.
- Procedures for controlling changes to the knowledge assets.
- A mechanism for knowledge asset storage and retrieval.

• In developing an understanding of the domain, it is important to identify and manage both the commonalities (such as features, capabilities, or functions) and the differences or variations of the system elements (including where a common system element has variations in parameters depending on the system instance). The domain representations should include:
  - Definition of the boundaries.
  - Relationships of the domains to other domains.
  - Domain models that incorporate the commonalities and differences allowing for sensitivity analysis.
  - An architecture for a system family or product line within the domain, including their commonalities and variations (see Section 4.2.4).

Elaboration

General KM Implementation. KM focuses on capturing the organizational, project, and individual knowledge for use throughout the organization in the future. It is important to capture end-of-project lessons learned prior to the project personnel moving on to new assignment. However, an effective Knowledge Management process has the knowledge capture mechanisms in place to capture the relevant information throughout the life of the project, rather than trying to piece it together at the end.

KM for Product Lines and Reuse. KM also includes identification of systems that are part of a product line or system family (see Section 4.2.4) and system elements that are designed for reuse. For the first instance of these systems and system elements, the KM system needs to capture the domain engineering artifacts in a way to facilitate their use in the future. For subsequent instances, the KM system needs to provide the domain engineering information and capture any variations, updates of technology, and lessons learned. Issues important to the organization include:

• Definition and planning of KM activities for domain engineering and asset preservation, including tasks dedicated to domain engineering of product lines or system families and to the preservation of reusable assets.

• Integration of architecture management into the KM system including frameworks, architecture reuse, architecture reference models, architecture patterns, platform-based engineering, and product line architecture.

• Characterization of the types of assets to be collected and maintained including an effective means for users to find the applicable assets.

• Determination of the quality and validity of the assets.

Potential Reuse Issues. There are serious traps in reuse, especially with respect to commercial off-the-shelf (COTS) (see Section 4.3.3) and non-developmental item (NDI) elements:

• Do the new system or system element requirements and operational characteristics closely match the prior one? Trap: the prior solution was intended for a different use, environment, or performance level, or it was only a prototype.

• How did the prior system or system element perform? Trap: it worked perfectly, but the new application is outside the qualified range (e.g., using a standard car for a high-speed track race).

• Is the new system or system element going to operate in the same environment as the prior one? Trap: it is not certain, but there is no time to study it. One NASA Mars probe was lost because the development team used a radiator design exactly as was used on a successful satellite in Earth orbit. When the Mars mission failed, the team then realized that Earth orbiting environment, while in space, is different from a deep space mission.

• Is the system/system element definition defined and understood (i.e., requirements, constraints, operating scenarios, etc.)? Trap: too often, the development team assumes that if a reuse solution will be applied (especially for COTS), there is no need for well-defined system definition. The issues may not show up until systems integration, causing major cost and schedule perturbations.

- Is the solution likely to have emergent requirements/behaviors where the reuse is being considered? Trap: a solution that worked in the past was used without consideration for the evolution of the solution. If COTS is used, there may be no way to adapt or modify it for emergent requirements.

A properly functioning KM system paired with well-defined processes and engineering discipline can help avoid these problems.

2.3.4   Technical Management Processes

The engineering of new or existing systems is managed by the conduct of projects. For this reason, it is important to understand the contribution of SE to the management of the project. This contribution is provided through the Technical Management Processes, which ensure the successful management of the SE effort within the project.

The Technical Management Processes are defined in ISO/IEC/IEEE 15288 as follows:

> [5.7.4] The Technical Management Processes are concerned with managing the resources and assets allocated by organization management and with applying them to fulfill the agreements into which the organization or organizations enter. The Technical Management Processes relate to the technical effort of projects, in particular to planning in terms of cost, timescales and achievements, to the checking of actions to help ensure that they comply with plans and performance criteria, and to the identification and selection of corrective actions that recover shortfalls in progress and achievement. They are used to establish and perform technical plans for the project, manage information across the technical team, assess technical progress against the plans for the system products or services, control technical tasks through to completion, and to aid in the decision-making process.

Technical management, which is the application of technical and administrative resources to plan, organize and control engineering functions, consists of the following eight processes: Project Planning, Project Assessment and Control, Decision Management, Risk Management, Configuration Management, Information Management, Measurement, and Quality Assurance. The Technical Management Processes are used consistently throughout the system life cycle so that system-specific Technical Processes can be conducted effectively. They work with the project management processes to establish and perform technical plans, manage information across the technical teams, assess technical progress against the plans, control technical tasks and risks through to completion, and aid in the decision-making process.

SE practitioners continually interact with project management practitioners. Both contribute to the project with unique professional competences. A life cycle from the project management practitioner's point of view (project start–project end) is defined differently than from the SE practitioner's point of view (system concept to system retirement). But there is a "shared space" where both must collaborate to drive the team's performance and success (Langley, et al., 2011). See Section 5.3.3 for treatment of the integration between SE and project management.

2.3.4.1   *Project Planning Process*

Overview

Purpose As stated in ISO/IEC/IEEE 15288,

> [6.3.1.1] The purpose of the Project Planning process is to produce and coordinate effective and workable plans.

Description Project planning starts with the identification of a new potential project and continues after the authorization and activation of the project until its termination. The Project Planning process is performed in the context of the organization, and in compliance with the Life Cycle Model Management process (see Section 2.3.3.1) that identifies and establishes relevant policies and procedures applicable to all projects owned by the organization.

The Project Planning process identifies the project objectives, technical activities, interdependencies, resource requirements, risks and opportunities, and management approach for the technical effort. The planning includes the estimates of needed resources and budgets and the determination of the need for project enablers, including specialized equipment, facilities, and specialists during the project to improve efficiency and effectiveness and decrease cost overruns. This requires coordination across the set of processes to develop a set of consistent planning for all activities. For example, different disciplines work together in the performance of the System Requirements Definition, System Architecture Definition, and Design Definition processes to evaluate the parameters such as producibility, testability, operability, maintainability, and sustainability against product performance. Project tasking may be concurrent to achieve the best results.

Project planning establishes the direction necessary to enable execution of the project and the assessment and control of the project progress. It identifies the details of the work and the right set of personnel, skills, infrastructure, and facilities with a schedule for needed resources from within and outside the organization.

Inputs/Outputs Inputs and outputs for the Project Planning process are listed in Figure 2.23. Descriptions of each input and output are provided in Appendix E.

Process Activities The Project Planning process includes the following activities:

- *Define the project.*
  - Analyze the project supply response and related agreements to define the project objectives, assumptions, constraints, and scope.
  - Identify or establish tailoring of organization procedures and practices to carry out planned effort (see Section 4.1).

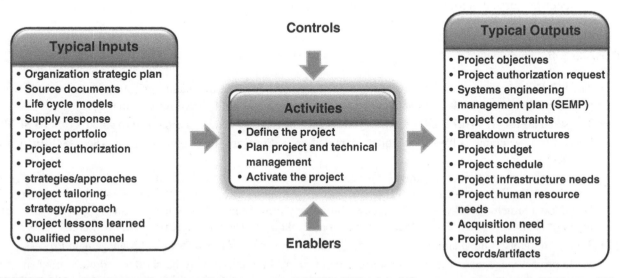

FIGURE 2.23   IPO diagram for Project Planning process. INCOSE SEH original figure created by Shortell, Walden, and Yip. Usage per the INCOSE Notices page. All other rights reserved.

- Develop or select Breakdown Structures based on the evolving system architecture (see paragraph on Breakdown Structures hereafter) and the constraints on the resources.
- Define and maintain a life cycle model that could be tailored from the defined life cycle models of the organization. This includes the identification of major milestones, decision gates, and project reviews.

- *Plan project and technical management.*
  - Establish the roles and responsibilities for project authority.
  - Define top-level work packages for each activity identified. Each work package should be tied to required resources including procurement strategies.
  - Develop a project schedule (e.g., an integrated project schedule, a SE Master Schedule (SEMS)) based on objectives and work estimates.
  - Determine the infrastructure and services needed for the project.
  - Estimate the costs and establish a project budget.
  - Plan the acquisition of materials, goods, and enabling systems.
  - Generate and communicate a Systems Engineering Management Plan (SEMP), also called a Systems Engineering Plan (SEP), for project and technical management/execution, including the technical reviews and audits (see Section 2.1.4).
  - Contribute to the quality management, configuration management, risk management, information management, and measurement plans to meet the needs of the project with regard to SE efforts (may be the SEMP for smaller projects).
  - Establish the achievement criteria to be used for major milestones, decision gates, and internal reviews.
  - Establish criteria for project performance.
- *Activate the project.*
  - Obtain project authorization and resources. The Portfolio Management process provides this authorization (see section 2.3.3.3).
  - Obtain authorization for the necessary project resources.
  - Commence execution of the project plans.

Common approaches and tips:

- The SEMP (or equivalent technical planning) is an important outcome that identifies activities, key events, work packages, and resources. It references other planning artifacts that are tailored for use on the project.
- The standard ISO/IEC/IEEE 24748–4 on Systems Engineering Planning is a reference to aid in writing a SEMP.
- Plans for developing software are often captured in a Software Development Plan. (See ISO/IEC/IEEE 24748–5.)
- The creation of the Work Breakdown Structure (WBS) and other breakdown structures (e.g., Function Tree/Functional Breakdown Structure (FBS), Product Tree/Product Breakdown Structure (PBS), Organizational Breakdown Structure (OBS), Cost Breakdown Structure (CBS)) is an activity where SE and Project Management intersect (Forsberg, et al., 2005). (See paragraph on Breakdown Structures hereafter and Section 5.3.3.)
- Taking shortcuts in the planning process reduces the effectiveness of other Technical Management Processes.
- Agile project management methods also include planning—the cycles may be shorter and more frequent, but planning is an essential process. Agile planning process is not related to the entire project but addressing only the next already known iterations while applying learning from the previous iterations.
- Defining project objectives, value, and the criteria for success are critical to guide project decision making. The project value should be expressed in technical performance measures (TPMs) (Roedler and Jones, 2006) (see Section 2.3.4.7).

- Incorporate risk assessment early in the planning process to identify areas that need special attention or contingencies (see Section 2.3.4.4). Always attend to the technical risks (PMI, 2013).
- If a Project Management Plan (PMP) already exists or is in preparation (in accordance with practices as defined by the Project Management Body of Knowledge (PMBOK®) (2021) from the Project Management Institute (PMI), for example), then it is important to coordinate in order to have a global consistency between these artifacts. The SEMP should reference, or provide a link to, the PMP for direction on how the SEMP will be updated and controlled on the project.

Elaboration

Project Planning Concepts. Project planning estimates the project budget and schedule against which project progress will be assessed and controlled. SE practitioners and PM practitioners must collaborate in project planning. SE practitioners perform technical management activities consistent with project objectives (see Section 5.3.3). Technical management activities include planning, scheduling, reviewing, and auditing the SE process as defined in the SEMP.

Systems Engineering Management Plan (SEMP). The SEMP is the key technical management plan that integrates the SE effort. It defines how the total set of engineering processes will be organized, structured, and conducted and how it will be controlled to provide a product that satisfies stakeholder requirements. The SEMP typically includes the following content (a complete outline can be found in ISO/IEC/IEEE 24748–4 (2016), which is aligned with ISO/IEC/IEEE 15288 and this handbook):

- organization of how SE interfaces with the other parts of the organization
- responsibilities and authority of the key engineering roles
- clear system boundaries and scope of the system
- key, technical objectives, assumptions, and constraints (or link to them)
- infrastructure support and resource management (i.e., facilities, tools, IT, personnel)
- technical schedule, including key milestones, decision gates, and associated criteria
- definition of the SE processes, including interaction with other engineering and project processes
- approach and methods for planning and executing the Technical Processes (see Section 2.3.5)
- approach and methods for planning and executing the Technical Management Processes (see Section 2.3.4)
- approach and methods for planning and executing applicable quality characteristic (QC) approaches (see Section 3.1)
- major technical deliverables of the project

A SEMP should be prepared early in the project, submitted to the customer (or to management for in-house projects), and used in technical management for the concept and development stages of the project. The format of the SEMP can be tailored to fit project, customer, or company standards. In addition to being a stand-alone artifact, the SEMP can be a part of an integrated project plan, be a distributed set of plans, or be in a format other than a document (e.g., it may be composed of different models, management tools, or other artifacts).

The SEMS is an essential part of the SEMP and a tool for project control because it identifies the critical path of technical activities in the project. The schedule of tasks and dependencies helps prioritize the effort and justify requests for personnel and resources needed throughout the development life cycle.

Breakdown Structures. The purpose of the breakdown structures is to hierarchically decompose constructs in manageable and understandable elements. In projects, breakdown structures provide:

- a framework for ensuring that all requirements, functions, and products of the system design are identified and arranged in a logical relationship that can be traced to, and satisfy, the business and stakeholder needs;
- an identification of all activities and resources needed to the product;
- a cost relationship to the activities being performed;
- an organizational context for the project to perform the activities needed to the product;

- an identification, by name, within the organization of the responsible person for performing each activity;
- a basis for configuration control once a particular project breakdown structure is baselined, and a basis for effective management of changes;
- a framework to help identify risks and subsequent risk management;
- a basis for financial control and interface responsibilities resulting from business agreements.

The SE practice is to derive system functions from requirements and then allocate these functions into products or services, usually through the development of a functional and physical architecture (see Section 2.3.5.4). Functions and products are organized in breakdown structures that have the organizational framework of a tree, such as Function Tree and Product Tree. The Function Tree also, called Functional Breakdown Structure (FBS), is a breakdown of the functions of the required SoI into successively lower levels of its functional architecture. The Function Tree includes the technical characteristics of each function. The Product Tree, also called Product Breakdown Structure (PBS), is a breakdown of the SoI into successively lower-level details of its physical architecture (see Section 1.3.5).

The work to be carried out to reach the project objectives can be organized in a breakdown structure, as a hierarchical tree, where the lower-level activities provide more details. This is the Work Breakdown Structure (WBS), which is based on the FBS in the initial stages of system maturity (e.g., feasibility, conceptual design) and the PBS in the later stages.

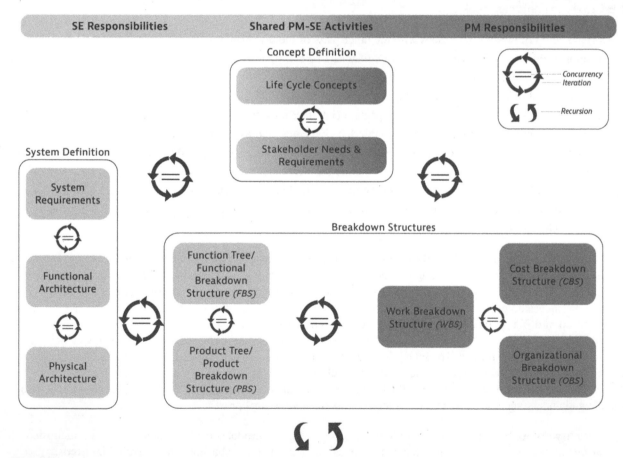

FIGURE 2.24   The breakdown structures. INCOSE SEH original figure created by Roussel and Dazzi on behalf of the INCOSE PM-SE Integration Working Group. Usage per the INCOSE Notices page. All other rights reserved.

The WBS includes all activities needed to develop the product. Each branch of the WBS is used to define a work package (WP). Each WP describes the work to be performed, related input and output, who is doing the work, the related interfaces with other WPs, the related cost and deliverables and the key dates and milestones. The WBS serves as a reference for the identification of cost elements arranged into a Cost Breakdown Structure (CBS). Along with the overall management organization, the WBS is also used to determine who does what. This is represented in an Organizational Breakdown Structure (OBS), which is a hierarchical tree of the organizational elements. Figure 2.24 illustrates these different project breakdown structures with their relationships. See Section 5.3.3 for the relationship between PM and SE.

2.3.4.2 *Project Assessment and Control Process*

Overview

Purpose As stated in ISO/IEC/IEEE 15288,

> [6.3.2.1] The purpose of the Project Assessment and Control process is to assess if the plans are aligned and feasible; determine the status of the project, technical and process performance; and direct execution to help ensure that the performance is according to plans and schedules, within projected budgets, to satisfy technical objectives.

Assessments are scheduled periodically and for all milestones and decision gates. The intention is to maintain good communications within the project team and with the stakeholders, especially when deviations are encountered. The Project Assessment and Control process uses these assessments to direct the efforts of the project, including redirecting the project when the project does not reflect the anticipated maturity.

Description The Project Planning process (see Section 2.3.4.1) identified details of the work effort and expected results. The Project Assessment and Control process collects data to evaluate the adequacy of the project infrastructure, the availability of necessary resources, and the compliance with project performance measures. Assessments also monitor the technical progress of the project and may identify new risks or areas that require additional investigation. A discussion of the creation and assessment of measures is found in Section 2.3.4.7—Measurement process.

The rigor of the Project Assessment and Control process is directly dependent on intrinsic characteristics of the project and the SoI, such as the complexity, urgency, and consequence of failure to deliver or failure of the SoI. Project control involves both preventive and corrective actions taken to ensure that the project is performing according to plans and schedules and within projected budgets. The Project Assessment and Control process may trigger activities within Technical Management Processes.

Inputs/Outputs Inputs and outputs for the Project Assessment and Control process are listed in Figure 2.25. Descriptions of each input and output are provided in Appendix E.

Process Activities The Project Assessment and Control process includes the following activities:

- *Plan for project assessment and control.*
  - Develop a strategy/approach for assessment and control for the project.
- *Assess the project.*
  - Determine whether project objectives and plans are aligned with the project context.
  - Determine cost, schedule, and performance variances for the project and technical effort through assessment of status versus plans.
  - Evaluate the effectiveness and efficiency of the performance of project activities.
  - Determine if the project roles, responsibilities, accountabilities, and authorities are adequate.
  - Assess the adequacy and the availability of the project infrastructure and resources.

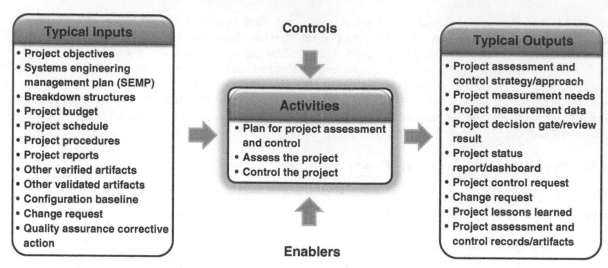

FIGURE 2.25   IPO diagram for Project Assessment and Control process. INCOSE SEH original figure created by Shortell, Walden, and Yip. Usage per the INCOSE Notices page. All other rights reserved.

- Using project measures and milestone status, assess the progress of the project.
- Conduct required reviews, audits, and inspections to determine readiness to proceed to the next milestone (see paragraph on reviews and audits in the elaboration and Section 2.1.4 for more details).
- Monitor areas of high risk, such as critical tasks and new technologies/applications (see Section 2.3.4.4).
- Recommend actions that are warranted by measurement results and other project information.
- Make recommendations for adjustments to project plans—these are input to the project control process and other decision-making processes.
- Record and provide the project status report, gathering all needed measures on technical progress aspects (e.g., performance, requirements compliance, verification and validation progress) and technical management aspects (e.g., schedule, cost, risk, configuration status).
- Communicate status in the project status report as designated in agreements, policies, and procedures.

- *Control the project.*
  - Initiate preventive actions when assessments indicate a trend toward deviation.
  - Initiate problem resolution when assessments indicate nonconformance with performance success criteria.
  - Initiate corrective actions when assessments indicate deviation from approved plans.
  - Update project planning as needed based on the project control or corrective actions.
  - Implement change actions to reflect contractual changes to cost, time, or quality. This is usually due to the impact of acquirer or supplier request.
  - Authorize the project to proceed when assessments support a decision gate or milestone event.

Common approaches and tips:

- One way to remain updated on project status is to conduct regular team meetings. Short stand-up meetings on a daily or weekly schedule are effective for smaller groups.
- Prevailing wisdom suggests that "what gets measured gets done," but projects should avoid the collection of measures that are not used in decision making.
- Good practices show that status should be concise and visual (e.g., usage of Red/Yellow/Green "traffic lights") in order to quickly and easily identify the critical issues on which urgent actions for recovery are required. Another useful tool is a project dashboard that provides a timely and easy summary of status.
- A template for the project status report is a good practice. This template may be included in the SEMP (or PMP).
- Methods and techniques for Project Assessment and Control should be formally described in the SEMP (or PMP) and agreed with the project team.
- The Project Management Institute (PMI) provides industry-wide guidelines for project assessment, including Earned Value Management techniques.
- Project teams need to identify critical areas and control them through measurement, risk management, analysis, configuration management, and information management.
- The Project Assessment and Control process requires close cooperation between the PM practitioner and SE practitioner, with PM being accountable for the overall results of the project and SE being accountable for the achievement of the technical activities.
- The typical common responsibilities between PM and SE practitioners are risk management, external supplier relations, quality management and life cycle planning.
- An effective feedback control process is an essential element to enable the improvement of project performance.
- Incremental and evolutionary models typically schedule frequent assessments and make project control adjustments on tighter feedback cycles than sequential development models (see Section 2.2).
- Tailoring of organization processes and procedures (see Section 4.1) should not jeopardize any certifications. Processes must be established with effective reviews, assessments, audits, and improvements.
- Standard ISO/IEC 24748–8 / IEEE 15288.2 (2014) is a useful reference on how to define and manage technical reviews and establish requirements for the related milestones.

Elaboration

Integration of Technical Management Artifacts. Each of the Technical Management Processes provides essential insight into the health and progress of the project through the life cycle with respect to the specific focus of the particular process. However, it is important to look at the results of these processes in an integrated view, especially since there are relationships between these processes and their artifacts. For example, the results of the Measurement process provide useful insights into risks, technical reviews and audits, and quality assurance, as well as many other things. Similarly, other processes may identify new information needs for which new measures should be initiated. Mechanisms should be put in place to provide an integrated view of the results or artifacts in a way that the decision makers can interpret quickly and see trends and trigger points to aid decisions. Two such mechanisms are the project status report and the project dashboard. Both organize and provide a summary of similar information about the project; the status report usually presents the information in report form and the dashboard is usually a digital representation that uses gauges, graphs, indicators, or other visual representations of the information. In both mechanisms relationships are shown and trends or areas needing attention are highlighted.

Technical Reviews and Audits. Technical reviews and audits are a foundational element of an effective SE approach and form the backbone of robust technical assessment. Technical reviews and audits provide a venue for baselining stakeholder and system requirements, evaluating the system's technical maturity, and identifying and assessing risks to system performance, cost, and schedule. In order for a project's technical management to have a balanced information basis on which to base any required project control actions, each technical review or audit should be conducted from an integrated project viewpoint, including technical status and progress, cost and schedule status, and impacts and risk assessment, to help ensure that technical review decisions do not create unrecognized and unacceptable future project impacts. See Section 2.1.4 Technical Reviews and Audits and ISO/IEC 24748–8 / IEEE 15288.2 (2014) for more information.

2.3.4.3 *Decision Management Process*

Overview

Purpose As defined by ISO/IEC/IEEE 15288,

> [6.3.3.1] The purpose of the Decision Management process is to provide a structured, analytical framework for objectively identifying, characterizing and evaluating a set of alternatives for a decision at any point in the life cycle and select the most beneficial course of action.

Table 2.4 provides a partial list of decision situations (opportunities) that are commonly encountered throughout a system's life cycle. Buede and Miller (2009) provide a much larger list.

Decision management as a critical SE activity. Consider the number of decisions involved in identifying a business/mission need, crafting a technology development strategy, defining the stakeholder and system requirements, selecting a system architecture, converging on a detailed design, developing verification and validation plans, determining make-or-buy decisions, creating production ramp-up plans, crafting maintenance and logistics plans, and selecting disposal approaches. New product developments entail an array of interrelated decisions throughout the system life cycle.

Description The Decision Management process transforms a broadly stated decision situation into a recommended course of action and associated implementation plan. The process requires a decision maker with full responsibility, authority, and accountability for the decision, a decision analyst with a suite of decision tools, subject matter experts with performance models, and a representative set of end users and other stakeholders (Parnell, et al., 2013). The decision process is executed within the policy and guidelines established by the system sponsor. A well-structured decision process will capture and communicate the impact that different value judgments have on the overall decisions and facilitate the search for alternatives that remain attractive across a wide range of value schemes.

TABLE 2.4   Partial list of decision situations (opportunities) throughout the life cycle

Life cycle stage	Decision situation (opportunity)
Concept	Assess technology opportunity/initial business case
	Craft a technology development strategy
	Inform, generate, and refine a capability artifact
	Conduct analysis of alternatives
	Supporting program initiation decision
	Select system architecture
Development	Select system element
	Select lower-level elements
	Select verification and validation methods
	Perform make-or-buy decision
Production	Select production process and location
Utilization, support	Select maintenance approach
Retirement	Select disposal approach

Inputs/Outputs Inputs and outputs for the Decision Management process are listed in Figure 2.26. Descriptions of each input and output are provided in Appendix E.

Process Activities The Decision Management process includes the following activities:

- *Prepare for decisions.*
  - Develop the decision management strategy/approach for system or project decisions.
  - Establish and challenge the decision statement and clarify the decision to be made.
  - Determine the analyses methods, other processes, and tools required to support decision activities. (Note that the System Analysis process (see Section 2.3.5.6) is often applied to perform analyses to provide input for the decisions.)
  - Provide resources to implement the strategy.
- *Analyze the decision information.*
  - Frame, tailor, and structure each decision.
  - Develop objectives and measures.
  - Generate creative alternatives.
  - Assess alternatives via deterministic analysis.
  - Synthesize results.
  - Identify uncertainties and conduct probabilistic analysis.
  - Assess impact of the uncertainties.
  - Improve alternatives.
  - Communicate trade-offs.
  - Present recommendation and implement action plan.
- *Make and manage decisions.*
  - Record the decision with relevant data, models, and supporting documentation (i.e., the decision authority, source, and rationale)
  - Describe analyses methods, other processes, and tools actually used to support decision activities.
  - Communicate new directions from the decision

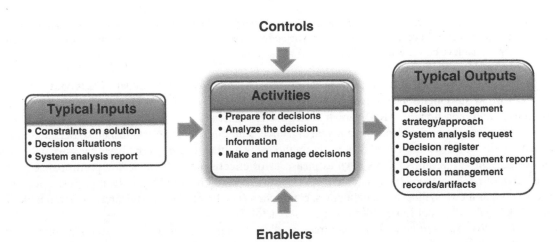

FIGURE 2.26   IPO diagram for the Decision Management process. INCOSE SEH original figure created by Walden, Shortell, and Yip. Usage per the INCOSE Notices page. All other rights reserved.

Common Approaches and Tips:

- Since there are many decisions across the spectrum of project management, system definition, and life cycle activities, the Decision Management process is applied in conjunction with most of the Technical Management and Technical Processes.
- It is important to verify and validate the data and assumptions used in the decision analyses, since the validity of the analysis results depends on the use of valid data and assumptions, and the application of appropriate analytic methods.

Elaboration

SE practitioners face many decision situations throughout the life cycle of a project. They must choose the analytical approach that best fits the frame and structure of each decision problem. For instance, when there are "clear, important, and discrete events that stand between the implementation of the alternatives and the eventual consequences" (Edwards, et al., 2007), a decision tree is often a well-suited analytical approach, especially when the decision structure has only a few decision nodes and chance nodes. As the number of decision nodes and chance nodes grows, the decision tree quickly becomes unwieldy and loses some of its communicative power. Furthermore, decision trees require end node consequences be expressed in terms of a single number.

The decision management method most employed by SE practitioners is a multiple objective decision approach (MODA) (Parnell, 2016), in which an objective function is formulated to synthesize an alternative's response across multiple, often competing, objectives. The aim is to define, measure, and assess stakeholder value and then synthesize this information to facilitate the decision maker's search for alternatives that represent the best balance with often competing objectives. If time and funding allow, SE practitioners may want to conduct trade studies using several techniques, compare results, and reconcile any differences to ensure findings are robust.

The following are a summary of decision management good practices.

Framing, Tailoring, and Structuring Decisions. Capturing a description of the system baseline, as well as the concept of operations with some indication of system boundaries and anticipated interfaces, helps ensure the understanding of the decision context. This includes such details as the time frame allotted for the decisions, an explicit list of stakeholders, a discussion regarding available resources, and expectations regarding the type of action to be taken as a result of the decision at hand. It may also include decisions anticipated in the future (Edwards, et al., 2007).

Developing Objectives and Measures. Defining the decision to be made may require balancing a large number of ambiguous and potentially conflicting stakeholder need statements, engaging in uncomfortable discussions regarding the relative priority of each requirement, and establishing walkaway points and stretch goals. Per Keeney (2002):

> "Most important decisions involve multiple objectives, and usually with multiple-objective decisions, you cannot have it all. You will have to accept less achievement in terms of some objectives to achieve more on other objectives. But how much less would you accept to achieve how much more?"

Use the information obtained from the Business or Mission Analysis, Stakeholder Needs and Requirements Definition, System Requirements Definition, System Architecture Definition, and Design Definition processes to develop objectives and measures for MODA models that use fundamental objectives (why, what, where, and when), but not means objectives (how). For each fundamental objective, a measure must be established so that alternatives that more fully satisfy the objective receive a better score on the measure than those alternatives that satisfy the objective to a lesser degree. These measures (also known as measures of effectiveness (MOEs), key performance parameters (KPPs), measures of performance (MOPs), technical performance measures (TPMs), critical performance measures, attributes, criterion, or metrics) must be unambiguous, comprehensive, direct, operational, and understandable (Keeney and Gregory, 2005) (Roedler and Jones, 2005) (see Section 2.3.4.7).

Generating Creative Alternatives. For many trade studies, the alternatives will be systems composed of many interrelated system elements. It is important to establish a meaningful product structure for the SoI and to apply this product structure consistently throughout the decision analysis. The product structure should be a useful decomposition of the elements of the SoI that explores the trade space. Each alternative is composed of specific design choices for each element. The ability to communicate the differentiating design features of the alternatives is essential. An alternative to a finite number of alternatives is Set-Based Design (SBD). SBD has been shown to effectively and efficiently explore the trade space (Specking, et al., 2018).

Assessing Alternatives via Deterministic Analysis. The decision team should engage subject matter experts by creating models using operational and test data along with the defined objectives, measures, and alternatives to assess performance and using structured scoring sheets. Each score sheet contains a summary description of the alternative and the scoring criteria. Ideally, the models and simulations should be integrated with the performance, value, and cost models so a design change impacts all models.

Synthesizing Results. Using the data summarized in the objective measure consequence table, explore, understand, aggregate the data, and display results in a way that facilitates stakeholder understanding.

Identifying Uncertainty and Conducting Probabilistic Analysis. It is important to identify potential uncertainty surrounding the assessed score and variables that could impact one or more scores (see Section 1.4.1). One example of uncertainty is that system concepts are described as a collection of system element design choices, but knowledge of the system element performance during system design is often incomplete. Subject matter experts can often assess an upper, nominal, and lower bound score by making three separate assessments: (i) assuming a low performance, (ii) assuming moderate performance, and (iii) assuming high performance.

Accessing Impact of Uncertainty. Decision analysis uses many forms of sensitivity analysis including line diagrams, tornado diagrams, waterfall diagrams, and several uncertainty analyses, including Monte Carlo simulation, decision trees, and influence diagrams (Parnell, et al., 2013). Monte Carlo simulations are used to identify the relative impact of each source of uncertainty on the performance, value, and cost of each alterative. Risks should be identified when significant uncertainty is present.

Improving Alternatives. One could be tempted to end the decision analysis here, highlight the alternative that has the highest total value, and claim success. Such a premature ending would not be considered good practice. Good practice includes further analysis to mine the data generated for the first set of alternatives to reveal opportunities to modify some system element design choices to identify untapped value and reduce risk.

Communicating Trade-Offs. The decision team should identify key observations regarding what stakeholders seem to want and what they may be willing to give up to achieve it. The decision team highlights the design decisions that are least significant and/or most influential and provide the best stakeholder value. In addition, the important uncertainties and risks should also be identified. Observations regarding combinatorial effects of various design decisions are also important products of this process step. Finally, competing objectives that are driving the trade-offs should be highlighted as well.

Presenting Recommendations and Implementing the Action Plan. It is helpful to clearly describe the recommendation as an actionable task list to increase the likelihood of the decision analysis leading to some form of action showing tangible value. Decisions should be documented using digital engineering artifacts. Reports that include the analysis, decisions, and rationale are important for historical traceability and future decisions.

2.3.4.4 Risk Management Process

Overview

Purpose As stated in ISO/IEC/IEEE 15288,

> [6.3.4.1] The purpose of the Risk Management process is to identify, analyze, treat and monitor the risks continually.

Description Risk Management is a disciplined approach to dealing with the uncertainty that is present throughout the entire system life cycle (see Section 1.4.1). Opportunity management may be performed in conjunction with or as part of risk management. A primary objective of risk management is to identify and manage uncertainties that threaten or reduce the value provided by a business enterprise or organization. A primary objective of opportunity management is to identify and manage uncertainties that enhance or increase the value provided by a business enterprise or organization. Since risk cannot be reduced to zero, another objective is to achieve a proper balance between risk and opportunity.

Risk management, as it relates to SE, is defined in ISO/IEC/IEEE 15288 and elaborated upon in ISO/IEC/IEEE 16085 (2021). As stated in ISO/IEC/IEEE 16085,

> [6.1] The Risk Management process is a continual process for systematically addressing risk throughout the life cycle of a system, product, or service. It can be applied to risks related to the acquisition, development, maintenance, or operation of a system.

When using this process for opportunity management, the above statement, with the term "opportunity" substituted for the term "risk," is also true.

Inputs/Outputs Inputs and outputs for the Risk Management process are listed in Figure 2.27. Descriptions of each input and output are provided in Appendix E.

Process Activities The Risk Management process includes the following activities:

- *Plan risk management.*
  - Develop the risk management strategy/approach.
  - Capture the Risk Management process context, including risk categories.
- *Maintain the risk profile.*
  - Capture the thresholds and conditions of the risks.
  - Establish and maintain a risk profile to include context of the risk and its likelihood of occurrence, severity of consequences, risk thresholds, and priority and the risk action requests along with the status of their treatment.
  - Ensure updates of the risk profile are available to relevant stakeholders.

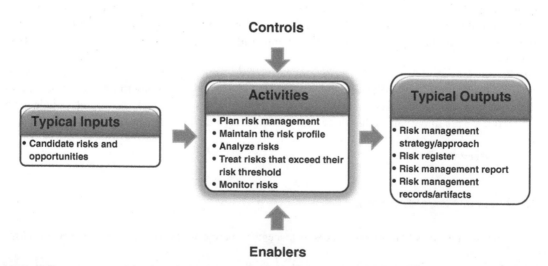

FIGURE 2.27   IPO diagram for Risk Management process. INCOSE SEH original figure created by Shortell, Walden, and Yip. Usage per the INCOSE Notices page. All other rights reserved.

- *Analyze risks.*
  - Identify risks consistent with the risk management strategy/approach.
  - For each risk, estimate its likelihood and consequence of occurrence.
  - For each risk, use the risk thresholds to evaluate the risk for potential treatment.
  - For risks that exceed the threshold, capture recommended treatment strategies and measures.
- *Treat risks that exceed their risk threshold.*
  - Identify a set of feasible alternatives for the treatment of risks.
  - Establish measures to provide insight into the risk treatment effectiveness.
  - Execute the treatments for the risks.
  - When management action is needed for risk treatments, ensure effective coordination.
- *Monitor risks.*
  - Maintain the record of risk items and how they were treated.
  - Monitor high priority risks.
  - Monitor risks and the risk management context to capture changes and update priorities and actions.
  - Throughout the life cycle, monitor for new risks and sources of risk.

When using this process for opportunity management, the above process activity description, with a few adjustments in terminology, and the term "opportunity" substituted for the term "risk," is reasonable.

Common approaches and tips:

- In the Project Planning process, a risk management plan (RMP) is tailored to satisfy the policies, procedures, standards, and regulations related to and affecting the management of risks for the project.
- Process Enablers—It has been found that an organization's structure and culture can have a significant effect on the performance of the Risk Management process. ISO 31000 (2018), outlines a model that advocates the establishment of principles for managing risk and a framework for managing risk that work in concert with the process for managing risk.
- Typical strategies for coping with risk include transference, avoidance, acceptance, or taking action to reduce the potential negative effects of the situation.
- Most Risk Management processes include a prioritization scheme whereby risks with the greatest potential negative consequence and the highest likelihood are treated before those deemed to have lower potential negative consequences and lower likelihood. The objective of risk management is to balance the allocation of resources such that a minimum amount of resources achieves the greatest risk mitigation (or opportunity realization) benefits.
- Communication errors and misunderstandings can be prevented by defining and communicating the risk terminology to be used by the project and including with the project's risk management plan (RMP).
- Experience has shown that terms such as "positive risk" and concept models that define opportunity as a subset of risks serve only to confuse. Take care to define the terminology and concepts to be used by the project team and provide training to reinforce a common understanding.
- Practices used for writing good requirements help with risk statements. For example, one good practice for identifying and clarifying risks is to use an "if <situation>, then <consequence>, for <stakeholder >" pattern. This pattern helps to determine the validity of a risk and assess its magnitude or importance.
- Risk management is most successful when risk-based thinking is embraced and integrated into the culture. All personnel are responsible for identifying risks early and continuously throughout the project life cycle.

- Negative feedback toward personnel who identify a potential problem will discourage the full cooperation of engaged stakeholders and could result in failure to identify and address serious risk-laden situations. Conduct a transparent Risk Management process to encourage all stakeholders to assist in risk mitigation efforts.
- Some situations can be difficult to categorize in terms of probability and consequences; involve all relevant stakeholders in this evaluation to capture the maximum variety in viewpoints.
- Risk measurement is not an exact science. Variation in stakeholder perspectives, perceptions, and tolerance levels, along with high uncertainty in available data, can make reliance on quantitative measures of risk insufficient. For example, some low-likelihood/high-severity risks might require treatment and monitoring regardless of the estimated likelihood of occurrence (Taleb, 2018) (Siegel, 2019).
- External risks are often neglected in project management. External risks are risks caused by or originating from the surrounding environment of the project (Fossnes, 2005). Project participants often have no control or influence over external risk factors, but they can learn to observe the external environment and eventually take proactive steps to minimize the impact of external risks on the project. The typical issues are time-dependent processes, rigid sequence of activities, one dominant path for success, and little slack.

Elaboration

Definitions of Risk. Few terms used in engineering have as many different published definitions as the term "risk." In practice, risk terminology and concepts vary considerably across industries; however, most published definitions of risk align with one of two concept models. Below are two prominent definitions of risk that capture the essence of both concepts:

- The effect of uncertainty on objectives [see ISO/IEC/IEEE 15288, ISO Guide 73, ISO/IEC/IEEE 16085, ISO/IEC 31000, ISO 27000]
- The combination of the probability of occurrence of harm and the severity of that harm [see ISO Guide 51, ISO 22367, ISO 14971]

Both definitions may be used in an SE project. The first definition includes the concept that effects may be negative or positive. In this respect the first definition accommodates use of the second definition. In SE it is common to use the term "risk" when referring to scenarios with a negative effect, and the term "opportunity" when referring to scenarios with a positive effect. The second definition (which accommodates only negative effects) is commonly used in safety engineering, and its use may be required in order to demonstrate compliance to risk management standards and regulations applicable to products and systems that impact public health, safety, and security. For example, in the medical industry (see Section 4.4.2), particularly for medical devices, risk management is often centered on product (patient and user) safety risk (referred to as system safety in this handbook, see Section 3.1.11).

Evolving Risk and Opportunity Management Concepts. According to Conrow (2003), "Traditionally, risk has been defined as the likelihood of an event occurring coupled with a negative consequence of the event occurring. In other words, a risk is a potential problem—something to be avoided if possible, or its likelihood and/or consequences reduced if not." As a corollary to risk, Conrow (2003) defines opportunity as "the potential for the realization of wanted, positive consequences of an event." The idea of considering opportunities and positive outcomes (in addition to negative outcomes) as an integral part of a Risk Management process has gained favor with some experts and practitioners. New risk and risk management concepts intended to support this broadened scope for risk management are evolving.

The measurement of risk has two components (see Figure 2.28):

- The likelihood that an event will occur
- The undesirable consequence of the event if it does occur

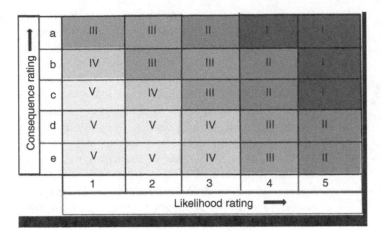

FIGURE 2.28 Level of risk depends upon both likelihood and consequence. From ISO/IEC 31010 (2019). Used with permission. All other rights reserved.

The generic consequence/likelihood matrix in Figure 2.28 is a way to display risks according to their consequence (illustrated with the generic a-e, high-to-low, consequence rating scale), and their likelihood (illustrated with the generic 1-5, low-to-high, likelihood rating scale), and to combine these characteristics to display a rating for risk level (illustrated with the generic Roman numeral I-V, high-to-low, risk significance scale). The combination of low likelihood and low undesirable consequences gives low risk, while high risk is produced by high likelihood and highly undesirable consequences. Risk prioritization and decision rules (such as the level of management attention or the urgency of response) can be linked to the matrix cells. Note that this generic matrix is conceptual and cannot be applied without careful customization to address the specific project. Detailed guidance and examples for designing rating scales and matrices suitable for use on specific projects, products and systems are provided in IEC 31010 (2019).

A positive consequence scale may be used in the matrix shown in Figure 2.28, thereby changing the outcome adjective from undesirable to desirable, and the cells in the matrix from risks to opportunities. Note that the foundational concept and structure of the matrix diagram remains the same.

SE and project management are all about pursuing an opportunity to solve a problem or fulfill a need. Opportunities enable creativity in resolving concepts, architectures, designs, and strategic and tactical approaches, as well as the many administrative issues within the project. It is the selection and pursuit of these strategic and tactical opportunities that determine just how successful the project and system will be. Of course, opportunities usually carry risks, and each opportunity will have its own set of risks that must be intelligently judged and properly managed to achieve the full value (Forsberg, et al., 2005). These are the risks that must be managed to enhance the opportunity value and the overall value of the project (see Figure 2.29). Opportunity management and risk management are therefore essential to—and performed concurrently with—the planning process but require the application of separate and unique techniques that justify this distinct technical management element.

Balancing Project, Risk, and Opportunity Management for SE. No realistic project can be planned without risk. The challenge is to define the system and the project that best meet overall requirements, allow for risk, and achieve the highest chances of project success. Figure 2.30 illustrates the major interactions between the four risk categories: technical, cost, schedule, and programmatic. The arrow labels indicate typical risk relationships, others are possible.

The Risk Management process is used to understand the potential cost, schedule, and performance (i.e., technical) risks associated with a system, and then take a (proactive) structured approach to anticipate negative outcomes and respond to them before they occur. With respect to opportunities, this process is used to understand the potential cost, schedule, and performance (i.e., technical) improvement opportunities associated with a system, and then take a (proactive) structured approach to defining potential positive outcomes and responding to them by adopting the best candidate improvements before the "window of opportunity" is missed. Care is taken to consider new and increased risk created as a result of pursuing a new opportunity. This practice can help identify unintended negative consequences that might be introduced by the proposed change.

Integrating Risk Management. Per ISO/IEC 31000, "integrating risk management with all organizational processes improves the performance of risk management while gaining efficiencies." Section 7 of ISO/IEC/IEEE 16085 "Risk management in life cycle processes" provides a methodical approach for the integration of risk management and "risk-based thinking" into all SE life cycle processes. Organizations typically manage risks and opportunities of many types,

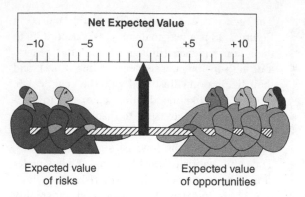

FIGURE 2.29 Intelligent management of risks and opportunities. From Forsberg, et al. (2005) with permission from John Wiley & Sons. All other rights reserved.

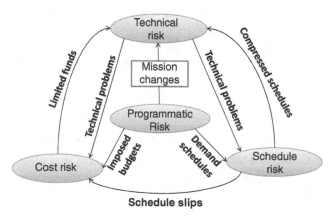

FIGURE 2.30 Typical relationship among the risk categories. INCOSE SEH original figure from INCOSE SEH v1 Figure 4.5-7. Usage per the INCOSE Notices page. All other rights reserved.

across and throughout the organization. Risks and opportunities associated with system development should be managed in a manner consistent with the organization's overall risk and opportunity management strategies.

Risk Management and the System Life Cycle. Once the scope and context of a system have been established from a hierarchical standpoint, it is possible to define and model the system (and its associated risks) in relation to its life cycle, i.e., the differences in the risks in different life cycle stages. For example, risks in the concept stage are quite different than the risks in the retirement stage. It is often necessary to consider risks in other stages while performing activities in the current stage.

Risk Assessment Techniques. ISO/IEC 31010, *Risk management—Risk assessment techniques*, provides detailed descriptions and application guidance for over 30 assessment techniques ranging from brainstorming and checklists to Failure Mode and Effects Analysis (FMEA), Fault Tree Analysis (FTA), Monte Carlo simulation, and Bayesian statistics and Bayes nets. Although a comparable set of (published) techniques for opportunity management is not available, it is notable that ISO/IEC 31010 is not without mention of opportunity, and contains the Strength, Weakness, Opportunity, and Threat (SWOT) Analysis technique. In addition, many of the techniques in ISO/IEC 31010 can be used to assess positive outcomes as well as negative outcomes. For example, FTA can be used to perform Success Tree Analysis, and techniques such as brainstorming, checklists, Monte Carlo simulation, and Bayesian statistical analysis are broadly used for most any purpose, including the assessment of opportunities. A variant of SWOT analysis that is not mentioned in ISO/IEC 31010 is Threats, Opportunities, Weaknesses, and Strengths (TOWS), which puts the emphasis on the external environment (threats and opportunities) rather than on the internal environment (strengths and weaknesses).

Risk Treatment Approaches. Risk treatment approaches (also referred to as risk handling approaches) are often established for the moderate- and high-risk items identified in the risk analysis effort. These activities are formalized in the RMP. There are four basic approaches to treat risks:

1. *Acceptance*: Accept the risk and do no more.
2. *Avoidance*: Avoid the risk through change of requirements or redesign.
3. *Control (or Mitigation)*: Taking actions to reduce the risk by expending budget and/or other resources to reduce likelihood and/or consequence over time.
4. *Transference*: Transfer the risk by agreement with another party that it is in their scope to treat. Look for a partner that has experience in the dedicated risk area.

The following are some of the steps that can be taken to avoid or control unnecessary risks:

- *Requirements scrubbing*—Requirements that significantly complicate the system can be scrutinized to ensure that they deliver value equivalent to their investment. Find alternative solutions that deliver the same or comparable capability.
- *Selection of most promising options*—In most situations, several options are available. A trade study can include project risk as a criterion when selecting the most promising alternative.
- *Staffing and team building*—Projects accomplish work through people. Attention to training, teamwork, and employee morale can help avoid risks introduced by human errors.

For high-risk technical tasks, risk avoidance is insufficient and can be supplemented by the following approaches:

- Early procurement
- Initiation of parallel developments
- Implementation of extensive analysis and testing
- Contingency planning

For each risk that is determined credible after analysis, a Risk Treatment Plan should be created that identifies the risk treatment strategy, the trigger points for action, and any other information to ensuring the treatment is effectively executed. The Risk Treatment Plan can be part of the risk record on the risk profile. For risks that have significant consequences, a contingency plan should be created in case the risk treatment is not successful. It should include the triggers for enacting a contingency plan.

Risk Monitoring. Project management uses measures to simplify and illuminate the Risk Management process (see Figure 2.34). Measures can help identify new risks, as well as provide insight into the effectiveness of the risk treatments.

Each risk category has certain indicators that may be used to monitor project status for signs of risk. Tracking the progress of key system technical parameters can be used as an indicator of technical risk. The typical format in tracking technical performance is a graph of a planned value of a key parameter plotted against calendar time. A second contour showing the actual value achieved is included in the same graph for comparative purposes. Cost and schedule risk are monitored using the products of the cost/schedule control system or some equivalent technique. Normally, cost and schedule variances are used along with a comparison of tasks planned to tasks accomplished.

2.3.4.5 Configuration Management Process

Overview

Purpose As stated in ISO/IEC/IEEE 15288,

> [6.3.5.1] The purpose of the Configuration Management (CM) process is to manage system and system element configurations over their life cycle.

CM establishes and maintains consistency, integrity, traceability, and control of a product's configuration. CM provides enduring truth, trust and traceability across the full life cycle of the product. Appropriate CM across the enterprise and its supply chain, provides efficient, effective, lean, resilient, financially responsible, mature, need realization and sustainment through quantified knowledge and insight. Inadequate CM increases risk to the product—see the example in Section 6.1 about the case of the Therac-25.

Description: Configuration Management (CM) is a Technical Management Process applying appropriate processes, resources, and controls, to establish and maintain consistency between product configuration information, and the product (SAE-EIA 649C).

Evolving system requirements, technology and the operating environment are a reality that must be addressed over the life of a system development effort and throughout the utilization and support stages. Furthermore, CM extended to the enterprise level supports the internal goals needed to achieve an efficient, effective, lean, and resilient enterprise.

Configuration management helps ensure:

- that product functional, performance, and physical characteristics are properly identified, documented, controlled, validated, and verified to establish product integrity;
- that changes to these product characteristics are properly identified, reviewed, approved, documented, and implemented;
- that the products produced against a given set of data are known, verified and validated.

Inputs/Outputs The functional model for the Configuration Management process is listed in Figure 2.31. Descriptions of each input and output are provided in Appendix E.

Process Activities The Configuration Management process includes the following activities:

- *Prepare for configuration management.*
  - Similar to other SE processes, configuration management needs to be planned as early as possible in the product life cycle. The result of CM planning could be a standalone configuration management strategy, could be incorporated in the SEMP, or could be part of the digital implementation of these principles throughout the development platform (e.g., in software development where we have an integrated platform).
  - Planning and managing configuration management is accomplished in conjunction with and integrated through other SE activities and should include the following:
    1) Identify the context and environment of the system that we want to apply this to
    2) Applying adequate configuration management resources and assigning responsibility
    3) Establishing performance and status measurements
    4) Establish, implement and maintain procedures

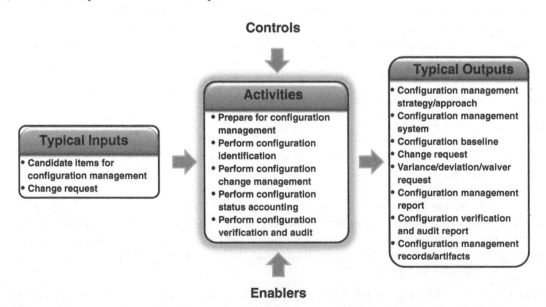

FIGURE 2.31   IPO diagram for Configuration Management process. INCOSE SEH original figure created by Shortell, Walden, and Yip. Usage per the INCOSE Notices page. All other rights reserved.

    5) Configuration management training

    6) Assessing compliance and effectiveness

    7) Supplier configuration management

    8) Product configuration information processes establishment (inc. collection and processing, controlling status, providing inter-operability and exchange, long-term preservation)

    9) Planning for configuration identification, configuration change management, configuration status accounting, configuration verification, and audit.

- *Perform configuration identification.*
  - Identify the items or elements of a system and associated data which should be under configuration management.
  - Establish unique identifiers for the items and data under configuration management.
  - Structure the items and information under configuration management.
  - Validate and release items and information under configuration management.
  - Establish and identify baselines at appropriate points throughout the life cycle. Baselines may coincide with a project milestone or decision gate as shown in Figure 2.4.
  - Manage interfaces and the constraints they impose.
  - Identify where distinct configuration control is required (governance) and designate Configuration Items (CIs).
- *Perform configuration change management.*
  - Manage changes and variances (i.e., non-conformances) throughout the system life cycle to ensure the integrity of the product/system.
  - This includes the identification, recording, review (incl. impact analysis), approve/disapprove, tracking, processing, implementing and closing of requests for change/variances, including relevant supporting documentation, whatever its origin.
  - Disposition of changes/variances are often performed by change boards (the names used in various industries like Configuration Control Board [CCB], Configuration Review Boards [CRB], etc.).
  - An important practice for the change/variance process is to track and manage implementation activities and close the loop ensuring that both the product and its associated information have been evolved to the current approved configuration.
- *Perform configuration status accounting.*
  - Communicate and maintain the status of controlled events, items, and data, as well as performance of CM processes across the life cycle of the product/system to the appropriate stakeholders.
  - Measures and means of measuring performance are established by the Project Assessment and Control process and the configuration status accounting supports these metrics and performance assessments (e.g., Performing reconciliation of the As-Designed data with the As-Built data).
- *Perform configuration verification and audit.*
  - Perform verification of CM processes, in conjunction with the Verification and Quality Assurance processes. Verification includes: review of CM processes; verifying to ascertain that the system has achieved specified requirements and the design of the system is accurately and completely documented in configuration information; verify physical, functional, and interface requirements defined in the approved product definition information, are achieved by the product; verify approved changes to its configuration. Auditing supports the verification process by validating traceability and status between the product to its design, product design to its requirements, and the implementation of changes. Auditing at events like a functional configuration audit (FCA) or a physical configuration audit (PCA), is often accomplished at the end of the development effort and/or testing.
  - Furthermore, the acquirer may have the requirement or wish to perform surveillance and, where necessary, audits to ensure the correct application of CM processes in their supply chain.

Common approaches and tips:

- Begin the Configuration Management process at the beginning of the system life cycle and continue through until retirement of the system. Tailoring of the configuration management approach is key for its successful applications across various domains; this includes an appropriate understanding of the information and processes that need to be in place to fulfill all CM requirements.

Elaboration

Additional guidance regarding configuration management can be found in the current versions of SAE-EIA 649C, ISO 10007, and IEEE 828. Application domain-specific practices, such as SAE ARP 4754A, GEIA HB 649, MIL HDBK 61 B, NIST 800–53, NIST 800–128 provide additional application details.

Configuration management must account for horizontal and vertical integration (see Section 2.3.5.8), in addition to other factors that can affect the system definition over time. Change is a fundamental characteristic of every large-scale system during its life cycle; baselines are set, design fidelity and completeness are improved, and problems are resolved as analyses are performed, impacts are assessed, and trade studies result in decisions that change the system definition. This constancy of change as the design matures makes it imperative to understand the impact of change across all interacting elements and to ensure the complete incorporation of change decisions. Consequently, configuration management, including change management, coordinates maturation of the system.

In Model-Based Systems Engineering (MBSE), CM is required to assure and ensure that the product/system and its product configuration information (i.e., the configuration) are appropriately captured, organized, managed, and communicated for the benefit of the model's stakeholders and participants (see Section 4.2.1).

The corresponding testing and deployment provisions need to be considered in terms of checks against validation rules, interface compatibility, flow time alignment, technical performance measure evaluation, physical clashing, and other domain-specific characteristics.

Moreover, although cyber security is traditionally thought of as a software engineering problem, it needs to be taken into account in a wider system's engineering thought process. Hardware components on which the software is deployed as well as system interfaces can be just as susceptible to cyber-attacks as software itself. That is why proper configuration management needs to also include continuous auditing of potential cyber vulnerabilities. CM processes that originated in agile software engineering (SWE), are now widely used in other Engineering disciplines, including MBSE where the most challenging aspect is the constant need to maintain the relationships between the appropriate configurations of each domain while ensuring accountability and consistency. Several well-established CM practices in agile SWE help with addressing those pain points:

- revisions are managed as a stream of commits;
- baselines are established by tagging specific commits;
- concurrent changes are managed through branching and merging;
- testing, evaluation and/or deployment are automated through a Continuous Integration and Continuous Delivery (CI/CD) process
- security is ensured through the DevSecOps life cycle by integrating security tools into DevOps (see Figure 2.8)

The digital thread establishes communication paths between the individually configured domains. It is also responsible for correctly tying together the appropriate configurations in each domain and to form a consistent configuration for a specific system/product and their elements. More details on traceability can be found in Section 3.2.3.

2.3.4.6   Information Management Process

Overview

Purpose As stated in ISO/IEC/IEEE 15288,

> [6.3.6.1] The purpose of the Information Management process is to generate, obtain, confirm, transform, retain, retrieve, disseminate, and dispose of information to designated stakeholders.

Information management plans, executes, and controls the provision of information to designated stakeholders that is unambiguous, complete, verifiable, consistent, traceable, and presentable. Information includes technical, project, organizational, integration, contractual, agreement, and user information. Information is often derived from data artifacts of the organization, system, process, or project.

Information management needs to provide relevant, timely, complete, valid, and, if required, protected information to designated parties during and, as appropriate, after the product/system life cycle. It manages all defined information, including technical, project, organizational, integration, contractual, agreement, and user information.

Information management ensures that data is properly defined, stored, structured, maintained, secured, exchanged and accessible to those who need it, thereby establishing/maintaining integrity of relevant system life cycle artifacts.

Description Information exists in many forms, and different types of information have different values within an organization. Information assets, whether tangible or intangible, have become so widespread in contemporary organizations that they are indispensable. Information Security has become a fundamental requirement for every industry to work within digital environments with confidence. The following are important terms in information management:

- Information is what an organization has compiled or its employees know. It can be stored and communicated, and it may include classified or unclassified, export restrictive, proprietary, and/or protected (e.g., by copyright, trademark, or patent) and unprotected (e.g., business intelligence) intellectual property. Specific domain classification may apply as well (e.g., further classifications, like Controlled Unclassified Information (CUI) protections in the US defense domain).
- Information assets are intangible information and any tangible form of its representation, including drawings, models of all flavors (systems, software, design, simulation, manufacturing, etc.), specifications, memos, email, computer files, and databases.
- Information security generally refers to the protection, confidentiality, integrity, and availability of the information assets (ISO 17799, 2005).
- Information security management includes the controls used to achieve information security and is accomplished by implementing a suitable set of formalized controls, which could be policies, practices, procedures, organizational structures, and software.
- Information Security Management System is the life cycle approach to implementing, maintaining, and improving the interrelated set of policies, controls, and procedures that ensure the security of an organization's information assets in a manner appropriate for its strategic objectives.

Information management must be associated very closely with configuration management to ensure the integrity, initial release and change control of the information and data. Information management provides the basis for the management of and access to information throughout the system life cycle from ideation through disposal. Designated information may include organizational, project, integration, contractual, agreement, technical, and user information. The mechanisms for maintaining historical knowledge in the prior processes—decision making, risk, and configuration management—are under the responsibility of configuration management working in concert with information management. Figure 2.32 is the IPO diagram for the Information Management process.

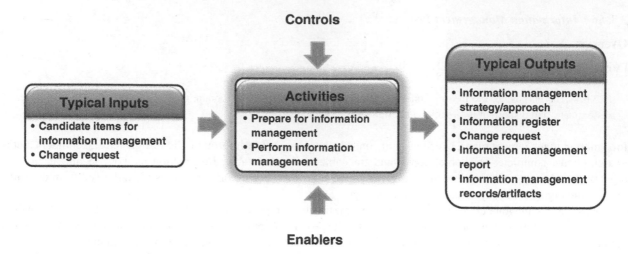

FIGURE 2.32 IPO diagram for Information Management process. INCOSE SEH original figure created by Shortell, Walden, and Yip. Usage per the INCOSE Notices page. All other rights reserved.

Knowledge management is also closely linked to information management. One of the motivations for putting knowledge management in place is for "Information sharing across the organization" thus information management is key for knowledge management. Whereas "Share knowledge and skills throughout the organization" as one of the process-activities for knowledge management draws an identifiable relationship between the information management and Knowledge Management processes via the feedback loop created with inputs and outputs between the processes if mapped out.

Inputs/Outputs Inputs and outputs for the Information Management process are listed in Figure 2.32. Descriptions of each input and output are provided in Appendix E.

Process Activities The Information Management process includes the following activities:

- *Prepare for Information Management.*
  - Support establishing and maintaining a system data dictionary—see project planning outputs.
  - Define system-relevant information, revisioning scheme, storage requirements, access privileges, and the duration of maintenance.
  - Define formats and media for capture, retention, transmission, and retrieval of information.
  - Identify valid sources of information (e.g., business processes) and designate authorities and responsibilities regarding the origination, generation, capture, release, archival, and disposal of information in accordance with the Configuration Management process.
- *Perform Information Management.*
  - Periodically obtain or transform artifacts of information. This is not necessarily specific to waterfall approaches. "Periodically" can also mean at each increment or each iteration/sprint.
  - Maintain information according to integrity, security, and privacy requirements.
  - Retrieve and distribute information in an appropriate form to designated parties, as required by agreed schedules, definitions, or defined circumstances.
  - Archive designated information for compliance with legal, audit, knowledge retention, and project closure requirements.

– Dispose of unwanted, invalid, or unverifiable information according to organizational policy, security, privacy, and legal requirements applicable to the data.

Common approaches and tips:

- Identify information-rich artifacts and store them for later use even if the information is informal, such as a design engineer's notebook (in any media or format).
- Identify the information set at the start of a project if you are going to follow a digital engineering approach.
- In the Project Planning process (see Section 2.3.4.1), an information management plan is tailored to satisfy the individual project procedures for information management. An information management plan identifies the system-relevant information to be collected, retained, controlled, secured, and disseminated, with a schedule for disposal.

Elaboration

The initial planning efforts for information management are defined in the information management plan (and should align with the Configuration Management Plan), which establishes the scope of information that is maintained; identifies the resources and personnel skill level required against the defined tasks to be performed; defines the rights, obligations, and commitments of parties for generation, management, and access; and identifies information management tools and processes, as well as methodologies, standards, and procedures that will be used on the project and managed by appropriate configuration management.

Effective information management provides readily accessible information and management means to authorized project and organization personnel. Database management, security, and revision of data, sharing data across multiple platforms and organizations are facilitated by information management. With all emphasis on knowledge management, organizational learning, and information as competitive advantage, these activities are gaining increased attention.

2.3.4.7 *Measurement Process*

Overview

Purpose As stated in ISO/IEC/IEEE 15288 (and ISO/IEC/IEEE 15939),

[6.3.7.1] The purpose of the Measurement process is to collect, analyze, and report objective data and information to support effective management and *address information needs about* the products, services, and processes.

Description The Measurement process defines the types of information needed to support project and technical management decisions and implement actions to manage and improve performance. The key SE measurement objective is to assess the SE processes and work products with respect to project and organization needs, including timeliness, meeting performance requirements and quality characteristics, product conformance to standards, effective use of resources, and continuous process improvement in reducing cost and cycle time.

The *Practical Software and Systems Measurement (PSM) Guide* (2003), Section 1.1, states:

Measurement provides objective information to help the project manager.

Specific measures are based on information needs and how that information will be used to make decisions and take action. Measurement thus exists as part of an integrated set of management processes and includes not just the project manager, but also SE practitioners, analysts, quality management/assurance, and nearly all other technical and management functions/roles. The decisions to be made drive the information needs and the information needs drive the data to be collected, analyzed, and reported. As a result, numerous benefits are realized from effective measurement (see Table 2.5).

TABLE 2.5   Measurement benefits

Benefit to Project Manager/Technical Lead
Communicate effectively throughout the project organization
Identify and correct problems early
Support making key trade-offs
Track specific project objectives
Defend and justify decisions
Enable continuous process improvement

Successful measurement communicates meaningful information to the decision makers. The presentation of the information must be relevant and unambiguous to those using it, ensuring the intended interpretation.

Inputs/Outputs Typical inputs and outputs for the Measurement process are listed in Figure 2.33. Descriptions of each input and output are provided in Appendix E.

Process Activities The Measurement process includes the following activities:

- *Prepare for measurement.*
  - Identify the measurement stakeholders and their measurement information needs and develop a strategy to meet them.
  - Identify and select relevant prioritized measures that aid with the management and technical performance of the project.
  - Define the base measures, derived measures, indicators, data collection, measurement frequency, measurement repository, reporting method and frequency, trigger points or thresholds, and review authority.
- *Perform measurement.*
  - Gather, process, store, verify, and analyze the data to obtain measurement results (information products).
  - Record and review the measurement information products with the measurement stakeholders and recommend action, as warranted by the results.

Common approaches and tips:

- Measurement for measurement sake is a waste of time and effort. Collecting data without an information need and an intended use is not effective use of limited resources.
- Each measure should be regularly reviewed by the measurement stakeholders. The frequency of review is determined by a number of factors, including frequency of data availability/change, level of risk, maturity of the organization, and cycle times.

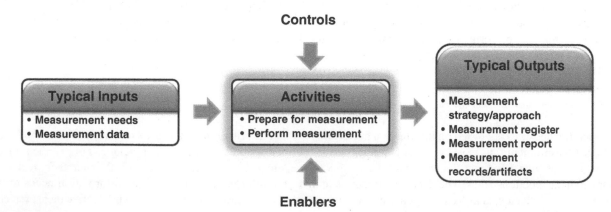

FIGURE 2.33   IPO diagram for Measurement process. INCOSE SEH original figure created by Shortell, Walden, and Yip. Usage per the INCOSE Notices page. All other rights reserved.

SYSTEM LIFE CYCLE PROCESSES   **95**

- Some agreements identify measures of effectiveness (MOEs) that must be met. The derived measures of performance (MOPs) and Technical Performance Measures (TPMs) that provide the necessary insight into meeting the MOEs are default measures to be included within the measurement plan. Other measures to consider should provide insight into technical and programmatic execution of the project (Roedler and Jones, 2005).

- The best measures are repeatable, can be implemented with automated data collection or require minimal effort for data collection, are straightforward to understand, and are presented in a consistent format on a regular basis (with trend data and, where applicable, projections).

- Many methods are available to present the data to the measurement stakeholders. Line graphs, control charts, and Red/Yellow/Green "traffic lights" are some of the more frequently used. Tools are available to help with measurement.

- If a need for corrective action is perceived, further investigation into the measures may be necessary to identify the root cause of the issue to ensure that corrective actions address the cause instead of a symptom.

- Measurement by itself does not control or improve process performance, project success, or product quality. Measurement results must be provided to decision makers in a manner that provides the needed insight for the right decisions to be made. Action must be taken, to realize any benefit.

Elaboration

Measurement Concepts. Measurement concepts have been expanded upon in the previous works shown in Table 2.6 that the SE measurement practitioner should reference for further insights.

Measurement Approach. As discussed in the INCOSE Measurement Primer (2010), measurement may be thought of as a feedback control system. Value is obtained from measurement when the data analysis provides insight for assessment or action by decision makers (e.g., action is taken due to a variance from a target value or the need to improve current performance to a more desirable level). Comparing the target value and the allowable difference between the target and actual values enables decisions based upon evaluation of risk to the project or product performance meeting their required goals.

Relationship of Measurement to Risk Management and Decision Management. The measures for a project are driven by the information needs of the project and its decision makers. One source of the information needs are the objectives of the project, which can be related to resources, technical performance of the system, product or process

TABLE 2.6 Measurement references for specific measurement focuses

Reference Focus	Reference
General Reference	*Systems and Software Engineering -Measurement Process* (ISO/IEC/IEEE 15939, 2017)
	Guide to the Systems Engineering Body of Knowledge (SEBoK), Part 3: SE and Management/Systems Engineering Management/Measurement (SEBoK, 2023)
	Practical Software and Systems Measurement (PSM) Guide V4.0c, (PSM, 2003)
	Capability Maturity Model Integration (CMMI®) for Development V2.0, Measurement and Quantitative Management Process Areas (CMMI, 2018)
Guidance for New Practitioners	*INCOSE Systems Engineering Measurement Primer*, Version 2.0 (INCOSE Measurement Primer, 2010)
Technical Measurement / Performance	*Technical Measurement Guide* (Roedler and Jones, 2005)
System Development	*System Development Performance Measurement Report* (NDIA, et al., 2011)
Project Management	*Project Manager's Guide to Systems Engineering Measurement for Project Success* (INCOSE PMGtSEMfPS, 2015)
Continuous Iterative Development	*Continuous Iterative Development Measurement Framework* (PSM, et al., 2021)
Digital Engineering	*Practical Software and Systems Measurement (PSM) Digital Engineering Measurement Framework* (INCOSE, et al., 2022)
Leading Indicators	*Systems Engineering Leading Indicators Guide*, Version 2.0 (Roedler, et al., 2010)

INCOSE SEH original table created by Roedler. Usage per the INCOSE Notices page. All other rights reserved.

quality, or other aspects of the project that are considered essential. Another key source of information needs are the key risks of the project. As shown in Figure 2.34, the Risk Management process identifies risks that need to be monitored, thus creating information needs that drive new measures. The Measurement process helps characterize and quantify the risks. In turn, the results of the measurement analysis may uncover new risks that need to be considered by the Risk Management process. The results of both risk management and measurement provide essential insight to decision makers that is essential to the Decision Management process. Measurement also provides insight to all other processes, especially Project Planning, Project Assessment and Control, Quality Assurance, Life Cycle Model Management, and the Technical Processes.

Digital Engineering (DE) Measurement. DE has three interrelated concerns: the transformation of engineering activities to a fully digital infrastructure, artifacts, and processes; the use of authoritative sources of truth (ASOTs) to improve the efficiency and productivity of engineering practice; and the use of MBSE practice to fully integrate system data and models with engineering, project management, and other domains and disciplines. Measurement in DE focuses on the implementation of DE transformations on projects and in enterprises, including the realization of measurable benefits in performance, effectiveness, and product quality relative to traditional engineering methods. DE measures can also serve as useful leading indicators for other product related measures. For more information, see INCOSE, et al. (2022) and Section 5.4.

Continuous Iterative Development. As organizations and projects move toward incremental and evolutionary approaches for acquisition and life cycle models, measurement is key to understanding progress and quality (see Section 2.2). Measures are needed to address team, product, and enterprise perspectives. Measures are needed that balance both speed and quality that delivers a best value solution based on project objectives. For more information, see PSM, et al. (2021).

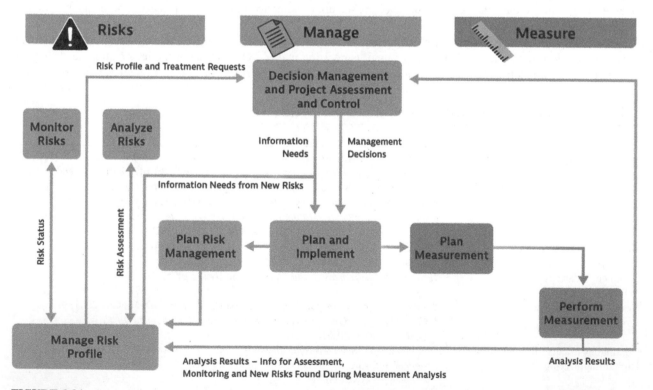

FIGURE 2.34 Integration of Measurement, Risk Management, and Decision Management processes. INCOSE SEH original figure created by Roedler. Usage per the INCOSE Notices page. All other rights reserved.

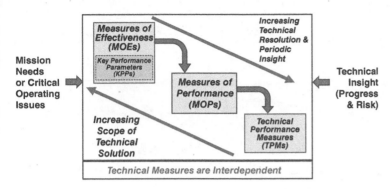

Leading Indicators. Per Roedler, et al. (2010), a leading indicator is a measure for evaluating the effectiveness of how a specific activity is applied on a project in a manner that provides information about impacts that are likely to affect the system performance or SE effectiveness objectives.

A leading indicator may be an individual measure, or collection of measures, that is predictive of future system performance before the performance is realized. Leading indicators aid leadership in delivering value to stakeholders and end users while assisting in taking interventions and actions to avoid rework and wasted effort.

Rather than provide status and historical information, leading indicators use trend information to facilitate predictive analysis (forward looking). By analyzing the trends, quantitative relationships of key factors can be developed with known correlations and predictions can be forecast on the outcomes of certain activities. Trends are analyzed for insight into both the entity being measured and potential impacts to other entities. This enables proactive decisions and actions (preventive and corrective).

For a more detailed treatment of this topic, including measurement examples, refer to Roedler, et al. (2010). In addition, NDIA, et al. (2011) provides specific leading indicators developed from the previously referenced guide for the defense and aerospace domains. However, most of the indicators have a broader application.

Product-Oriented Measures. As shown in Roedler and Jones (2005), product measures can be thought of as an interdependent hierarchy (see Figure 2.35).

Measures of Effectiveness (MOEs), which are stated from the acquirer (customer/user) viewpoint, are the acquirer's key indicators of achieving the mission needs for performance, suitability, and affordability across the life cycle. Although they are independent of any particular solution, MOEs are the overall operational success criteria (mission performance, safety, operability, operational availability, etc.) to be used by the acquirer for the delivered system, services, and/or processes.

Key Performance Parameters (KPPs) are used in some domains to indicate the minimum number of performance parameters needed to characterize the major drivers of operational performance, supportability, and interoperability. Each KPP has a threshold and objective value. The acquirer defines the KPPs at the time the operational concepts and requirements are defined.

Measures of Performance (MOPs) measure attributes considered as important to ensure that the system has the capability to achieve operational objectives. MOPs are used to assess whether the system meets design or performance requirements that are necessary to satisfy the MOEs. MOPs should be derived from or provide insight for MOEs or other user needs.

Technical Performance Measures (TPMs) are used to assess design progress, show compliance to performance requirements, and track technical risks. They provide visibility into the status of important project technical parameters to enable effective management, thus enhancing the likelihood of achieving the technical objectives of the project. TPMs are derived from, or provide insight for, the MOPs and focus on the critical technical parameters of specific architectural elements of the system as it is designed and implemented. Selection of TPMs should be limited to critical technical thresholds or parameters that, if not met, put the project at cost, schedule, or performance risk. The TPMs are not a full listing of the requirements of the system or system element. The SEMP should define the approach to TPMs (Roedler and Jones, 2005).

Figure 2.36 illustrates a sample TPM. Values are established to provide limits that give early indications if a TPM is out of tolerance. The tolerance band is generally wider earlier in the life cycle and gets tighter as the system

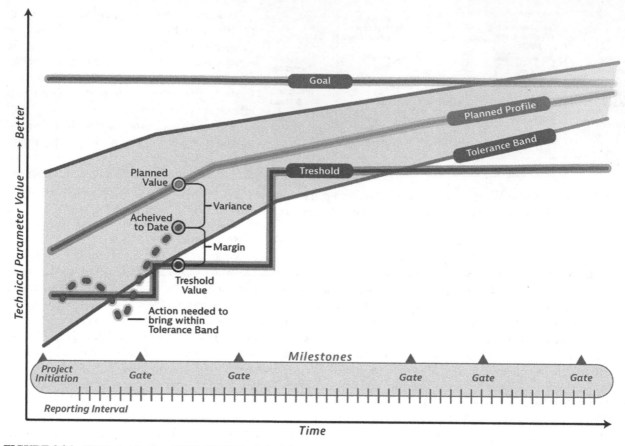

FIGURE 2.36   TPM monitoring. INCOSE SEH original figure created by Roedler and Walden. Usage per the INCOSE Notices page. All other rights reserved.

development progresses. Measured values that fall outside an established tolerance band alert management that it may be necessary to take corrective action.

The progress of some TPMs relies on maturing a particular technology. Thus, it may be necessary to have a technology plan and technology readiness level (TRL) assessment as part of the input associated with a TPM.

2.3.4.8   *Quality Assurance Process*

Overview

Purpose As stated in ISO/IEC/IEEE 15288,

> [6.3.8.1] The purpose of the Quality Assurance process is to help ensure the effective application of the organization's Quality Management process to the project.

Description Quality Assurance (QA) is broadly defined as the set of activities throughout the entire project life cycle necessary to provide adequate confidence that a product or service conforms to stakeholder requirements or that a process adheres to established methodology (ASQ, 2007). SE practitioners adopt and use QA processes as key contributors to project and systems development success. QA is a key aspect in QM from the project level to the individual processes. It involves the resourcing and improved performance of processes with built-in quality that is designed to prevent defects from occurring in delivered products and services.

Inputs/Outputs Inputs and outputs for the QA process are listed in Figure 2.37. Descriptions of each input and output are provided in Appendix E.

Process Activities The Quality Assurance process includes the following activities:

- *Prepare for quality assurance.*
  - Establish and maintain the QA strategy (often captured in a QA plan).
  - Establish and maintain QA guidelines, policies, standards, and procedures.
  - Define responsibilities and authorities.
- *Perform product or service evaluations.*
  - Perform the evaluations at appropriate times in the life cycle as defined by the QA plan, ensuring V&V of the outputs of the life cycle processes. Ensure that QA perspectives are appropriately represented during design, development, verification, validation, and production activities.
  - Evaluate product verification results as evidence of QA effectiveness.
- *Perform process evaluations.*
  - Implement prescribed surveillance on processes to provide an independent evaluation of whether the developing organization is in compliance with established procedures.
  - Evaluate enabling tools and environments for conformance and effectiveness.
  - Flow applicable procedural and surveillance requirements throughout the project supply chain and evaluate subcontractor processes for conformance to allocated requirements.
- *Manage QA records and reports.*
  - Create, maintain, and store records and reports in accordance with applicable requirements.
  - Identify incidents and problems associated with product and process evaluations.
- *Treat incidents and problems.*

FIGURE 2.37   IPO diagram for the Quality Assurance process. INCOSE SEH original figure created by Shortell, Walden, and Yip. Usage per the INCOSE Notices page. All other rights reserved.

Note: Incidents are short-term anomalies or observations that require immediate attention, and problems are confirmed nonconformities that would cause the project to fail to meet requirements.

– Document, classify, report, and analyze all anomalies.

– Perform root cause analysis and note trends.

– Recommend appropriate actions to resolve anomalies and errors, when indicated.

– Track all incidents and problems to closure.

Common Approaches and Tips

- Management's commitment to QA is reflected in the integration of QM principles in the strategic planning and budgeting of the organization, and the allocation of educational resources to achieve and sustain a reliable QM culture (see Section 2.3.3.5).

- A quality policy, mission, strategies, goals, and objectives provide essential inputs along with a description of an organization's fundamental values for quality assurance and the support of a growing QM culture.

Elaboration

QA Generally accepted theory and practice. QA is one of the two Quality Management (QM) defect prevention methods. The second is Quality Control which is described and contrasted in Section 2.3.3.5. QA can be described as "putting good things into our processes" so that they perform as designed and conform to our stakeholder's requirements. Like QM, QA was born in the aerospace industry and was originally referred to as "reliability engineering." It is generally associated with activities such as failure testing and pre-inspecting batches of materials and system elements that are then certified for use, thus preventing errors and defects from occurring by building-in quality. QA also includes infusing processes with reliable human resources and the appropriate policies, procedures, and training. W. Edwards Deming noted that "Quality comes not from inspection, but from improvement of the production process" (Deming, 1986).

QA Culture. "Ultimately, it is the people in an organization who can create a work culture in which quality is promoted and value is delivered to stakeholders" (Kennedy, 2005). An effective QA methodology defines competent, well-prepared humans as the major asset within processes that are then supported by the appropriate corporate environment, resources, and technologies to improve outcomes. It supports a high-performing work culture that diligently defines and fulfills stakeholder requirements with a Zero Defects Attitude (ZDA) (see Section 2.3.3.5) and is focused on continuous improvement. Philip Crosby noted that "Quality is the result of a carefully crafted cultural environment. It has to be the fabric of the organization, not part of the fabric" (Crosby, 1979).

The fabric of a QA-strengthened work culture is defined by fundamental skills and supporting values that create a sense of ownership by all participants. Workers who identify with an organization's core values have a stronger sense of psychological ownership and higher job satisfaction. At its core, psychological ownership is about an employee's possession and stewardship of an organization's core values and the pride they have about their enterprise/mission (Journal of Organizational Behavior, 2004). The workforce must have skills and experience that are directly related to the output objectives, and when skills are supported by shared values it creates a reliable work culture. (See Section 2.3.3.5.) This strengthening of the work culture leads to greater employee engagement and naturally results in products and services with higher quality, along with other benefits to both the workforce and the corporation (Gallup, 2017, 2020). QM is an educational technology with systems, methods and language that help us reach our business goals and QA performs an essential resourcing, educational and process improvement role in ensuring that all elements of an organization execute its activities in accordance with its plans, and procedures as a means of building quality into products or services. While QA is focused on improving processes to prevent errors from occurring, QC provides an essential feedback loop to QA by providing defect rates and identifying their source in processes. By applying Work Process Analysis (WPA) to the defect data, QA can define and initiate input and process improvements to produce lean outcomes.

As the complexity of a project increases, the challenges to effectiveness and risk management also increase. These factors further emphasize the need for a coordinated QM culture with the proper balance of QC and QA along with the skills, experience, and values that align with the requirements of the project. Kennedy calls this properly configured alignment "Vocational Certainty," and that a high-performing work culture is measured by identifiable professional and personal attributes or values within an organization's workforce (Kennedy, 2005). Professional values for an effective QA educational initiative must build upon personal vocational certainty, and on administrative consistency that extends our attention to process details beyond the initial documentation of requirements and progress reports. We must continue to interact with and challenge the stakeholders to mature their requirements so that stakeholder satisfaction can be assured.

2.3.5 Technical Processes

The ISO/IEC/IEEE 15288 includes 14 Technical Processes that are invoked concurrently, iteratively, and recursively throughout the system life cycle in conjunction with supporting agreement and technical management process activities. The Technical Processes are defined in ISO/IEC/IEEE 15288 as follows:

> [5.7.5] The Technical Processes are used to define the requirements for a system, to transform the requirements into an effective product, to permit consistent reproduction of the product where necessary, to use the product, to provide the required services, to sustain the provision of those services and to dispose of the product when it is retired from service.

Technical Processes enable SE practitioners to coordinate the interactions between engineering specialists, other engineering disciplines, acquirers, operators, manufacturing/production and other system stakeholders. They also address conformance with the expectations and legislated requirements of society. These processes lead to the creation of a necessary and sufficient set of needs and requirements as well as resulting system solutions that address the needed capabilities within the bounds of performance, environment, external interfaces, ethical norms, societal expectations, regulations, and design constraints. Without the Technical Processes, the risk of project failure would be unacceptably high. Figure 2.38 provides a graphical representation of the Technical Processes in context.

As shown in Figure 2.38, at the beginning of the system life cycle are stakeholder real-world expectations for a SoI. The SoI could be the integrated system, a set of system elements, or a system element within the system architecture.

For each SoI, through a series of transformational actions across the life cycle, the technical processes transform input artifacts into output artifacts that are inputs into other technical processes, which in-turn transform those artifacts into additional artifacts. This series of transformations results in an SoI that addresses the capabilities needed by the stakeholders.

It is important to understand several key points for Figure 2.38.

1. While the figure depicts the series of transformations in a linear fashion, in practice the Technical Processes are intended to be practiced concurrently, iteratively, and recursively as the project team moves down the layers of the system architecture. As such, the figure applies to each system element within the system architecture.

2. The Integration process is applied from the beginning of the project, managing the integrated system as the project team traverses the system architecture. In doing so, the project team is continuously addressing interactions of the parts that make up the integrated system as well as interactions with the macro system of which it is a part. In addition, the project team is assessing the behavior of the system as a function of these interactions and looking for emerging properties—both good and bad—which is a key activity involved in Interface Management (see Section 3.2.4).

3. Following each transformation, the output artifacts are verified against the system requirements via the Verification process to ensure the output artifacts' transformation was "right" as defined by their requirements.

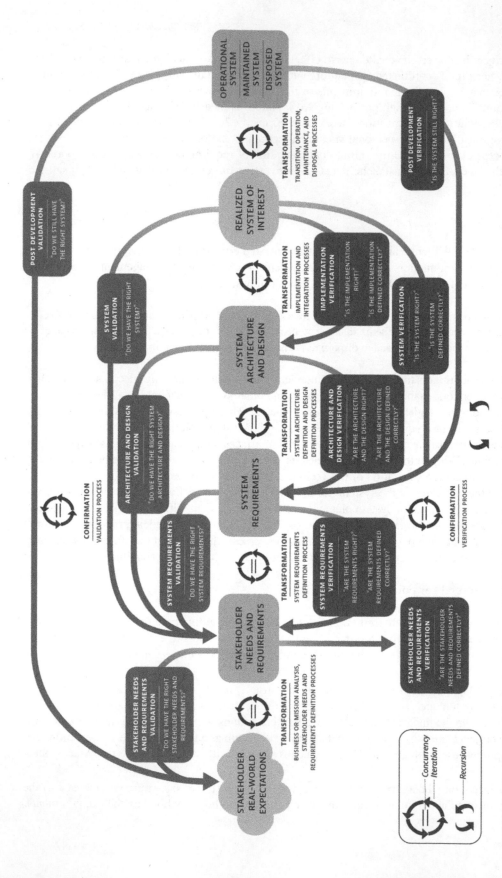

FIGURE 2.38 Technical Processes in context. INCOSE SEH original figure created by Roedler, Walden, and Wheatcraft derived from INCOSE NRM (2022). Usage per the INCOSE Notices page.

4. Following each transformation, the output artifacts are validated against the stakeholder needs and requirements via the Validation process to ensure the output artifacts are the "right" artifacts as defined by the stakeholder needs and requirements.

5. Once the SoI has been deployed and has entered into operations by its intended users, post-deployment validation is performed to help ensure the SoI remains the right SoI that meets the stakeholder real-world expectations—is the SoI still the right system? In addition, post-deployment verification is performed to help ensure the SoI is still meeting its requirements over time—is it still "right"?

Further elaboration of these key points is included within the following Technical Process sections.

New requirements can be placed on the SoI by the SoS configurations in which the SoI will participate. The SoS technical considerations apply to all the system life cycle processes across the life cycle stages, especially the Technical Processes that provide the concept and system definition (see Section 4.3.6).

2.3.5.1 Business or Mission Analysis Process

Overview

Purpose As stated in ISO/IEC/IEEE 15288,

[6.4.1.1] The purpose of the Business or Mission Analysis process is to define the overall strategic problem or opportunity, characterize the solution space, and determine potential solution class(es) that can address a problem or take advantage of an opportunity.

Description The Business or Mission Analysis process initiates the life cycle of the SoI by defining the problem or opportunity space; defining the mission, business, or operational problems or opportunities; identifying major stakeholders; characterizing the solution space by identifying environmental conditions and business constraints that bound the solution domain; identifying and prioritizing business needs; identifying and prioritizing business requirements, defining critical business success measures; developing preliminary life cycle concepts from the organizational perspectives including operations, acquisition, deployment, support, and retirement; and evaluating alternative solution classes and selecting a preferred solution class.

Inputs/Outputs Inputs and outputs for the Business or Mission Analysis process are listed in Figure 2.39. Descriptions of each input and output are provided in Appendix E. Note that, as with all processes, the Business or Mission Analysis process is applied concurrently and iteratively evolving throughout the life cycle so that all SE artifacts mature as a result of the iterative application of the processes.

Process Activities The Business or Mission Analysis process includes the following activities:

- *Prepare for business or mission analysis.*
  - Identify potential problems and opportunities resulting from changes in the organization's strategy and Concept of Operations, while considering desired organization mission(s), goals, objectives, and other organizational business needs and business requirements. This may involve the development of concepts for a new solution but may also involve identifying gaps or deficiencies in existing capabilities, systems, products, or services and concepts for addressing those gaps or deficiencies.
  - Establish the strategy/approach for business or mission analysis. This involves the organizational approach(es) to defining the problem space, the characterization of the solution space, and the identification of an appropriate alternative solution classes.
  - Plan for the necessary enabling systems or services needed through the life cycle for business or mission analysis. This includes interfaces to organizational enabling systems or services such as business, acquisition,

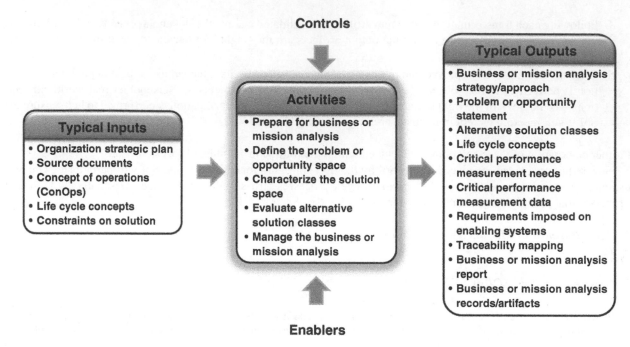

FIGURE 2.39   IPO diagram for Business or Mission Analysis process. INCOSE SEH original figure created by Shortell, Walden, and Yip. Usage per the INCOSE Notices page. All other rights reserved.

operations, production, project management, and SE tools and applications, financial systems, information technology, databases, security controls, and other data and information repositories.

– Ensure enabling system or service access needed to support business or mission analysis.

• *Define the problem or opportunity space.*

– Considering relevant trade-space factors as part of the problem and opportunity analysis. This analysis is focused on understanding the scope, drivers, constraints, risks, needs, and requirements associated with the problem or opportunity. The analysis includes changes in business needs, business requirements, opportunities, capabilities, performance improvement, security risks, safety risks, factors such as cost and effectiveness, value, regulation changes, standard changes, user dissatisfaction, lack of existing systems, and PESTEL factors (Political, Economic, Social, Technological, Environmental, and Legal). This may be accomplished through external, internal, or SWOT (Strengths, Weaknesses, Opportunities, and Threats) analysis.

– Define the problem or opportunity, mission, goals, objectives, and associated business needs and requirements to be addressed by a solution class. This definition is solution-class independent, since the solution could be an operational change, a change to an existing system or service, or a new system.

– Prioritize the problem or opportunity, mission, goals, objectives, needs, and requirements against other business needs and business requirements.

– Define critical business success measures. The business must define how it will know that the solution provided will meet its needs. Validation criteria establish critical and desired system performance—thresholds and objectives for system performance parameters that are critical for system success and those that are desired but may be subject to compromise to meet the critical parameters.

– Obtain agreement on the problem or opportunity, mission, goals, objectives, business needs, business requirements, and success measures.

- *Characterize the solution space.*
  - Define preliminary life cycle concepts for acquisition, development, deployment, operations, support, and retirement of the solution. Business stakeholders identify the stakeholders (individuals or groups) who are to be involved in any of the life cycle concepts. The life cycle concepts define what the system needs to do and how well from the business stakeholder's perspective of the intended use in the intended operational environment, when operated by the intended users in the context of all its life cycle activities, the required interactions with external systems, drivers and constraints, security, risks, business needs, and business requirements at the strategic level.
  - Establish a set of alternative classes spanning the potential solution space.
- *Evaluate alternative solution classes that span the potential solution space.*
  - Evaluate the set of alternative solution concepts and select the preferred solution concepts against the organization's business needs, business requirements, and critical business success measures. Appropriate modeling, simulation, and analytical analysis will help determine the feasibility, value, and appropriateness of the alternative solution classes.
  - Select the preferred solution class(es) and ensure each has been validated in the context of the proposed strategic level life cycle concepts. Feedback on feasibility, value, market factors, and alternatives is also provided for use in completing the definition of the organization's level life cycle concepts.
  - Provide feedback to organization level life cycle concepts in terms of the selected solution class(es).
  - Obtain agreement on the problem or opportunity statement, mission, goals, objectives, critical business success measures, life cycle concepts, business needs, and business requirements.
- *Manage the Business or Mission Analysis.*
  - As key decisions are made, record the decision along with supporting information and rationale.
  - Establish and sustain traceability (analysis, rationale, and alternative solution classes).
  - Give CM the information items, work products, or other artifacts needed for baselines.

Common approaches and tips.

- Identify the enabling systems and materials needed for transition early in the life cycle to allow for the necessary lead time to obtain or access them.

Elaboration

Identify Major Stakeholders. Although the identification of stakeholders is undertaken at each stage of system development, during the Business or Mission Analysis process, business managers are responsible for nominating key stakeholders and are often responsible for establishing a stakeholder register and means of exchanging information. It is fundamentally a business management function to ensure stakeholders are available and able to contribute to the system development activities for the SoI—stakeholders are often occupied in other business operations activities and must be authorized in terms of both budget and time to expend the needed effort and resources on other than their current operational tasks.

Identify Business Needs and Requirements. For each problem or opportunity, it is important to identify the business needs and business requirements associated with needed capabilities, functionality, performance, and security as well as risk and compliance with standards and regulations. Business needs exist at several levels of abstraction, consist of identification of "what is needed" by the business to address the problem or opportunity, and can be communicated in several forms, such as the mission statement, goals, objectives, critical success measures, use cases, user stories, and individual need statements. The business requirements communicate what the business requires of the solution to address their needs without stating a specific solution. The life cycle concepts are developed in response to the business needs and business requirements. Together, the business needs, business requirements, and critical business success measures communicate what is "necessary for acceptance" at the business level.

Life Cycle Concepts. Life cycle concepts address not only the concepts for the SoI during operations by the intended users in the operational environment, but also includes the concepts required to address the business needs, business requirements, critical business success measures, and higher-level stakeholder needs and stakeholder-owned system requirements across the system life cycle. Preliminary life cycle concepts are established and assured through the Business or Mission Analysis process to the extent needed to define the problem or opportunity space and characterize the solution space. Principal life cycle concepts include:

Concept of operations (ConOps)—Describes the way the organization will operate to achieve its missions, goals, and objectives. The ConOps captures how the system will potentially impact the acquiring and other organizations. "The ConOps describes the organization's assumptions or intent in regard to an overall operation or series of operations of the business with using the system to be developed, existing systems and possible future systems. The ConOps serves as a basis for the organization to direct the overall characteristics of the future business and systems, for the project to understand its background, and for [its] users … to implement the stakeholder requirements elicitation" (ISO/IEC/IEEE 29148, 2018) Ideally, the enterprise level ConOps should be an input to the Business or Mission Analysis process, but if it does not exist, it may need to be jointly developed and maintained. The ConOps also describes the higher-level system in which the SoI must operate.

Operational concept (OpsCon)—Describes the way the system will be used during operations, for what purpose, in its operational environment by its intended users and does not enable unintended users to negatively impact the intended use of the system nor allow unintended users from using the system in unintended ways. Also addressed are the needed capabilities, functionality, performance, quality, safety, security, compliance with standards and regulations, interactions with external systems, and operational risks. An OpsCon provides a user-oriented perspective that describes system characteristics of the to-be-delivered system. The OpsCon is used to communicate overall quantitative and qualitative system characteristics to the acquirer, user, supplier and other organizational elements.

Acquisition concept—For solutions that will be procured from a supplier, the acquisition concept describes the way the system will be acquired including aspects such as stakeholder engagement, needs definition, requirements definition, design, production, verification, validation, and contract deliverables. The supplier enterprise(s) may need to develop more detailed concepts for production, assembly, verification, validation, transport of system, and/or system elements. For solutions that will be provided internal to the organization, the acquisition concept will include a production concept that describes the way the system will be developed and produced including aspects such as stakeholder engagement, needs definition, requirements definition, design, production, integration, verification, and validation.

Deployment concept—Describes the way the system will be delivered, integrated into its operational environment, and introduced into operations, including deployment considerations when the system will be integrated with other systems that are in operation and/or replace any systems in operation.

Support concept—Describes the logistics, desired support infrastructure and staffing considerations for supporting the system after it is deployed. A support concept would address operating support, engineering support, maintenance support, supply support, training support, and post-deployment verification and validation.

Retirement concept—Describes the way the system will be removed from operation and retired, including the disposal of any hazardous materials used in or resulting from the process and any legal obligations—for example, regarding IP rights protection, any external financial/ownership interests, sustainability, environmental impacts, and security concerns.

These preliminary life cycle concepts are defined first at the organizational level, to the extent required at that level, for the identified solution classes that address the problem or opportunity. The preliminary life cycle concepts are then elaborated and refined through the Stakeholder Needs and Requirements Definition process (Section 2.3.5.2). Through iteration, the life cycle concepts are refined throughout the life cycle as required as a result of feedback obtained through the conduct of the rest of the Technical Processes.

Uncertainties and risk. There will be uncertainties (see Section 1.4.1) in the preliminary life cycle concepts. Uncertainties can be related to differing stakeholder perspectives, business factors, market, management, technical performance, schedule, development and production costs, operations and support costs, security, and sustainability. These uncertainties are a source of risk. Each of these uncertainties need to be addressed using the Risk Management process (Section 2.3.4.4) in conjunction with the rest of the Technical Processes, especially the Stakeholder Needs and Requirements Definition process (Section 2.3.5.2), the System Requirements Definition process (Section 2.3.5.3), and the System Architecture Definition process (Section 2.3.5.4).

2.3.5.2 *Stakeholder Needs and Requirements Definition Process*

Overview

Purpose As stated in ISO/IEC/IEEE 15288,

> [6.4.2.1] The purpose of the Stakeholder Needs and Requirements Definition process is to define the stakeholder needs and requirements for a system that can provide the capabilities needed by users and other stakeholders in a defined environment.

Description Successful projects depend on meeting the stakeholder real-world expectations as communicated by the needs and requirements of the stakeholders throughout the system life cycle. A stakeholder is any entity (individual or organization) with a legitimate interest in the system. Stakeholders exist at each of the levels of an organization and system architecture. The focus of the Stakeholder Needs and Requirements Definition process is on elaboration of the preliminary the life cycle concepts, on the stakeholder needs transformed from those concepts, and on the stakeholder requirements transformed from those needs. The activities during the process are constrained and driven by the preliminary life cycle concepts, business needs, business requirements, and critical business success measures developed during the Business or Mission Analysis process (see Section 2.3.5.1).

In addition to identifying the stakeholders, this process elicits the operational use cases, scenarios, and life cycle concepts from stakeholders, identifies drivers and constraints, determines interactions with the operational and enabling systems, determines interactions with users and operators, characterizes the operational environment, and assesses risks associated with the development of a new or changed capability or new opportunities addressed by a solution class. The life cycle concepts are analyzed, matured, and transformed into a set of stakeholder needs. These needs are analyzed and transformed into a set of stakeholder requirements for the SoI. These stakeholder requirements communicate what the stakeholders expect from the SoI that will result in their needs being met using their terminology.

The stakeholder requirements drive and constrain the solution space by addressing stakeholder expectations for the SoI, characterizing the operational environment, and identifying external interface boundaries between the SoI and external systems across which there is an interaction. Traceability between the life cycle concepts, stakeholder needs, and stakeholder requirements is established as part of this process.

Stakeholder requirements govern the SoI's development and are an essential factor in further defining or clarifying the scope of the development project and elaborating on what is "necessary for acceptance." If an organization is acquiring the system, this process provides the basis for the technical description of the deliverables in an agreement—typically in the form of a set of system requirements for a SoI and defined interfaces at the SoI boundaries.

Inputs/Outputs Inputs and outputs for the Stakeholder Needs and Requirements Definition process are shown in Figure 2.40. Descriptions of each input and output are provided in Appendix E.

Process Activities The Stakeholder Needs and Requirements Definition process includes the following activities:

- *Prepare for stakeholder needs and requirements definition.*
    - Identify the stakeholders with an interest in the solution. Resolve differing interests. These stakeholders or classes of stakeholders will help identify constraints and define operational-level life cycle concepts, transform those concepts into operational-level stakeholder needs, which are then transformed into operational-level stakeholder requirements.
    - Establish the strategy/approach for stakeholder needs and requirements definition. Understand the role and perspective of each stakeholder and identify any potential conflicts with other stakeholders to develop a stakeholder management plan and a strategy for defining life cycle concepts, stakeholder needs, and stakeholder requirements.

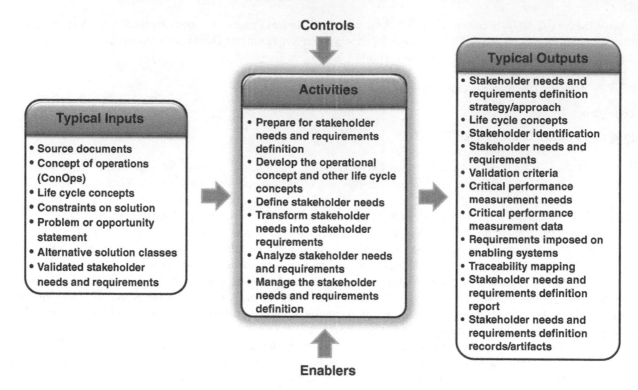

FIGURE 2.40   IPO diagram for Stakeholder Needs and Requirements Definition process. INCOSE SEH original figure created by Shortell, Walden, and Yip. Usage per the INCOSE Notices page. All other rights reserved.

- Plan for the necessary enabling systems or services needed through the life cycle for stakeholder needs and requirements definition. Enabling systems may include tools for elicitation of stakeholder life cycle concepts, recording drivers and constraints, defining risks, analysis, recording needs, recording requirements, and providing traceability between SE artifacts generated during this process, and managing those artifacts.
- Ensure enabling system or service access needed to support stakeholder needs and requirement definition.

- *Develop the operational concept and other life cycle concepts.*
  - Elaborate the operational concept (OpsCon) within the context of the concept of operations (ConOps), other life cycle concepts, and the preferred solution class(es). Preliminary life cycle concepts are developed in the Business or Mission Analysis process (see Section 2.3.5.1)—in this process they are elaborated in more detail.
  - Define a set of operational scenarios (or use cases) to identify required capabilities that correspond to anticipated operational and other life cycle concepts. Identify use cases and operational scenarios and associated capabilities, behaviors, and responses of the system or solution and environments across the SoI life cycle (concept, development, production, utilization, support, and retirement). The use cases and operational scenarios provide the information from the stakeholders needed to define the life cycle concepts; the range of intended uses of the system; the intended operational environment and the system's impact on the environment; intended users, and interfacing systems, platforms, or products.
  - Capture the characterization of the SoI's intended operational environment and users.
  - Considering usability, identify user interactions. Capture factors (e.g., skills) that can affect the interactions.
  - Identify external interface boundaries across which the SoI interacts.
  - Considering the stakeholder and technical objectives and limitations, identify constraints on the solution.

- Identify risks associated with management, development, operations, and disposal, including misuse and loss scenarios (see Sections 3.1.12 and 3.1.13).

- *Define stakeholder needs.*
  - Identify stakeholder needs that reflect the intended life cycle concepts and associated constraints (see the INCOSE NRM [2022] and INCOSE GtNR [2022] for more details). Specify quality, health, safety, security, environment, compliance, and other stakeholder needs and functions that relate to critical qualities.
  - In conjunction with the Decision Management process, prioritize and select the essential needs. The System Analysis process (see Section 2.3.5.6) is used to analyze the life cycle concepts, resolve conflicts, and access feasibility of those concepts. Needs are assessed and priorities are assigned in terms of critically, value, completeness, correctness, consistency, security and feasibility.
  - As the stakeholder needs are selected, record the needs with their sources and rationale. Transform the life cycle concepts and other sources into a homogeneous, agreed-to integrated set of stakeholder needs. Establish traces from the needs to their sources. Although the stakeholder needs are not required to be as stated as rigorously as requirements, it is useful to follow the same rules as for quality requirements (see Section 2.3.5.3 and the INCOSE GtWR [2023] for more details) since better-formed needs and sets of needs will result in less ambiguity in the transformation of the needs into requirements. For each stakeholder need, define attributes such as source, rationale, priority, and criticality.

- *Transform stakeholder needs into stakeholder requirements.*
  - Transform the needs into stakeholder requirements. Define a set of stakeholder requirements consistent with the stakeholder needs.
  - Identify any additional stakeholder requirements. Define stakeholder requirements that relate to safety, security, sustainability, human systems integration, etc. (see Section 3.1). Define stakeholder requirements that relate to high priority and critical functionality, performance, the operational environment, interactions with users, interactions with external interfacing and enabling systems, and compliance with standards and regulations. Ensure the stakeholder requirements are consistent with the life cycle concepts, needs, scenarios, interactions, constraints, operational risks, and SoI considerations.
  - Ensure high quality stakeholder requirements. Each stakeholder requirement should follow the rules for quality requirements and possesses characteristics such as necessary, singular, correct, unambiguous, feasible, appropriate to level, complete, conforming, and can be validated (see Section 2.3.5.3 and the INCOSE GtWR (2023) for more details). For each stakeholder requirement, define attributes such as source, rationale, priority, and criticality.

- *Analyze stakeholder needs and requirements.*
  - Analyze the complete sets of stakeholder needs and stakeholder requirements. Analyze the sets of stakeholder needs and stakeholder requirements to ensure they are correct, complete, consistent, comprehensible, appropriate to level, and feasible (see Section 2.3.5.3 and the INCOSE GtWR [2023] for more details).
  - Enable technical achievement monitoring through the definition of critical performance measures and quality characteristics.
  - Define system validation criteria for each stakeholder need and requirement, the validation strategy, validation method, and responsible organization for providing evidence the stakeholder needs and requirements have been met.
  - Review the analyzed stakeholder requirements with the applicable stakeholders to validate that their needs and expectations have been adequately captured and expressed.
  - Resolve stakeholder needs and requirements issues. Negotiate changes, amendments, and modifications to resolve inconsistencies, conflicts, and unrealizable or impractical stakeholder needs and requirements.

- *Manage the stakeholder needs and requirements definition.*
  - Obtain explicit agreement on the stakeholder needs and requirements.
  - Establish and sustain traceability (stakeholder needs and requirements).
  - Give CM the information items, work products, or other artifacts needed for baselines.
  - Manage changes to the stakeholder needs and stakeholder requirements, as needed.

Common approaches and tips.

- Identify the enabling systems and materials needed for transition early in the life cycle to allow for the necessary lead time to obtain or access them.

Elaboration

This section elaborates and provides "how-to" information on the Stakeholder Needs and Requirements Definition process. Further guidance on elicitation, life cycle concepts, needs and requirements definition can be found in ISO/IEC/IEEE 29148 (2018), the INCOSE GtWR (2023), the INCOSE GtNR (2022), and the INCOSE NRM (2022).

Verified and validated stakeholder needs and stakeholder requirements are drivers and constraints for the majority of the system life cycle Technical Processes. Depending on the system development model, life cycle concepts definition, and maturation, the stakeholder needs and stakeholder requirements capture should be conducted at the beginning of the development cycle and assessed as a continuous, concurrent, and iterative activity as the project team moves recursively through the system architecture and across all life cycle activities. The reason for eliciting and analyzing the life cycle concepts, stakeholder needs, and stakeholder requirements is to understand the expectations of stakeholders well enough to support the System Requirements Definition processes.

Identify Stakeholders. One of the biggest challenges in system development is the identification of the set of stakeholders from whom life cycle concepts, needs, and requirements are elicited. When identifying stakeholders, take into account those who may be affected by, are able to influence, or will support the life cycle stages of the SoI, Typically, stakeholders include customers, users, operators, maintainers, procurement, organization decision makers, approving authorities, regulatory bodies, developing organizations, verifiers, validators, support organizations, and society at large (within the context of the business and proposed solution). This can include the stakeholders of external systems (e.g., interoperating, interfacing, other constituent systems in a system of systems) and enabling systems, as these will usually impose constraints that need to be identified and considered in the SoI or could have impacts on those systems or the environment. In sustainable development, this includes identifying representation for future generations. When direct contact is not possible, agents are identified, such as marketing or user groups to represent the concerns of classes of stakeholders such as consumers or future generations. There also may be stakeholders who oppose the system. These detractors of the system are first considered in establishing consensus needs. Beyond this, they are addressed through the Risk Management process, the threat analysis of the system, or the system requirements for security, adaptability, agility, or resilience.

Elicit or Derive Stakeholder Needs and Stakeholder Requirements. Determining stakeholder needs and requirements requires the integration of a number of disparate views, which may not necessarily be harmonious. It is important to have a "reconcile" path in the establishment of stakeholder needs and stakeholder requirements, since the stakeholder expectations and the life cycle concepts may be in conflict, incomplete, ambiguous, infeasible, or unable to be satisfied collectively within project constraints. This circumstance illustrates an aspect of "horizontal integration" (see Section 2.3.5.8), recognizing that there will often be prioritization of competing concerns, or even outright rejection of some stakeholder concerns because of inconsistencies with other stakeholders' needs and requirements or a lack of feasibility.

As the SE processes are applied, a common paradigm for examining and prioritizing available information and determining the value of added information should be created. Each of the stakeholder's views of the needed systems can be translated to a common system description that is understood by all participants, and all decision-making activities recorded for future examination. The stakeholder views will be influenced by cognitive biases (see Section 1.4.2) based on their specific role, education, work experiences, culture, etc. Stakeholder views are framed in the context of these biases. It is important for the project team to understand this during elicitation to better understand the perspective of each stakeholder.

SE practitioners support project management in defining what must be done and gathering the information, personnel, and analysis tools to elaborate the life cycle concepts, needs, and requirements. This includes eliciting or deriving stakeholder needs, stakeholder requirements, system/project constraints (e.g., cost and schedule constraints, technology limitations, applicable specifications, and requirements), "drivers" (e.g., capabilities of the competition, external threats, and critical environments), and risks.

The output of the Stakeholder Needs and Requirements Definition process should be sufficient definition of the life cycle concepts, stakeholder needs, and stakeholder requirements to gain authorization and continuing funding for through the Portfolio Management process (see Section 2.3.3.3). The output should also provide necessary technical definition to the Acquisition process (see Section 2.3.2.1) to generate a request for supply if the system is to be acquired through an acquisition or to gain authorization to develop and market the system if the SoI is to be developed within the organization.

Since stakeholder needs and requirements come from multiple sources, eliciting and capturing them constitutes a significant effort on the part of the project. The life cycle concepts help the project team understand the context within which the needs and requirements are captured and defined. Modeling, analysis, and simulation tools can also be used to evaluate candidate solutions and select a desired solution (see Section 3.2.1).

It is essential to establish a database of the data and information which represents the artifacts generated during this process. The database also includes traces between the stakeholder needs, stakeholder requirements, and system requirements. They serve as a foundation for later refinement and/or revision by subsequent activities across the life cycle. Tools for capturing and managing requirements can be used.

Refine Life Cycle Concepts. Stakeholder needs and requirements result from obtaining an understanding of stakeholder expectations through the definition, analysis, and maturation of in a series of life cycle concepts (e.g., acquisition concept, deployment concept, operations concept, support concept, and retirement concept). Development of preliminary life cycle concepts were introduced in the Business or Mission Analysis process (see Section 2.3.5.1). These life cycle concepts need to be refined as part of the Stakeholder Needs and Requirements Definition process.

The primary objective of the development of life cycle concepts is to ensure that stakeholder needs and requirements are clearly understood and the rationale for each is incorporated into the decision mechanism for later transformation into the system requirements. Interviews with manufacturing/coding stakeholders, operators, maintainers, and disposers of current/similar systems, potential users, owners of interoperating, interfacing, and enabling systems (see Section 1.3.3), and site visits provide valuable stakeholder input toward establishing life cycle concepts. Other objectives are as follows:

- To provide traceability between stakeholder needs and stakeholder requirements and their source.
- To establish a holistic understanding of the capabilities needed to address the problem or opportunity in terms of people, process, and products.
- To establish a basis for needs and requirements to support the system over its life, such as personnel requirements, enabling systems, and support requirements.
- To establish a basis for design, system verification, and system validation planning across the life cycle and resulting artifacts and requirements for enabling systems needed as part of the validation and verification activities.
- To assess interactions of the SoI with users and its operating environment including interactions across interface boundaries with external and enabling systems.
- To provide the basis for analysis of system performance, behavior under (over)-load, and mission-effectiveness calculations.
- To validate needs and requirements at all levels and to discover implicit requirements overlooked from other sources.

The life cycle concepts are used to define an integrated set of stakeholder needs which are transformed into the set of stakeholder requirements.

Uncertainties and Risk. During the development of the preliminary life cycle concepts as part of the Business or Mission Analysis process (see Section 2.3.4.3), there may have been uncertainties (see Section 1.4.1) from several perspectives including business, market, management, technical performance, schedule, development and production costs, operations and support costs, security, and sustainability. These uncertainties are a source of risk. Each of these uncertainties must be addressed during the Stakeholder Needs and Requirements Definition process and further elaborated during the Systems Requirements Definition Process (see Section 2.3.4.3).

Record and manage the life cycle concepts, needs, and requirements. The life cycle concepts, stakeholder needs, and stakeholder requirements should be recorded and managed within the project database in a form that allows traceability between the life cycle concepts and the resulting stakeholder needs and requirements (see the INCOSE GtNR [2022] and the INCOSE NRM [2022] for more details).

2.3.5.3   *System Requirements Definition Process*

Overview

Purpose As stated in ISO/IEC/IEEE 15288,

> [6.4.3.1] The purpose of the System Requirements Definition process is to transform the stakeholder, user-oriented view of desired capabilities into a technical view of a solution that *meets the operational needs of the user.*

Description. System requirements are the foundation of system definition and form the basis for the System Architecture Definition, Design Definition, Integration, and Verification processes. Each requirement carries a cost, so the system requirements should be the minimum set necessary and sufficient to realize the intent of the stakeholder needs and requirements. Typically, the later in the project that changes are introduced to the system requirements, the greater the impact is to cost and schedule. Where there is more uncertainty in the requirements, the uncertainty should be managed until the requirements mature.

The System Requirements Definition process generates system requirements from a technical perspective using the stakeholder needs and requirements that reflect the stakeholders' perspectives. As such, the stakeholder needs and requirements drive and constrain the SoI being developed. The quality of the resulting system requirements is dependent on the quality of the agreed-to stakeholder needs and requirements.

System requirements definition is concurrent, iterative, and recursive. Thus, the System Requirements Definition process is done concurrently and iteratively with the other Technical Processes, particularly the Stakeholder Needs and Requirements Definition and the System Architecture Definition processes. With each iteration, more detailed information is discovered and defined based on the analysis and maturation of the life cycle concepts and the system solution. In addition, the System Requirements Definition processes is recursively applied to define the requirements for each lower-level system element within the SoI architecture. The allocation of the system requirements is performed concurrently with the System Architecture Definition process. Lower-level system elements are defined via the System Architecture Definition process, and then the SoI level requirements are allocated to the system elements at the next level. For each lower-level system element, the Stakeholder Needs and Requirements Definition and System Requirements Definition processes are repeated recursively until all system elements have their system requirements defined. The outputs of System Requirements Definition process must be traceable and consistent with the life cycle concepts and stakeholder needs and stakeholder requirements, without introducing unnecessary implementation biases. The System Requirements Definition process adds the verification criteria to each system requirement as it is derived.

Inputs/Outputs. Inputs and outputs for the System Requirements Definition process are listed in the IPO diagram in Figure 2.41. Descriptions of each input and output are provided in Appendix E.

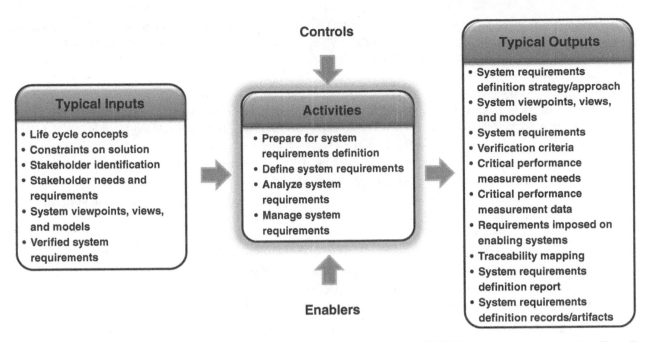

FIGURE 2.41   IPO diagram for System Requirements Definition process. INCOSE SEH original figure created by Shortell, Walden, and Yip. Usage per the INCOSE Notices page. All other rights reserved.

Process Activities. The System Requirements Definition process includes the following activities:

- *Prepare for system requirements definition.*
  - Establish the strategy/approach for system requirements definition.
  - Plan for the necessary enabling systems or services needed through the life cycle for system requirements definition. Enabling systems include tools for elicitation of requirements, life cycle concepts, recording drivers and constraints, defining risks, analysis, recording system needs, recording system requirements, and providing traceability.
  - Ensure enabling system or service access needed to support system requirements definition.
- *Define system requirements.*
  - Define the functional boundary of the system in terms of the behavior and properties to be provided.
  - Identify the life cycle concepts and stakeholder requirements from which the system requirements will be transformed and then define each function and associated performance.
  - Define each expected system function, including the associated performance. Include both primary functions and enabling functions.
  - Define necessary constraints. These include higher-level requirements allocated to the SoI, operational conditions, and interactions with external systems. Define interactions with users, operators, maintainers, and disposers.
  - Identify system requirements that relate to risks, criticality of the system, critical quality characteristics, and compliance with standards and regulations.

- Define verification success criteria for each system requirement, the verification strategy, verification method, and responsible organization for providing proof the system requirements have been met (see Section 2.3.5.9).
- Capture the system requirements and their attributes.

- *Analyze system requirements.*
  - Analyze the system requirements for characteristics of individual requirements and of the set of requirements (can be the set of requirements for the current increment, build, or sprint). Analyze the set of requirements to ensure they are correct, complete, consistent, comprehensible, appropriate to level, and feasible (see the elaboration below and the INCOSE GtWR [2023] for more details).
  - Enable technical achievement monitoring through the definition of critical performance measures.
  - Review the analyzed requirements with the applicable stakeholders.
  - Perform issue resolution for the system requirements. Negotiate changes, amendments, and modifications to resolve inconsistencies, conflicts, and unrealizable or impractical requirements.

- *Manage system requirements.*
  - Confirm agreement that the system requirements meet the stakeholder needs and requirements.
  - Capture key system requirements decisions, rationale, alternatives, and enablers.
  - Establish and sustain traceability (system requirements).
  - Manage system requirements change.
  - Give CM the information items, work products, or other artifacts needed for baselines.

Common approaches and tips.

- Identify the enabling systems and materials needed for transition early in the life cycle to allow for the necessary lead time to obtain or access them.

Elaboration

This section elaborates and provides "how-to" information on the System Requirements Definition process. Additional guidance on needs and requirements definition can be found in ISO/IEC/IEEE 29148 (2018), the INCOSE GtWR (2023), the INCOSE GtNR (2022) and the INCOSE NRM (2022).

Stakeholder Requirements versus System Requirements. The set of stakeholder requirements are SoI requirements written from the stakeholders' perspectives to represent what they require of the SoI in order to meet their needs. The set of system requirements represent the technical perspective of what the SoI must meet during the System Architecture Definition and Design Definition processes that will result in a system that meets the stakeholder needs and stakeholder requirements. Another key distinction is that the focus of the stakeholder requirements is often on high-priority and critical functions, performance, quality, compliance, etc., while the system requirements are more encompassing and detailed including enabling functions, performance, quality, compliance, etc. that will result in the stakeholder requirements to be implemented.

In some cases, the stakeholder requirements can be copied directly into the set of system requirements "as is" and additional requirements added as needed. For smaller, internal projects, the set of stakeholder requirements could be used as the set of system requirements, depending on how much analysis went into the definition of the set of stakeholder requirements such that their implementation will result in the stakeholder needs to be met.

When a set of stakeholder requirements is provided to a supplier by an acquirer, the supplier uses these as inputs to their SE processes to develop the set of system requirements. When defining the system requirements, rather than treating the supplied stakeholder requirements as the *only* source of requirements, the supplier has an obligation to do an assessment for derived system requirements and as well as requirements from other "non-acquirer" stakeholders. For example, the supplier's production team needs the product to be manufacturable, their test team needs the product to be testable, the supplier and public need the product to be safe and secure from a cybersecurity perspective, the users and operators need the product to be easy and safe to interface with from a human perspective, and the organization has regulatory compliance considerations. For products to be developed by an outside supplier, the supplier's company may need the product to conform to a strategic development effort aligning with other products produced, internal standards, and technology maturation. The acquirer may not have included all of these considerations when developing their set of requirements. If the supplier blindly follows only the acquirer supplied requirements specified in their contract, they are likely to generate a SoI that may not work in the integrated system or operational environment, resulting in a system that fails system validation.

Plan for system requirements definition. The System Requirements Definition process should begin with a review of the problem, threat, or opportunity for which the SoI is to address, and the mission, goals, objectives and critical success measures defined by the Business or Mission Analysis process (see Section 2.3.5.1) and the set of stakeholder needs, stakeholder requirements, and life cycle concepts defined by the Stakeholder Needs and Requirement Definition process (see Section 2.3.5.2). For contracted development efforts, mission, goals, objectives, and critical success measures can come from both the acquirer and supplier organizations. Before the System Requirements Definition process, the project team will need to define the strategy to be used to transform the stakeholder needs and stakeholder requirements, define drivers and constraints, assess risks, define, analyze, and mature life cycle concepts, and derive an integrated set of system requirements resulting from these activities.

Requirements Definition. The integrated set of stakeholder requirements is transformed into system requirements to address what the system must do to meet those needs. The transformation process involves additional analysis and further elaboration of the models developed during life cycle concept analysis and maturation. The system requirements must address function, fit, form, quality, and compliance with stakeholder and business needs. System requirements must also address interactions with external systems, users, operators, maintainers, disposers, and the operational environment. SE practitioners collaborate with the stakeholders of the external systems to define each of the interactions and record an agreement of those definitions in some configuration managed form, as well as any constraints or interface requirements (see Section 3.2.4).

Definition of the system requirements is a complex process that includes function and performance analysis; trade studies; constraint evaluation; inclusion of (or reference to) specific requirements from relevant standards and regulations; risk assessment; technology assessment; detailed characterization of the operational environment; detailed assessment of the interactions of the parts that make up the SoI, detailed assessment of the interactions between the SoI and users, operators, maintains, disposers, and external systems; and cost–benefit analysis. System requirements cannot be established without determining their impact (achievability) on lower-level system elements, especially in terms of cost, schedule, and technology. Therefore, system requirements definition is a concurrent, iterative, and recursive balancing process that works both "top-down" (called allocation, derivation, and flow-down) and "bottom-up" (called compliance analysis and flow-up).

The system requirements are inputs to the System Architecture Definition and Design Definition processes, in some domains these requirements are referred to as "design-to" or "design input" requirements. When the requirements are defined, it is important that they are expressed at a level of abstraction that is appropriate to the SoI and systems hierarchy level to which they apply. Although it is good practice to avoid implementation when defining the system requirements, it is not always possible.

In defining system requirements, care should be exercised to ensure each requirement statement is appropriately crafted. The characteristics shown in Table 2.7 should be considered for each individual requirement statement (INCOSE GtWR, 2023). In addition to the characteristics of individual requirement statements, the characteristics shown in Table 2.8 should be considered for requirement sets (INCOSE GtWR, 2023).

System requirement statements may have a number of attributes attached to them (either as fields in a database or through relationships with other artifacts) shown in Table 2.9. *The attributes annotated with an asterisk ("*") represent a proposed minimum set.* See the INCOSE NRM (2022) for the definition and description of these attributes.

Allocation, derivation, and flow-down. The next level of the system hierarchy is defined in conjunction with the System Architecture Definition and Design Definition processes. System requirements are allocated to the system elements at the next level of the system hierarchy. Once the allocation has been determined, the system requirements are derived (assigned) for the next system elements at the level of system hierarchy such that the intent of the allocated parent requirement is met.

TABLE 2.7   Requirement statement characteristics

Requirement Statement Characteristic	Definition
Necessary	The requirement statement defines a capability, characteristic, constraint, or quality factor needed to satisfy a life cycle concept, need, source, or parent requirement.
Appropriate	The specific intent and amount of detail of the requirement statement is appropriate to the level (e.g., the level of abstraction, organization, or system architecture) of the entity to which it refers.
Unambiguous	The requirement statement is stated such that the intent is clear and the requirement can be interpreted in only one way by all the intended stakeholders.
Complete	The requirement statement sufficiently describes the necessary capability, characteristic, constraint, conditions, or quality factor to meet the need, source, or higher-level requirement from which it was transformed.
Singular	The requirement statement states a single capability, characteristic, constraint, or quality factor.
Feasible	The requirement statement can be realized within entity constraints (e.g., cost, schedule, technical, legal, ethical, safety) with acceptable risk.
Verifiable	The requirement statement is structured and worded such that its realization can be verified to the approving authority's satisfaction.
Correct	The requirement statement is an accurate representation of the need, source, or higher-level requirement from which it was transformed.
Conforming	The requirement statement conforms to an approved standard pattern and style guide or standard for writing and managing requirements.

TABLE 2.8   Requirement set characteristics

Requirement Set Characteristic	Definition
Complete	The requirement set for a given SOI should stand alone such that it sufficiently describes the necessary capabilities, characteristics, functionality, performance, drivers, constraints, conditions, interactions, standards, regulations, and/or quality characteristics without requiring other sets of requirements at the appropriate level of abstraction.
Consistent	The requirement set contains individual requirements that are unique, do not conflict with or overlap with others in the set, and the units and measurement systems they use are homogeneous. The language used within the sets is consistent (i.e., the same words are used throughout the set to mean the same thing). All terms used within the requirement statements are consistent with the architectural model, project glossary, and project data dictionary.
Feasible	The requirement set can be realized within entity constraints (e.g., cost, schedule, technical) with acceptable risk.
Comprehensible	The requirement set is written such that it is clear as to what is expected of the entity and its relation to the macro system of which it is a part.
Able to be validated	The requirement set will lead to the achievement of the set of needs and higher-level requirements within the constraints (such as cost, schedule, technical, and regulatory compliance) with acceptable risk.
Correct	The requirement set is an accurate representation of the needs, sources, or higher-level requirements from which it was transformed.

TABLE 2.9   Requirement attributes

Attributes to Help Define Needs and Requirement and Their Intent	A24—Approval Date
A1—Rationale*	A25—Date of Last Change
A2—Trace to Parent*	A26—Stability/Volatility
A3—Trace to Source*	A27—Responsible Person
A4—States and Modes	A28—Need or Requirement Verification Status*
A5—Allocation/Budgeting*	A29—Need or Requirement Validation Status*
	A30—Status of the Need or Requirement
Attributes Associated with System Verification and System Validation	A31—Status (of Implementation)
A6—System Verification or System Validation Success Criteria*	A32—Trace to Interface Definition
A7—System Verification or System Validation Strategy*	A33—Trace to Dependent Peer Requirements*
A8—System Verification or System Validation Method*	A34—Priority*
A9—System Verification or System Validation Responsible Organization*	A35—Criticality or Essentiality*
A10—System Verification or System Validation Level	A36—Risk (of Implementation)*
A11—System Verification or System Validation Phase	A37—Risk (Mitigation)*
A12—Condition of Use	A38—Key Driving Need or Requirement
A13—System Verification or System Validation Results	A39—Additional Comments
A14—System Verification or System Validation Status	A40—Type/Category
	Attributes to Show Applicability and Enable Reuse
Attributes to Help Manage the Requirements	A41—Applicability
A15—Unique Identifier*	A42—Region
A16—Unique Name	A43—Country
A17—Originator/Author*	A44—State/Province
A18—Date Requirement Entered	A45—Market Segment
A19—Owner*	A46—Business Unit
A20—Stakeholders	**Attributes to Aid in Product Line Management**
A21—Change Control Board	A47—Product Line
A22—Change Proposed	A48—Product Line Common Needs and Requirements
A23—Version Number	A49—Product Line Variant Needs and Requirements

From INCOSE NRM (2022). Usage per the INCOSE Notices page. All other rights reserved.

The System Requirements Definition process is repeated recursively for each level of the system hierarchy until the system elements are to the level of detail needed to be realized via a make (e.g., build, code), buy, or reuse decision. The resulting sets of system requirements for the system elements represent the allocated baseline of the SoI.

Requirements Management. According to ISO/IEC/IEEE 29148, requirements management encompasses those tasks that record and maintain the evolving requirements and associated context and historical information from the requirements engineering activities. Effective requirements management occurs within the context of an organization's project and Technical Processes. Requirements management also establishes procedures for defining, controlling, and publishing the baseline requirements for all levels of the SoI. The resulting sets of requirements are provided to the Configuration Management process (see Section 2.3.4.5) process for baselining at the appropriate time. The Configuration Management process is used to establish and maintain configuration items and baselines. Requirements management also ensures traceability is established between requirements and other artifacts (see Section 3.2.3), that appropriate requirements reviews occur, and requirements measures are established and used. See also the INCOSE GtNR (2022) and the INCOSE NRM (2022) for further elaboration concerning requirements management.

2.3.5.4 *System Architecture Definition Process*

Overview

Purpose As stated in ISO/IEC/IEEE 15288,

> [6.4.4.1] The purpose of the System Architecture Definition process is to generate system architecture alternatives, select one or more alternative(s) that address stakeholder concerns and system requirements, and express this in consistent views and models.

System Architecture Definition process transforms related architectures (e.g., strategic, enterprise, reference, and SoS architectures), organizational and project policies and directives, life cycle concepts and constraints, stakeholder concerns and requirements, and system requirements and constraints into the fundamental concepts and properties of the system and the governing principles for evolution of the system and its related life cycle processes. This process results in a system architecture description for use by the project, its organization, other organizations, and various stakeholders. The Project Management Plan (PMP) and Systems Engineering Management Plan (SEMP) in some cases will provide management directives on how to perform this process, but usually the programmatic view and other related views developed by the System Architecture Definition activities will guide the PMP and the SEMP. The architecture governance activities at the organization level will provide additional direction for the System Architecture Definition process through its issuance of architecture governance directives. Since the directives and stakeholder requirements can evolve throughout the system life cycle, the system architecture description should be treated as a living artifact reflecting both the changing expectations and the evolution of our understanding of what the system solution should be.

Development practices for architecture are specified by ISO 15704 for enterprises and the ISO/IEC/IEEE 42000 series of standards in software, systems, and enterprise fields of application. ISO 15704 specifies terms, concepts, and principles considered necessary to address stakeholder concerns, carry out enterprise creation programs and any incremental change projects required by the enterprise throughout its whole life. ISO/IEC/IEEE 42000 series of standards establishes processes, key principles, and concepts for conceptualization, evaluation, and description of architectures.

Description The System Architecture Definition process provides information and data useful and necessary for identifying and characterizing the fundamental concepts and properties of the system and its elements. These concepts and properties can be fundamentally human-centric, with individual, social, organizational, and political aspects, in human activity systems considering technical elements as enablement assets. The architecture information and data will be implementable through system and system element designs, which satisfy as far as possible the problem or opportunity expressed by models and views for a set of stakeholder and system requirements (traceable to business/mission requirements, as applicable) and life cycle concepts (e.g., Operational, Acquisition, Deployment, Support, and Retirement). During a stage in the system life cycle, the relevant enabling systems and the SoI are considered together as a solution but are distinguished from each other in the overall solution conceptualization.

System architecture definition focuses on achieving associated missions and characterizing the operational concepts of the system and performing market analysis to ensure viability of the SoI. It utilizes architectural principles and concepts to define the high-level structure of a system and its elements, and the intended properties and characteristics of the SoI. It highlights and supports trade-offs for the other System Definition processes. and possibly Portfolio Management and Project Planning. It incorporates incremental insights obtained about the emergent properties and behaviors of the SoI while achieving a balance for suitability, viability, effectiveness, and affordability. This process is iterative and requires participation of architects, SE practitioners, and specialists in relevant domains, subject matter experts and other stakeholders. The process continues recursively through the levels of the system and its system elements, with consistent feedback to ensure the system design continues to satisfy stakeholder needs and system requirements,

Inputs/Outputs Inputs and outputs for the System Architecture Definition process are listed in Figure 2.42. Descriptions of each input and output are provided in Appendix E.

Process Activities The System Architecture Definition process includes the following activities:
Prepare for system architecture definition.

- Identify and analyze relevant market, industry, stakeholder, organizational, business, operations, mission, legal, and other information and related perspectives that will guide the development of the system architecture.
- Identify key milestones and decisions to be informed by the system architecture effort. In particular, identify those key architecture artifacts and resources that guide the system architecture development.
- In conjunction with the System Requirements Definition process, determine the system context (i.e., how the SoI fits into the external environment) and system boundary are refined, that reflect operational scenarios and expected system behaviors. This task includes identification of expected interactions of the SoI with system elements, or other systems or entities.
- Establish the approach for architecting. This includes an architecture roadmap and strategy, methods, frameworks (see Section 3.2.5), patterns (see Section 3.2.6), modeling techniques, tools, and the need for any enabling systems (see Section 1.3.3), products, or services. The approach should also include the process requirements (e.g., measurement approach and methods), evaluation (e.g., reviews and criteria), and necessary coordination.
- Ensure the enabling items (registry, repository, library, competencies), services, resources and capabilities for executing the System Architecture Definition process are available. This includes planning for the need and identifying the requirements for the enabling items.

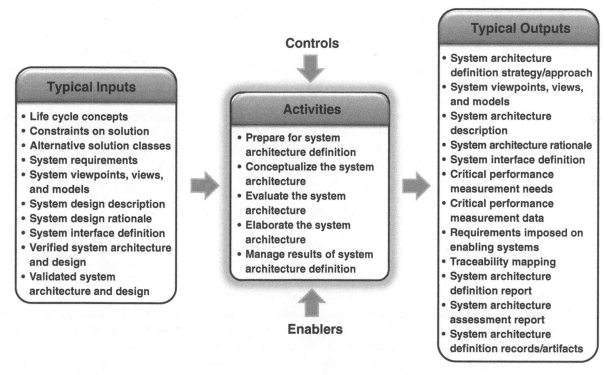

FIGURE 2.42 IPO diagram for System Architecture Definition process. INCOSE SEH original figure created by Shortell, Walden, and Yip. Usage per the INCOSE Notices page. All other rights reserved.

- Establish or identify associated architecture viewpoints and model kinds that facilitate analysis and understanding of the viewpoint. This task includes identifying expected uses and users of architecture information, identifying potential architecture framework(s), capturing rationale for selection of viewpoints, templates, metamodels and model kinds, selecting, modifying and developing relevant methods, techniques, and tools.

Conceptualize the system architecture.
Note: This activity is based on the Architecture Conceptualization process in ISO/IEC/IEEE 42020

- Characterize the problem space in conjunction with the BMA process and document it. The report focuses on architecture considerations that span one or more system life cycle stages
- In conjunction with the SNRD process, the system context and system boundary are refined, including identification of expected interactions of the SoI with system elements, or other systems or entities. This task includes determination of boundary conditions, quality measures, situation contexts, assumptions, degrees of freedom, constraints, conditions, and challenges.
- Define architecture objectives and critical success criteria that will be used to assess the extent to which the problems and opportunities will be addressed.
- Based on existing or previous solutions, and problem mitigation strategies, address the highest priority requirements and architecture considerations to synthesize a set of potential solutions. This task includes scanning for relevant technologies, problem patterns, solution patterns, naturally occurring solutions, enhancements to existing systems, heuristics, tactics, and discussion with experts.
- For each potential solution, identify strengths, weaknesses, gaps or shortfalls, required trade-offs, consequences, obligations, assumptions, critical success factors affecting critical success criteria and key performance indicators. Devise structural, behavioral, organizational and architectural entities (functions, input/output flows and flow items, states and modes, functional and physical interfaces, nodes and links, computational and communication resources, etc.) to formulate candidate architecture(s). Based on the set of candidate architecture(s), select the best architecture(s) for downstream use by using the Decision Management and Risk Management process. This task includes identifying and characterizing tradeoffs, defining context and scope, determining and mitigating risks, and identifying issues and areas for improvement.
- Select, adapt, or develop views and models of the best architecture(s), by capturing concepts, properties, decisions, processes, activities, tasks, characteristics, guidelines, and principles and utilizing architecture viewpoints to develop architecture descriptions. This task includes determining the scope, breadth and depth, use and users of each view and model, and expressing them in the specified form with sufficient level of detail.

Evaluate the system architecture.
Note: This activity is based on the Architecture Evaluation process in ISO/IEC/IEEE 42020.

- Determine evaluation objectives and criteria for value assessment and architecture analysis by identifying relevant mandates and imperatives, stakeholders and their concerns, policies and standards, value, and quality characteristics.
- Determine evaluation methods and integrate them with evaluation objectives and criteria.
- Collect and review evaluation related information including views and models, architecture concepts, properties, metrics and measures, sources of information, accuracy, errors, degrees of uncertainty, and qualification of correctness, completeness, and consistency of gathered information.
- Analyze, assess, and characterize architecture(s), by using evaluation methods and criteria, and applying the System Analysis and Measurement processes to produce architecture assessments. Architecture alternatives that are similar to each other or fail to meet identified mandates are eliminated and costs, risks, and opportunities are identified and characterized for appropriate actions.

- Formulate, capture, validate, and communicate the findings and recommendations, including implications, to relevant decision makers and stakeholders. The combined overall evaluation can be used to select a preferred system architecture solution.

Elaborate the system architecture.
Note: This activity is based on the Architecture Elaboration process in ISO/IEC/IEEE 42020.

- Based on the identified viewpoints, develop architecture models and views that adequately address stakeholder concerns, while, if applicable, conforming to selected architecture frameworks.
- Perform preliminary interface definition for interfaces with the level of detail necessary for understanding the architecture for decision making and risk management. The definition includes the internal interfaces between the system elements and the external interfaces with entities outside the system boundary.
- Analyze the architecture models and views for consistency and resolve any issues identified. ISO/IEC/IEEE 42010 correspondence rules from frameworks can aid in this analysis. This task includes relating architectural entities to elements of views and models, mapping related entities to relevant architecture and system concepts, properties, and principles, and assessing whether architecture views are consistent with corresponding viewpoints.
- In conjunction with the Verification and Validation processes (see Sections 2.3.5.9 and 2.3.5.11), verify and validate the models by execution or simulation, if modeling techniques and tools permit, and with traceability matrix of operational concepts. Where possible, use design tools to check their feasibility and validity. As needed, implement partial mock-ups or prototypes, or use executable architecture prototypes or simulators.
- Utilizing models and views, develop architecture descriptions by composing those views and models that adequately cover the uses and users of the architecture descriptions. Assess the architecture description against the intent of the architecture, as well as its suitability, correctness, completeness, and consistency.

Manage results of system architecture definition.

- Capture, maintain, and manage the rationale for selections among alternatives and decisions about the architecture, architecture framework(s), viewpoints, model kinds, views, and models. This task includes managing information for decisions, risks, constraints and assumptions and possible governance of upper-level architectures.
- Establish the means for the implementation of the directives of the governance of the architecture, including the roles, responsibilities, authorities, and other control functions. Monitor and assess whether governance directives and guidance are being followed.
- Establish a means for management of the architecture, including plans, measures, schedules, milestones, and other functional outcomes. Monitor and control the implementation of management instructions, provision of status reports, and corrective actions.
- Manage the maintenance and evolution of the architecture, including the architectural entities, their characteristics, and principles. Allocation and traceability matrices are useful to analyze impacts on the architecture.
- Manage the architecting effectiveness, including work performance tracking, reviewing, regulating the progress, dealing with management issues, dealing with resource allocation issues, dealing with methods and tools availability, and coordinating review of the architecture to achieve stakeholder agreement.
- Maintain bi-directional traceability of the system architecture including traceability between the architectural entities to the requirements, interface definitions, analysis results, related architectures, and stakeholder concerns.
- Manage the maintenance, evolution, and use of the architecture descriptions, including the architecture viewpoints, views, and models.

Common approaches and tips.

- Define the problem and the solution spaces with regard to the identified stakeholders
- Define the main principles governing the whole life cycle processes of a SoI, in the scope of the solution space.
- Identify the enabling systems and materials needed for transition early in the life cycle to allow for the necessary lead time to obtain or access them.
- Ensure that conflicting interests (e.g., performance vs. quality characteristics, distributed control vs. central control, new technologies vs. COTS) have been properly addressed.
- Use the Risk Management process to help ensure that the inherent risks associated with the use of new technologies are adequately assessed.

Elaboration

Architecture Processes ISO/IEC/IEEE 42020 (2019) provides a generic process reference model for architecture processes for enterprise, system, and software levels. The concept of architecture as considered in this standard is applicable for different kinds of entities being architected. It specifies 6 architecture processes for use by organizations and projects. As shown in Figure 2.43, the core architecture processes as outlined in the standard are: Architecture Conceptualization, Architecture Evaluation, and Architecture Elaboration. The Architecture Conceptualization process characterizes the problem space and determines suitable solutions that address stakeholder concerns, achieve architecture objectives, and meet relevant requirements. The Architecture Evaluation process determines the extent to which one or more architectures meet their objectives, address stakeholder concerns, and meet relevant requirements. The Architecture Elaboration process describes or documents an architecture in a sufficiently complete and correct manner for the intended uses of the architecture.

System Architecture The notion of a system is abstract, but it is a practical means to create, design, or redesign products, services, or enterprises. The SoI and the enabling systems that are necessary for development, utilization and support should be considered together in a solution to address a problem or an opportunity. Note that there may be several potential solutions to address the same problem or opportunity. System is represented with sets of interrelated entities—including human in socio-technical systems—achieving one or more stated purposes. These system entities may possess characteristics such as dimensions, environmental resilience, availability, robustness, learnability, execution efficiency, openness, modularity, scalability, and mission effectiveness.

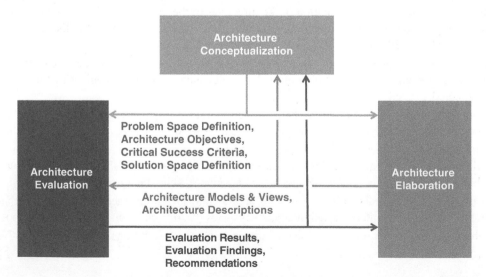

FIGURE 2.43 Core architecture processes. From ISO/IEC/IEEE 42020 (2019). Used with permission. All other rights reserved.

Architecture Description ISO/IEC/IEEE 42010 specifies the normative features of architecture frameworks, architecture description languages, and viewpoints and views as they pertain to architecture description. An architecture description expresses the architecture of a system and is composed of architecture views. A view is an information part comprising portion of an architecture description. It is composed of view components which are derived from models and non-model sources of information. A viewpoint is the set of conventions for the creation, interpretation and use of views to frame one or more concerns of stakeholders and specifies the ways in which the view components should be generated and used. An architecture framework contains standardized viewpoints, view templates, metamodels, model templates, etc. that aid in development of architecture views. An architecture description language contains syntax and semantics intended for describing the architecture and provides a way to create and understand view components.

Architecture Evaluation ISO/IEC/IEEE 42030 provides a generic, conceptual guiding framework that can be used for the planning, execution, and documentation of architecture evaluations. The elements described in this standard can be used to determine architecture value, determine architectural characteristics, validate whether the architecture addresses current and future stakeholder needs with architecture assessment against defined stakeholder acceptance criteria, and also provide inputs to decisions made at the business, operational and tactical levels.

Architecture Considerations Per ISO/IEC/IEEE 42010, Stakeholder concerns, architecture aspects and stakeholder perspectives are kinds of architecture considerations. Architecture frameworks help identify views and viewpoints to characterize the architectures with regard to these considerations.

Kinds of Architecture Entities Architecture is increasingly applied to systems and other entities that are not traditionally considered to be systems, such as enterprises, services, business functions, mission areas, product lines, families of systems, and software items. Corresponding to each of these entities, different kinds of architecture can be considered according to their purpose, domains of application, and roles within entity and architecture life cycles.

System Architecture vs System Design The System Architecture Definition process focuses on the essential concepts, properties' structure, behaviors, and features that apply to the system solution. It helps gain insights into the relation between the requirements for the system and the emergent properties and behaviors of the system that arise from the interactions and relations between the system elements. The Design Definition process focusses on developing an overall system design that is ultimately sufficiently detailed to allow its realization. An effective architecture is as design-agnostic as possible to allow for maximum flexibility in the design trade space. The Design Definition process provides feedback to the System Architecture Definition process to consolidate or confirm the allocation, partitioning, and alignment of architectural entities to system elements that comprise the system.

Architecting Styles Per (Evans, 2014), Architecting Styles provide a set of proven approaches for those who create, commission, use, and evaluate architecture products. These can help key decision makers to be better informed on the use and limitations of the architecture thereby ensuring that the different architecting activities consistently deliver value. These styles help to understand the architecting approach; architecture objectives; architectural entities; how value is created to make effective architecture-related decisions. The styles are driven by the purpose, culture, or reason for the architecture and reflect currently observed good practices. The four styles of architecting are: authoritative, directive, coordinative, and supportive.

Architecture Styles Per (Garlan, et al., 1994, 1996), an architecture style is a set of design elements or principles or properties or a generic pattern that provides guidance for the System Architecture Definition process. The set helps in identification and classification of architectures. Architecture styles can be understood as language, system of types and as theory. Architecture styles can be defined by architecture views, architectural elements and their relationships, architecture viewpoints, layouts, connections, interfaces, interaction mechanisms, communication factors, and applicable constraints.

Architecture Patterns Per (Bass, et al., 2012), an architecture pattern is a reusable, configurable architectural entity comprising a minimal set of elements that is complete under certain aspects and exhibits rules for instantiation that is applicable for different situations. It solves and delineates certain elements of the system architecture and can be used in many system architecture efforts. It has a fundamental structure of predefined elements and relationships, principles, rules and guidelines. Architecture patterns promote communication, streamline documentation, support high levels of reuse, improve architect's efficiency and productivity, and provide a starting point for additional ideas.

Value and Quality While the Systems Architecture Definition process creates a framework for addressing stakeholder concerns and requirements; the goal is to deliver value to all stakeholders, which might correlate to quality factors deemed important. It is essential that value is created over the life of the system so that the system remains satisfactorily in use. The perception of what is of value to stakeholder changes over time, and hence it is necessary to account for the different times at which value is being presented or reported. This requires that sources of stakeholder value are determined, system capabilities are defined to produce or influence value, and vulnerabilities that cause value degradation are identified. Per Kumar (2020), value-based approaches helps one to learn and understand the stakeholder's value system, their principles of behavior, expectations, ideals and belief systems, motivation, and the boundaries within which the stakeholder can be engaged.

Notion of Interface The notion of interface is one of the key items to consider when defining the architecture of a system. The term "interface" comes from Latin words "inter" and "facere" and means "to *do* something *between* things." Therefore, the fundamental aspects of an interface are functional and defined as inputs and outputs of functions. Interoperability is a stakeholder need and requirement, ensuring interfaces use open, well maintained and enduring standards is key to reduce future integration challenges.

Horizontal and vertical integration System Architectures ensure that requirements allocated throughout the system's design process account for system elements and interfaces as the design matures. The architecture establishes the significant operational and system development interfaces, both internal and external, that must be maintained through development and upgrades. The overall System Architecture is composed of system elements, which are integrated to form the entire system. It is essential to maintain cognizance of the end-to-end system performance expectations when evaluating integration of the system elements, so that those elements continue to perform as needed. When a dynamic relationship exists between one element in the system and another, there is an interdependency. This may involve relationships that are functional or physical in nature, or both. Depending on how tightly coupled these system elements are, the net effect on the system will vary. For example, there is often an interdependency between safety functional hazard conditions and certain function and physical system elements defined in the system architecture.

2.3.5.5   *Design Definition Process*

Overview

Purpose As stated in ISO/IEC/IEEE 15288,

> [6.4.5.1] The purpose of the Design Definition process is to provide sufficient detailed data and information about the system and its elements to realize the solution in accordance with the system requirements and architecture.

This process is driven by requirements that have been vetted through the architecture and more detailed analyses of feasibility.

Description The Design Definition process transforms architecture and requirements into a design of the system that can be realized. This process results in sufficiently detailed data and information about the system and its elements to enable implementation consistent with architectural entities defined in models and views of the system architecture, in conformance with applicable system requirements, and in alignment with design guidelines and standards adopted by the organization or project. Often these system elements are identified, and their fundamental concepts and properties are characterized, by the System Architecture Definition process. The design information and data will define the expected properties and characteristics allocated to each system element and enable transition toward their realization.

Inputs and Outputs Inputs and outputs for the Design Definition process are listed in Figure 2.44. Descriptions of each input and output are provided in Appendix E.

Process Activities The Design Definition process includes the following activities:
Prepare for design definition.

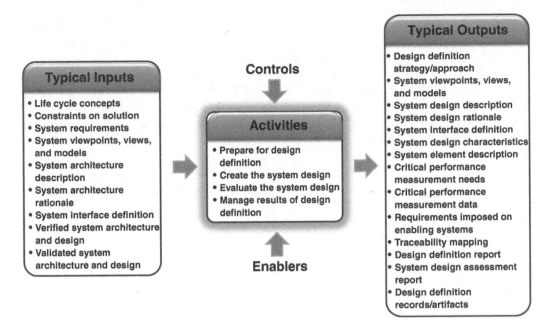

FIGURE 2.44 IPO diagram for Design Definition process. INCOSE SEH original figure created by Shortell, Walden, and Yip. Usage per the INCOSE Notices page. All other rights reserved.

- Determine design drivers for the system design and an appropriate design strategy and applicable approaches. Personnel (together with human factors), processes, products, and services intended to compose the system are among the many factors that will impact system design. Non-functional considerations and design constraints should be identified as these can also serve as design drivers.
- Determine the necessary technologies and the categories of system characteristics to be represented in the design. Capabilities, resources, and services should be identified as these can provide the necessary technologies. Quality models should be identified as these can categorize system characteristics.
- Examine the system architecture to determine the fundamental properties and concepts that apply to the system design, along with the principles that should govern the design and its evolution.
- Establish the approach for system design effort.
- Ensure the necessary system design-enabling elements, services, resources and capabilities are available.

Create the system design.

- Identify and assess design alternatives and create a system design using appropriate techniques such as adapting an existing design, composing a design from available system elements, creating a new design, or through a combination of these approaches. Section 3.2.7 briefly explains several design approaches.
- Allocate system requirements to system elements so that all the system requirements and architecture objectives are addressed. This task includes transformation of architectural entities and relationships to design elements and transformation of architectural characteristics into design characteristics.
- Where necessary, conceive a system design that does not already exist using synthesis techniques, e.g., brainstorming, analogical thinking, or using morphological charts. Synthesis can be applied at multiple levels within a system and focuses upon the solution space and the development of one or more potentially satisfactory design solutions.
- Transform architectural entities (e.g., enterprise or project goals, capabilities and effects, operational activities, resource functions) and relationships into design elements. Also transform architectural characteristics into

design characteristics (e.g., functionality, behavior, dimensions, shapes, materials, critical quality characteristics, data processing structures).

- Analyze the design as part of the design creation activity. A range of analysis techniques can be employed including (1) parametric design to explore within the potential solution space areas for further investigation, and (2) trade-off analysis to achieve balanced system design solutions. Analysis of the design which is systemic in nature (e.g., completeness and consistency analysis) is included in the Systems Analysis process (see Section 2.3.5.6).
- Analyze the interfaces that were identified and defined in the system architecture and refine or further define the interfaces to the level required for the design characteristics and interactions between the system elements and with external entities.
- Capture the system design expressing it as a work product, termed the system design description, addressing the composition, properties, and characteristics of the system design. Typically, methods, frameworks, diagrammatic forms, notations, and other forms of model are applied to express this work product. The system design description is developed into a specification which can be used to either procure or otherwise realize the system elements which comprise the design.

Evaluate the system design.

- Determine the overall suitability and "goodness" of potential design solutions in meeting the identified need or opportunity and in satisfaction of requirements and constraints. The design should be consistent with a governing architecture description.
- Analyze further the system design in support of such evaluation, for example to determine particular design properties and characteristics.
- Assess the value or worth the design will have for various stakeholders as well as potential negative consequences.
- Combine the analyses and assessments into an overall evaluation determination which can serve as the basis for selecting a preferred system design and serve as feedback to the System Architecture Definition process. This design evaluation activity can also provide useful information to the Verification process.

Manage results of design definition.

- Establish a means for the management of the system design, including agreement of the design.
- Maintain bidirectional traceability of the system design. Track the satisfaction of requirements, constraints, and objectives, as well as between the design characteristics and architecture entities, interfaces, and analysis results.
- Capture, maintain, and manage the rationale and supporting information for selections among alternatives and decisions about the design. Maintain the design configuration(s) and, in conjunction with the Configuration Management process, conduct change management.
- Manage certification of the design to qualify it as meeting specified quality standards, as applicable (e.g., for critical systems).
- Register the design to protect it and its constituent intellectual property, as applicable.
- Maintain design integrity during the development and evolution of the system design. A specific responsibility (e.g., a system design authority) is generally assigned to address this issue.
- Conduct design reviews to evaluate design progress, suitability, and quality, and record resulting actions and their satisfaction.

Common approaches and tips.

- Identify the enabling systems and materials needed for transition early in the life cycle to allow for the necessary lead time to obtain or access them.

- It is important to maintain good communication and coordination between the SE practitioners and the various practitioners from other disciplines during the design definition of the system elements, in order to ensure a holistic view as the elements evolve.

Elaboration

System Design versus System Architecture The Design Definition process focusses on developing an overall system design that is ultimately sufficiently detailed to allow its realization. It may be driven from stakeholder concerns, requirements, constraints, business opportunity, mission need, or from the architecture developed by the System Architecture Definition process. On the other hand, the system architecture focuses on the essential concepts, properties, structure, behaviors and features that apply to the system solution, the system design ultimately will capture a description of the proposed system solution which is sufficiently detailed to enable its implementation. System design focuses on system technical considerations such as the solution system elements, their interfaces and characteristics together with technological and other realization considerations such as materials, manufacture, software coding, and operator profiles.

Influence of Design Thinking Design Thinking focuses on design processes and reasoning, together with the resulting design concept and their development, specialization and realization (see Section 3.2.7). It has particular influence upon the creation of the system design. Design Thinking can have as important an influence on system design as Systems Thinking considerations (see Section 1.5) which focus on taking a holistic, systemic view.

Identification of Design Drivers Design drivers, often identified during the System Architecture Definition process, are those factors which should most heavily influence the system design. They may be identifiable from analysis of the business market for the system, operational considerations, through-life considerations, stakeholder requirements and constraints, human factors, technology characteristics and constraints, implementation factors, etc. The set of design drivers applicable to a particular system problem (and solution) determine the overall design approaches that should be employed, the applicable design principles, and the specific specialty considerations which should be employed and integrated into a coherent overall system design.

Usually, the functionality required of a system dominates the design considerations. However, in some situations, the non-functional attributes (e.g., safety, security, performance, dependability) will dominate the Design Definition process, for example, in propulsion systems (e.g., gas turbine engines), and for some kinds of systems (e.g., sensor systems) where quality attributes are key. Furthermore, in socio-technical systems, the relationships between system elements and things in the external world (e.g., financial, legal, social, economic, political), as well as human factors considerations, will often dominate the design of such systems. *(For example, see human-centric design).*

Design For X (DFX) Design drivers concerning considerations such as testability (DFT) or manufacturability (DFM), for example, are addressed by quality characteristics described further in Section 3.1. These drivers may apply to either the whole or just a part of the system and a commonly employed system design tactic is to concentrate their applicability just to a specific part of the system. The resulting system design elements or influences need to be integrated into an overall balanced, viable, and satisfactory design using holistic design practices.

Different Approaches used for developing a System Design A system design approach may originate from one or more sources:

- From requirements and constraints as identified in the Stakeholder Needs and Requirements Definition Process, and System Requirements Definition processes
- From the business or mission problem or opportunity and potential solution space characterization as determined by the Business or Mission Analysis process
- From the system architecture as defined by the System Architecture Definition process
- From the Design Strategy (e.g., maximum reuse).

Assurance of a System Design In certain circumstances, system designs are required with high levels of special characteristics such as integrity or dependability and this is likely to require either the more stringent application of design practices or the usage of specific practices (as part of the create the system design activity) to achieve design characteristics such as overall quality or some specific characteristics such as airworthiness, reliability, safety or security. Some of these characteristics may be externally regulated. Specific organizations (authorities) may have been delegated responsibility for identifying appropriate requirements and determining their satisfaction (often by independent means) by a system design. In addition, certification of the system design (as part of the Verification and Validation processes) may be undertaken to ensure that the system design complies with regulatory design certification requirements. This includes matters such as nuclear power plant operational certification, aircraft airworthiness certification, space launch range safety certification, and consumer product safety and emissions certifications.

Notions and Principles Used within Design The Design Definition process identifies the human activities and material system elements for realizing the system solution. Specialist technical disciplines will need to be harnessed and orchestrated to ensure achievement of a coherent overall solution that meets the specified or identified needs or opportunities. Every technical domain or discipline possesses its peculiar laws, rules, theories, and practices for developing solution parts. Designing an overall system entails identifying where and how specific practices should be employed and integrated. Common and coherent design descriptors should be employed across the system to ensure that overall required system properties are realized and that system elements are capable of interoperation.

Usage of Design Descriptors A design descriptor is the set of (1) design characteristics and (2) their possible values. System design entails the identification and quantification of relevant design descriptors for the system elements composing the system design. These descriptors may be determined through a combination of (1) top-down apportionment and allocation, and (2) bottom-up selection and measurement. Matching system elements together with their specific design descriptors to those of the overall system is a key part of system design.

- The following are examples of generic design characteristics that are specifically relevant to SE:
  - For overall system functional and structural aspects: Quality of service, modularity, openness, scalability, deployability, and degree of automation and autonomy
  - For overall system non-functional aspects: Aesthetics, commonality, balance, availability, reliability, affordability, and other relevant QCs (see Section 3.1)
- The following are examples of generic design characteristics that are related to other engineering disciplines, but are relevant to SE in ensuring system balance and supporting trade-offs across system elements:
  - For hardware elements: Weight/mass, power, geometry, volume, vibration, acoustic, and thermal
  - For software elements: Correctness, understandability, efficiency, maintainability, flexibility, and consistency
  - For data/information: Accuracy, completeness, reliability, relevance, and timeliness

Holistic Design System design starts with the system as a whole consisting of system elements and ends with a definition (i.e., design specification) for each of these system elements and how they are designed to work together as a complete system. The system architecture identifies the system elements, although the architecture description might only identify those elements that are architecturally significant with additional elements becoming evident as a result of more detailed design considerations.

Early Design Validation Techniques such as modeling, simulation, and prototyping can be employed to discover early problems in a system design with respect to stakeholder expectations before significant expenditure in effort and materials. Increasingly these make use of computer-based representations of the system design, including augmented and virtual reality. Such validation techniques can be employed as part of different design process cycles including staged design and as part of an iterative design cycle.

Maintenance of Design Integrity Design integrity concerns the completeness, consistency, and inherent quality characteristics of a system design. An important design responsibility concerns ensuring the integrity of a system design, and subsequently approving, or denying, any decisions concerning the design, its modification, and its implementation. This is important to the acquirer (and therefore also to the supplier) as a system design evolves throughout the Design Definition process, including any evolution following its realization and transition to operation.

Design Evolution The system design matures and evolves over time. Conceptual design focuses on the overall form of the design solution. Preliminary design (sometimes called embodiment) elaborates the system design(s) with more details than provided in the architecture, so that layout, technical, cost, and other realization issues can be addressed. Detailed design progresses the preliminary system design with specification such that the system can be realized unambiguously. Usually, a design continues to evolve for various reasons: to accommodate new technologies, address obsolescence, improve performance and functionality, account for a new threat to its operation and/or integrity, etc. In case the SoI is operated in an SoS environment, there can be continuous design evolution. Consequently, robustness and resilience of the system design need to be considered often and on many levels of the hierarchy. The system architecture can specify principles for evolution of the design. Where possible, features should be included in the design to accommodate change according to the specified architectural principles and design objectives.

2.3.5.6 *System Analysis Process*

Overview

Purpose As stated in ISO/IEC/IEEE 15288,

> [6.4.6.1] The purpose of the System Analysis process is to provide a rigorous basis of data and information for technical understanding to aid decision-making and technical assessments across the life cycle.

Description Analysis is "a detailed examination or study of something so as to determine its nature, structure, or essential features" (Oxford, 2020). System analysis uses models and simulations to assess the utility and integrity of system requirements, architecture, and design across the life cycle. Models are incomplete representations of reality that answer four types of questions: (1) descriptive—what has happened? (2) predictive—what could happen? (3) prescriptive—what should we do? And (4) definitive—how should an entity be defined? (Buede and Miller, 2016, p. 70; Lustig, et al., 2010; Watson IOT, 2017).

System models define quantitative (numerical) and qualitative (categorical) variables to represent system attributes along multiple dimensions that can be categorized according to Figure 2.45. By establishing the relationships among the selected variables, the models represent the emergent features of the system and provide answers to the four types of questions from a holistic perspective. In addition to addressing the complexity of many interacting variables, modeling approaches may be needed to address uncertainty, dynamic behavior, and feedback loops (Howard, 1968). Typical modeling approaches include MBSE with functional, structural, and behavioral modeling (Dennis, et al., 2020); mathematical analysis; probabilistic and statistical modeling; simulation; and other techniques to represent the relationships among the variables and to perform sensitivity analysis of the allowable range of values for the variables across all life cycle stages. In some cases, it may be necessary to employ a mixture of modeling approaches to obtain the necessary insight. For more information on models and simulation, see Section 3.2.1.

Inputs and Outputs Inputs and outputs for the System Analysis process are listed in Figure 2.46. Descriptions of each input and output are provided in Appendix E.

Process Activities The System Analysis process includes the following activities:

- *Prepare for system analysis.*
  - Establish the strategy/approach for system analysis.
  - Identify the situation (problem, opportunity, question, or decision) to be addressed by system analysis.
  - Identify the stakeholders and their perspectives for the system analysis.
  - Define the scope (including system level), objectives, and level of fidelity of the system analysis.
  - Choose analysis methods based on relevant drivers, such as time, cost, fidelity, and criticality.

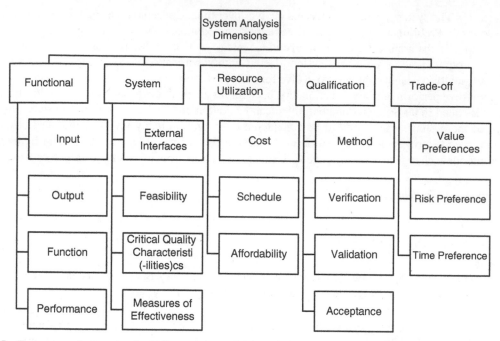

FIGURE 2.45   Taxonomy of system analysis dimensions. INCOSE SEH original figure created by Kenley. Usage per the INCOSE Notices page. All other rights reserved.

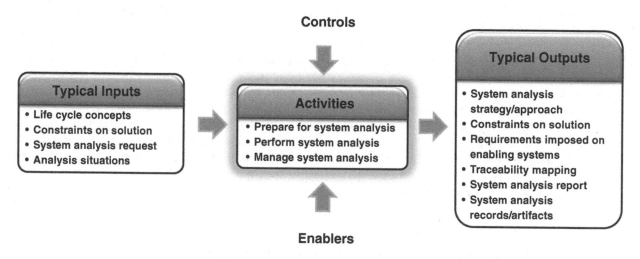

FIGURE 2.46   IPO diagram for System Analysis process. INCOSE SEH original figure created by Shortell, Walden, and Yip. Usage per the INCOSE Notices page. All other rights reserved.

- Plan for the necessary enabling systems or services needed through the life cycle for system analysis.
- Ensure enabling system or service access needed to support system analysis.
- Identify assumptions and ensure they are valid (correct and consistent).
- Ensure the data and inputs needed for the analysis are collected are timely and trustworthy.
- Establish criteria for trustworthiness of data needed for the analysis.
- Review the data and inputs for quality and validity (i.e., trustworthy data).

- *Perform system analysis.*
  - Execute the required system analysis using the selected analysis methods.
  - Perform uncertainty analysis and sensitivity analysis
  - Perform a quality and validity check on the analysis results.
  - Interpret the analysis results to develop relevant conclusions and recommendations.
  - Capture the system analysis results. This is often done in a system analysis report.
- *Manage system analysis.*
  - Establish and sustain traceability (system analysis results).
  - Give CM the information items, work products, or other artifacts needed for baselines.

Common approaches and tips:

- Models can never simulate all the behavior of a system: they operate in limited fields with a restricted number of variables. When a model is used, it is always necessary to make sure that the parameters and data inputs are part of the operation field; if not, irregular outputs are likely.
- Models evolve during the project: by modification of parameters, by entering new data, and by the use of new tools.
- It is recommended to concurrently use several types of models in order to compare the results and to take into account another characteristic or property of the system.
- Results of a simulation shall always be given in their modeling context: tool used, selected assumptions, parameters and data introduced, and variance of the outputs.

Elaboration

System Analysis Relationships to Other System Life Cycle Processes. Some of the SE processes leverage different dimensions of system analysis to answers different questions are as follows:

- The Business or Mission Analysis process to analyze and estimate candidate OpsCon and/or candidate business models related to a potential SoI in terms of effectiveness, feasibility, costs, risk preferences, and value preferences that address uncertainty, dynamic behavior, and feedback based on operator or end user behavior (Choi, et al., 2020).
- The Stakeholder Needs and Requirements Definition and System Requirements Definition processes to analyze issues relating to conflicts among the set of requirements, in particular those related to functionality, feasibility, performance, and effectiveness, including technical risks, costs, and externally imposed operational conditions and constraints (Beery and Paulo, 2019).
- The System Architecture Definition and Design Definition processes to analyze and estimate architectural and design space characteristics of candidate architectures and/or system element, providing information needed to select a design that provides the best value in terms of feasibility, and effectiveness, including technical risks, costs, and critical quality characteristics such as dependability, affordability, maintenance, and human-system interface considerations (Guariniello, et al., 2020).
- The Verification and Validation processes to understand and quantify the cost, schedule, information value, and the uncertainty characteristics inherent among the different choices of methods among inspection, analysis (including simulation), demonstration, and test (Salado and Kannan, 2018).
- The Project Planning and Project Assessment and Control processes to obtain estimates along with range of uncertainty of system metrics against established targets and thresholds, especially with respect to the technical measures (MOEs, MOPs, and TPMs) (Raz, et al., 2020).

The results of system analyses and estimations are provided to the Decision Management process as data, information, and arguments for selecting the alternatives or candidates that provide the best value to the decision maker(s) based on their value, risk, and time preferences. In some cases, the results may be provided to the Project Assessment and

Control process, if the information is needed to monitor the progress of the system or project against its system objectives, performance thresholds, or growth targets.

Cost Analysis. A cost analysis should consider the life cycle costs (LCC), which can be adapted according to the project and the system. The LCC may include labor and non-labor cost items; it may include development, manufacturing, service realization, sales, stakeholder utilization, supply chain, maintenance, and disposal costs (see Section 3.1.2).

Technical Risk Analysis. Technical risks address the operational system as opposed to the project risks, which are concerned with developing the system. The System Analysis process is often needed to perform the technical assessments that provide quantification and understanding of the probability or impact of a potential risk or opportunity (see Section 2.3.4.4).

Effectiveness Analysis. System effectiveness analysis is a term for a broad category of analyses that evaluate the degree or extent to which a system meets one or more criteria in its intended operational environment. The criteria may be derived from desired system characteristics, such as TPMs, MOPs, MOEs, or other attributes of the system (see Section 2.3.4.7). The system analysis assesses the probability the criteria are met and also the degree to which they are met (or fall short or exceed). This information is used to support evaluation of alternatives and trade-offs (such as which candidates to develop further, where improvements are needed or cost savings are possible).

2.3.5.7   Implementation Process

Overview

Purpose As stated in ISO/IEC/IEEE 15288,

[6.4.7.1] The purpose of the Implementation process is to realize a specified system element.

The Implementation process creates the system element per that element's description (concepts, requirements, architecture, design, including interfaces). Note that this process does not only occur during the production stage of the life cycle, but has activities in the other stages to ensure the element can be produced and to prepare for the production stage or other stages. For example: the supporting infrastructure may need to be defined or upgraded in preparation for the activities in a stage.

Description During the Implementation process, engineers follow the requirements allocated and derived to the system element to fabricate, code, or build each individual element outlined in system element descriptions. System element requirements are verified and system element stakeholder requirements are validated. If subsequent configuration audits reveal discrepancies, iterative and recursive interactions occur with predecessor activities or processes, as required, to correct them.

Inputs/Outputs Inputs and outputs for the Implementation process are listed in Figure 2.47. Descriptions of each input and output are provided in Appendix E.

Process Activities The Implementation process activities include the following:

- *Prepare for implementation.*
  - Define fabrication/coding procedures, tools and equipment to be used, implementation tolerances, and the means and criteria for auditing configuration of the resulting elements.
  - Elicit from stakeholders, developers, and teammates any constraints imposed by implementation technology, strategy, or impacted systems. Record the constraints for consideration in the definition of the requirements, architecture, design, and implementation.
  - Document the plan for acquiring or gaining access to resources needed during implementation. The planning includes the identification of requirements and interfaces for the enabling system.
  - Ensure enabling system or service access, and materials, needed to support implementation.

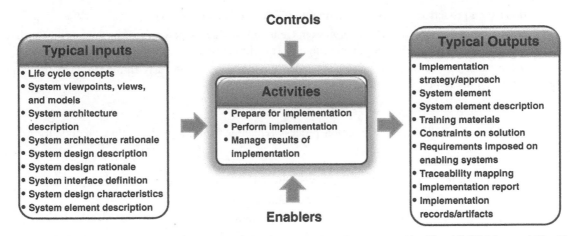

FIGURE 2.47   IPO diagram for Implementation process. IISE SEH original figure created by Shortell, Walden, and Yip. Usage per the INCOSE Notices page. All other rights reserved.

- Schedule the Implementation process, noting any critical path elements and ensuring that those critical element issues/constraints are addressed and understood by all stakeholders.

- *Perform implementation.*
  - Realize the system elements per the detailed product, process, and material specifications.
  - Produce documented evidence of implementation compliance:
  - Complete detailed product, process, material specifications, and system configurations.
  - Conduct peer reviews and testing—Inspect and verify software and hardware for correct functionality, hardware functional testing, etc.
  - Conduct conformation audits—Compare hardware and software elements to detailed drawings and design artifacts to ensure that each element meets its detailed specifications prior to integration with other elements.
  - Prepare initial training capability and draft training documentation—To provide the user community with the ability to operate, conduct failure detection and isolation, conduct contingency scenarios, and maintain the system as appropriate.
  - Prepare a hazardous materials log, if applicable.

- *Manage results of implementation.*
  - Identify and record implementation results. Provide baseline information for configuration management (see Section 2.3.4.5). Maintain the records per organizational policy (see Section 2.2.3.6).
  - Record any anomalies encountered during the Implementation process and resolve the anomalies (corrective actions or improvements) using the Quality Assurance process. (see Section 2.3.4.8)
  - Establish and sustain traceability of the implemented system elements with the system architecture, design, and system and interface requirements that are needed for implementation (see Section 3.2.3).

Common approaches and tips.

- Nearly all implementations have some issues requiring adjustments. Schedule pressures to get the system into operation can lead to lapses in tracking the needed adjustments. To assist consider the following:
- Keep the team (e.g., the Integrated Product Development Team (IPDT)) engaged to assist with configuration issues and redesign.

- Inspections are a proactive way to build in quality (Gilb and Graham, 1993).
- Conduct hardware conformation audits or system element level hardware verification; and ensure sufficient software configuration verification prior to entering the Transition process.
- Identify the enabling systems and materials needed for implementation early in the life cycle to allow for the necessary lead time to obtain or access them.

Elaboration

Implementation Concepts.   The implementation process typically focuses on the following four forms of system elements:

- Hardware—Output includes fabricated or adapted physical elements.
- Software—Output includes software code and executable images
- Operational resources—Output includes procedures and training.
- Services—Output includes specified services. These may be the result of one or more hardware, software, or operational elements resulting in the service.

The Implementation process can support either the creation (fabrication or development) or adaptation of system elements. For system elements that are reused or acquired (such as COTS), the Implementation process allows for adaption of the elements, if necessary, to satisfy the needs of the SoI. This may be accomplished via configuration settings provided within the element (e.g., hardware configuration switches and software configuration tables). Newly created products have more flexibility to be designed and developed to meet the needs of the SoI without modification (but at increased cost).

2.3.5.8   *Integration Process*

Overview

Purpose As stated in ISO/IEC/IEEE 15288,

[6.4.8.1] The purpose of the Integration process is to synthesize a set of system elements into a realized system that satisfies the system requirements.

Description   The focus of integration is the combination of system elements (hardware, software, and operational resources) that compose the SoI and verifying the correctness of the static and dynamic aspects of interfaces between, and interaction among, the implemented system elements. Integration also includes proactive activities to address potential integration issues early in the project such as modeling, analysis, simulation, prototyping, and early testing. Integration constraints and objectives are identified and considered during the definition of the system requirements, architecture, and design. The interaction of the Integration process with the system definition processes (i.e., System Requirements Definition, System Architecture Definition, and Design Definition) early in the development stage is essential for avoiding integration issues during the system realization.

The Integration process works closely with the Verification and Validation (V&V) processes. This process is iterated with the V&V processes, as appropriate, and includes an assessment of the integration maturity of elements to be integrated. As the integration of system elements occurs, the Verification process is invoked to check the correct implementation of system requirements, architectural characteristics, and design properties. The Validation process may be invoked to check that the individual system elements meet the stakeholder requirements and provide the function intended. The process checks that all boundaries between system elements have been correctly identified and described, including physical, logical, and human–system interfaces and interactions (physical, functional, sensory, and cognitive), and that all system and system element functional and performance requirements and constraints are satisfied.

Inputs/Outputs Inputs and outputs for the Integration process are listed in Figure 2.48. Descriptions of each input and output are provided in Appendix E.

Process Activities The Integration process includes the following activities:

- *Prepare for integration.*
  - Establish checkpoints for the correct implementation of the interfaces as the system elements are progressively aggregated.
  - Establish the strategy/approach for integration.
  - The strategy identifies risk mitigation approaches and sequences the order and levels for aggregating system elements while considering integration time and cost.
  - Identify system constraints and objectives necessary for successful integration to be addressed in the system requirements, architecture, or design such as those for accessibility, safety for integrators, required interfaces for enablers.
  - Plan for the necessary enabling systems or services needed to support integration such as integration facilities, training systems, or simulators.
  - Ensure enabling system or service access, and materials, needed to support integration.
- *Perform integration.*
  - Integrate system element configurations until the system is complete.
  - Manage interface availability as scheduled and track conformance of the interfaces to their requirements.
  - Address any conformance or availability issues.
  - Integrate the system elements in accordance with planned sequences.
  - Perform check of the interfaces, selected functions, and critical quality characteristics at different integration levels.

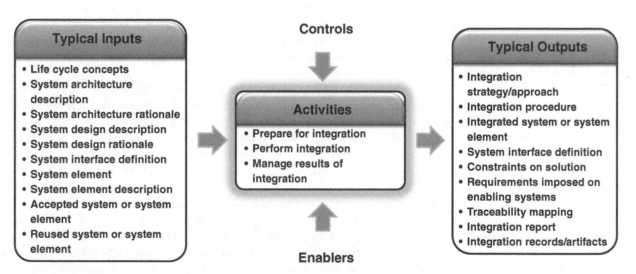

FIGURE 2.48   IPO diagram for Integration process. INCOSE SEH original figure created by Shortell, Walden, and Yip. Usage per the INCOSE Notices page. All other rights reserved.

- *Manage results of integration.*
  - Capture the integration results, including any anomalies or other issues identified. This includes anomalies due to the integration strategy, the integration enabling systems, execution of the integration, or incorrect system or element definition.
  - Where inconsistencies exist at the interface between the system, its specified operational environment, and any systems that enable the utilization stage the deviations lead to corrective actions or requirement changes. The Project Assessment and Control process (see Section 2.3.4.2) is used to analyze the data to identify the root cause, direct corrective or improvement actions, and to record lessons learned.
  - Maintain bidirectional traceability of the integrated system elements and the strategy, plans, and requirements (see Section 3.2.3).
  - Give CM the information items, work products, or other artifacts needed for baselines. The Configuration Management process (see Section 2.3.4.5) is used to establish and maintain baselines.

Common approaches and tips:

- The integration strategy should account for the schedule of availability of system elements and account for the personnel that will use, operate, maintain, and sustain the system). It should also be consistent with the defect/fault isolation and diagnosis practices.
- Development of integration enablers, such as tools and facilities, can take as long as the system itself and should be started early in the project.
- The Integration process of complex systems should use flexible approaches and techniques.
- Integrate aggregates in order to detect faults more easily. The use of the coupling matrix technique applies for all strategies and especially for the bottom-up integration strategy (see Section 3.2.4).

Elaboration

Integration occurs throughout the project from initial needs identification through utilization and support. The focus of integration evolves as the system evolves from concept definition to system definition to system realization to system deployment and use. As the system progresses, the emphasis of integration changes from its system definition, analysis, modeling, or prototypes to the deployed and operational system integrated into its intended environment, including interfacing systems. Integration should look proactively to mitigate risks and avoid integration issues, or discover them at the earliest point.

Concept of an "Aggregate." The integration of a system is based on the notion of an "aggregate." An aggregate is made up of several system elements and their physical and functional interfaces. Each aggregate is characterized by a configuration that specifies the system elements to be integrated and their configuration status. A set of verification actions is applied on each aggregate. To perform these verification actions, a verification configuration that includes the aggregate plus verification enabling systems is constituted. The verification enabling systems can be simulators (simulated system elements), emulators, stubs or caps, scaffolding, activators (launchers, drivers), harnesses, measuring devices, etc.

Integration Strategy and Approaches. The integration of evolving system elements is performed according to a predefined strategy. The strategy relies on the defined physical and functional architectures of the system and the organizational structure developing it. The detailed implementation of the strategy is described in an integration plan that defines the actions to be taken to mitigate integration risks and the configuration of expected aggregates of evolving system elements. It also defines the sequence these aggregates to carry out efficient verification actions and validation actions (e.g., inspections, analyses, demonstrations, or tests). The integration strategy is thus elaborated in coordination with the selected verification strategy and validation strategy (see Sections 2.3.5.9 and 2.3.5.11).

Several possible integration approaches and techniques can be used to define an integration strategy. Any of these may be used individually or in combination. The selection of integration approaches and techniques depends on several factors, in particular the type of system element, delivery time, order of delivery of system elements, risks, constraints, etc. Each integration approach has strengths and weaknesses, which should be considered in the context of the SoI.

Integration of the SoI and enabling systems occurs during development as well as utilization and support. Early in the life cycle, integration is concerned with concepts, requirements, architecture, and design. Approaches include models, analysis, simulations, and prototypes. In later life cycle stages, integration focuses on changes during utilization and support.

There are multiple options for the combination of system elements or aggregation of completed system elements or aggregates. Some common integration techniques are:

- *Global (or Big Bang) integration*—The simplest approach for low-risk, complicated, or simple systems is integration of the entire SoI. While the process is simplified, any issues or interface problems are difficult to find and resolve.
- *Bottom-up integration*—A common approach follows the reverse order of decomposition from lowest system element through levels of the architecture to the final system. Problems can be found at lower levels and more easily isolated to specific system elements. System level issues may not be discovered until late in the process.
- *Top-down integration*—This is a common variation of incremental integration (see below) that starts with the system elements that most closely reflect overall system performance with peripheral elements simulated and integrated later. The purpose is to detect system level issues, particularly with external interfaces, early.
- *Incremental integration*—In a predefined order, one or a small number of system elements are added to an already integrated increment of system elements. It can also include a portion of the system being integrated into a predefined increment. This approach can be effective for incremental and evolutionary development (see Section 2.2). For agile development, the order can be defined by features.
- *Subset integration*—System elements are assembled by subsets, and then subsets are assembled together. Subsets can be defined by functional chains or threads to perform specific tasks.
- *Criterion-driven integration*—The most critical system elements compared to the selected criterion are first integrated (e.g., dependability, complexity, technological innovation). The criteria are generally related to risks. This technique allows early integration and verification of intensively critical system elements.
- *Integration "with the stream"*—The delivered system elements are assembled as they become available.
- *Model-based integration*—The system elements are modeled physically or functionally and integrated in the model environment. Actual system elements can be inserted into the model environment as they developed.

Throughout the project, the integration strategy addresses management approaches to address risks such as communications issues. These include use of Integrated Product Teams (IPTs), Interface Control Working Groups (ICWGs), Systems Engineering Integration Teams (SEITs), or Technical Performance Measures (TPMs).

Horizontal & Vertical Integration. The Integration process needs to address the wide range of integration perspectives that apply across the life cycle. Horizontal integration typically refers to activities that are performed across elements that appear in a common hierarchy level of the system architecture. Structural aspects may be system elements that collectively constitute a system. Behavioral aspects include the sequence of discrete behaviors that together describe system functionality. Vertical integration typically refers to activities that are performed to help ensure that system elements at a given system hierarchy level are consistent with, and satisfy the expectations of, the system or higher-level system elements. The recursive nature of SE highlights how integration features span the levels of the system structure (see Section 2.3.1.2). As there is new information or learning on one level of the system structure, it is shared with both higher and lower levels. Other integration "directions" span additional viewpoints and stakeholder concerns, such as those relating to temporal or functional considerations, application of standards, satisfaction of regulatory expectations, or operational conditions and environments. Integration can also be viewed in relationship to the requirements concepts of horizontal traceability among parallel elements in the architecture and vertical traceability between system hierarchy levels (see Section 3.2.3).

2.3.5.9   *Verification Process*

Overview

Purpose As stated in ISO/IEC/IEEE 15288,

> [6.4.9.1] The purpose of the Verification process is to provide objective evidence that a system, system element, or artifact fulfils its specified requirements and characteristics.

Description The Verification process can be applied to any engineering artifact, entity, or information item that has contributed to the definition and realization of the SoI (e.g., verification of stakeholder needs, stakeholder requirements, system requirements, models, simulations, the system architecture, design characteristics, verification procedures, or a realized system or system element). The Verification process provides objective evidence with an acceptable degree of confidence to confirm:

1. The artifact or entity has been made "right" according to its specified requirements and characteristics,
2. No anomaly (error/defect/fault) has been introduced at the time of any transformation of inputs into outputs.
3. The selected verification strategy, method, and procedures will yield appropriate evidence that if an anomaly were introduced, it would be detected.

As is often stated, verification is intended to ensure that "the artifact or entity has been built right," while validation is intended to ensure that "the right artifact or entity will be or was built."

Inputs/Outputs Inputs and outputs for the Verification process are listed in Figure 2.49. Descriptions of each input and output are provided in Appendix E.

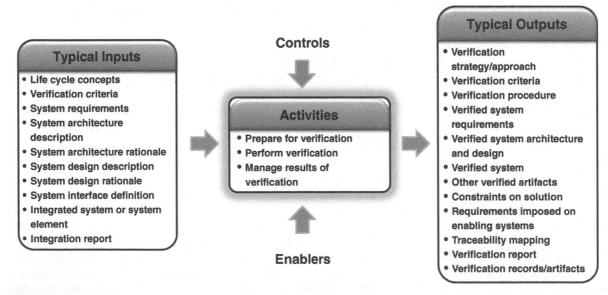

FIGURE 2.49   IPO diagram for Verification process. INCOSE SEH original figure created by Shortell, Walden, and Yip. Usage per the INCOSE Notices page. All other rights reserved.

Process Activities The Verification process includes the following activities:

- *Prepare for verification.*
  - Define the scope (what will be verified) and the verification actions (strategy, method, and success criteria). Verification activities consume resources: time, labor, facilities, and funds. The scope of the organization's verification strategy/approach should be documented within the project's SEMP and system integration, verification, and validation plans.
    - ° Establish a list of entities to be verified, including stakeholder needs, stakeholder requirements, system requirements, system architecture, prototypes, models, simulations, the system design, design characteristics, the system elements within the SoI architecture, and the integrated SoI itself.
    - ° Identify the specified requirements against which each entity will be verified.
  - Consider and capture constraints that could impact the feasibility or effectiveness of verification actions. The constraints could impact the implementation of the verification actions and include contractual constraints, limitations due to regulatory requirements, cost, schedule, feasibility to exercise a function, safety and security considerations, the laws of physics, physical configurations, accessibility, etc.
  - For each verification action, select one or more verification methods and associated success criteria. Verification methods include inspection, analysis, demonstration, and test (each of these methods are defined later in this section). The success criteria define what the verification actions must do that will result in sufficient objective evidence to show that the entity has fulfilled the requirement(s) against which it was verified against.
  - Establish the strategy/approach for verification, including trade-offs between scope and constraints. The verification strategy includes the method that will result in objective evidence that the verification success criteria has been met with an acceptable degree of confidence.
    - ° Define verification activities. For each verification instance, define a specific verification action that will result in objective evidence needed to verify the SoI meets one or more requirements per the defined verification strategy.
    - ° Define verification procedure requirements for each verification action. The verification procedure requirements are requirements that will drive the formulation of steps and actions for a given verification procedure.
  - Identify constraints and objectives from the verification strategy to be incorporated within the sets of system requirements, architecture, and design. These requirements are needed to support the defined strategy.
  - Plan for the necessary enabling systems or services needed through the life cycle for verification. Enabling systems include organizational support, verification equipment, simulators, emulators, test beds, test automation tools, facilities, etc.
  - Ensure enabling system or service access needed to support verification. This includes confirming everything required for the verification activities will be available, when needed and have passed their own verification and validation. The acquisition of the enablers can be done through several ways such as rental, procurement, development, reuse, and subcontracting.
- *Perform verification.*
  - Define the procedures for the verification actions. A procedure can support one action or a set of actions.
  - Execute the verification procedures for planned verification actions.
  - Schedule the execution of verification procedures. Each scheduled verification event represents a commitment of personnel, time, resources, and equipment that would ideally show up on a project's schedule.
  - Ensure readiness to conduct the verification procedures: availability and configuration status of the system/entity, the availability of the verification enablers, qualified personnel or operators, resources, etc.
- *Manage results of verification.*
  - Record the verification results and any defects identified. Maintain the results in verification reports and records per organizational policy as well as contractual and regulatory requirements.

- Analyze the verification results against the verification success criteria to determine whether the entity being verified meets those criteria with an acceptable degree of confidence.
- Throughout verification, capture operational incidents and problems and track them until final resolution. Problem resolution and any subsequent changes will be handled through the Project Assessment and Control process (see Section 2.3.4.2) and the Configuration Management process (see Section 2.3.4.5). Any changes to the SoI definition (e.g., stakeholder needs, stakeholder requirements, system requirements, system architecture, system design, design characteristics, or interfaces) and associated engineering artifacts are performed within other Technical Processes.
- Obtain agreement from the approval authority that the verification criteria have been met to their satisfaction. Combine the individual verification records into a verification approval package for the entity being verified and submit to the verification approval authority. The verification approval authority is the party authorized to determine whether sufficient evidence has been provided to show that the entity has passed verification with an acceptable degree of confidence.
- Establish and sustain traceability (verification). Establish and maintain bidirectional traceability of the verified entity and verification artifacts with the system architecture and design characteristics or requirements against which the entity is being verified.
- Give CM the information items, work products, or other artifacts needed for baselines. The Configuration Management process (see Section 2.3.4.5) is used to establish and maintain baselines. The Verification process identifies candidates for baseline and provides the items to the Configuration Management process.

Common approaches and tips.

- Identify the enabling systems and materials needed for verification early in the life cycle to allow for the necessary lead time to obtain or access them.
- Avoid conducting verification only late in the schedule or reducing the number of verification activities due to budget or schedule issues, since discrepancies and errors are more costly to correct later in the system life cycle.
- Review requirements as they are defined to ensure that the entities to which they apply can be verified against those requirements.

Elaboration

This section elaborates and provides "how-to" information on the Verification process. Additional guidance on verification can be found in the INCOSE NRM (2022) and INCOSE GtVV (2022).

Verification Planning. Planning for verification should begin when the system requirements are being defined. As the system requirements are defined, it is recommended to define the verification success criteria, method, and strategy and obtain acquirer and approval authority approval. Early planning helps drive cost and schedule estimates of the verification plan earlier in the project—maximizing the chance the full verification plan will be resourced.

Reduction of Verification Activities and Risk. If verification activities must be reduced due to cost and schedule concerns or other constraints, this should be done using a risk-based approach. The SE practitioner is urged to resist the temptation to blindly reduce the number of, or the costliest, verification activities due to budget or schedule concerns. Gaps and misses are more costly and time consuming to correct later in the life cycle—especially when these gaps show up at the integrated SoI level from reduced system element verification. If additional resources become available that allow an opportunity to verify to additional depth, the project should do so to reduce risk and increase the degree of confidence.

Notion of a Verification Action. A verification action describes verification in terms of an entity, the reference item against which the entity will be verified (e.g., a requirement, design characteristic, or standard), the expected result (success criteria deduced from the reference item the entity is being verified against), the verification strategy and method to be used, and on which level of integration of the system (e.g., system, system element). The performance of a verification action onto the submitted entity provides an obtained result which is compared with the expected result as defined by the verification success criteria. The comparison enables the determination of the acceptable conformance of the entity to the reference item with some degree of confidence. Figure 2.50 illustrates several common verification actions.

Examples of verification actions include:

- *Verification of a stakeholder requirement* (*requirement verification*)—(1) Verify the stakeholder requirement statement correctly transforms the source or stakeholder need from which it was transformed or derived and (2) verify the stakeholder requirement satisfies the characteristics of good requirement statements (see Section 2.3.5.3).
- *Verification of a system requirement* (*requirement verification*)—(1) Verify the system requirement statement correctly transforms the source, stakeholder requirement, or parent from which it was transformed or derived and (2) verify the system requirement satisfies the characteristics of good requirement statements (see Section 2.3.5.3)
- *Verification of a model or simulation* (*model or simulation verification*)—(1) Verify that the model/simulation meets its requirements consistent with its intended purpose, (2) verify the model/simulation against syntactic and grammatical rules, characteristics, and standards defined for the type of model/simulation, and (3) verify the correct application of the appropriate patterns and heuristics used and the correct usage of modeling/simulation techniques or methods as defined by the organization's guidelines and requirements concerning model/simulation development and use.

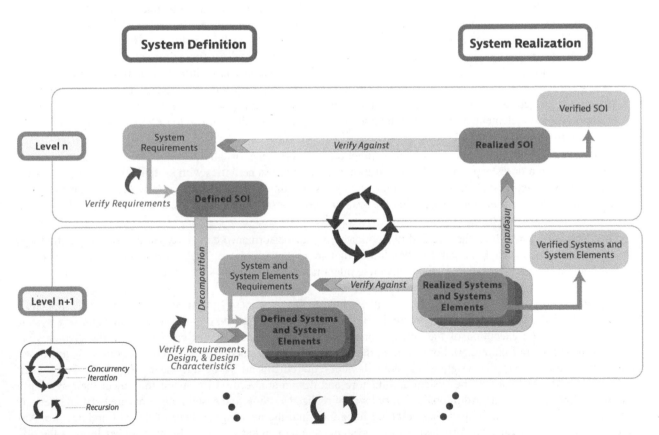

FIGURE 2.50   Verification per level. INCOSE SEH original figure created by Walden from Faisandier. Usage per the INCOSE Notices page. All other rights reserved.

- *Verification of the system architecture (architecture verification)*—(1) Verify that the SoI architecture, when realized by design, will result in a SoI that will pass system verification and (2) verify the correct application of the appropriate patterns and heuristics used and the correct usage of *architecture definition* techniques or methods as defined by the organization's guidelines and requirements concerning system architecture definition.
- *Verification of the system design (design verification)*—(1) Verify that the SoI design and associated design characteristics meets its system requirements and would result in a SoI that will pass system verification with an acceptable degree of confidence and (2) verify the correct usage of patterns, trade rules, or state of the art related to the concerned technology (e.g., software, mechanics, electronics, biology, chemistry) as defined by the organization's guidelines and requirements concerning system design.
- *Verification of a realized system (product, service, or enterprise) or system element (system verification)*—Verify the system or system element meets its system requirements and design characteristics with an acceptable degree of confidence.

Verification Methods. Basic verification methods are as follows (ISO/IEC/IEEE 29148, 2018):

- **Inspection**. An examination of the item against visual or other evidence to confirm compliance with requirements. Inspection is used to verify properties best determined by examination and observation (paint color, weight, etc.). Inspection is generally non-destructive and typically includes the use of sight, hearing, smell, touch and taste; simple physical manipulation; mechanical and electrical gauging; and measurement.
- **Analysis** *(including modeling and simulation)*. Use of analytical data or simulations under defined conditions to show theoretical compliance. Used where testing to realistic conditions cannot be achieved or is not cost-effective. Analysis (including simulation) may be used when such means establish that the appropriate requirement is met by the proposed solution. Analysis may also be based on "similarity" by reviewing a similar system or system element's prior verification and confirming that its verification status can legitimately be transferred to the present system or system element. Similarity can only be used if the systems or system elements are similar in design, manufacture, and use; equivalent or more stringent verification specifications were used for the similar system or system element; and the intended operational environment is identical to or less rigorous than the similar system or system element.
- **Demonstration**. A qualitative exhibition of functional performance, usually accomplished with no or minimal instrumentation or test equipment. Demonstration uses a set of test activities with system stimuli selected by the supplier to show that system or system element response to stimuli is suitable or to show that operators can perform their allocated functions when using the system. Often, observations are made and compared with predetermined responses.
- **Test**. An action by which the operability, supportability, or performance capability of an item is quantitatively verified when subjected to controlled conditions that are real or simulated. These verifications often use special equipment or instrumentation to obtain accurate quantitative data for analysis to determine verification.

Verification per Level. The SoI may have a number of hierarchical layers of system elements within its architecture. The planning of the verification is done recursively at each lower level as the definition of the system or a system element evolves. The execution of the verification actions occurs recursively for each layer as the elements are integrated as shown in Figure 2.50. For example, the stakeholder requirements are verified to ensure they meet their higher-level requirements, the system and system element requirements are verified to ensure they meet their higher-level system requirements and the system architecture and design are verified to ensure they meet their system or system element requirements. Additionally, every layer of realized systems and system elements are verified to ensure they meet their system requirements before being integrated into the next higher level of the SoI architecture. Any issues or discrepancies must be corrected before a system element is integrated into the next higher level of the SoI. Having passed verification at a given level, that set of elements are integrated into the next higher-level system as defined in the Integration process (see Section 2.3.5.8). System integration, system verification, and system validation continues until the integrated SoI has passed system verification.

Early Verification and MBSE. With the increased use of models and simulations as part of the design process, verification activities can be conducted earlier in the life cycle prior to implementation. Doing so will reduce the risk of issues and anomalies being discovered during system integration, system verification, and system validation activities with the actual physical hardware, mechanisms, and software and reduce the resulting expensive and time-consuming rework.

However, the SE practitioner is cautioned to resist substituting verification of the realized system with the verification results obtained using models and simulations, unless necessary. Doing so reduces the confidence level (as compared to verification against the actual realized system) and adds risk of the realized system failing system validation. As long as the realized system is not completely integrated and/or has not been validated to operate in the actual operational environment by the intended users, no result must be regarded as definitive until the acceptable degree of confidence is realized.

Managing the project's system verification program. In the progress of the project, it is important to know, at any time, the status of the verification activities, anomalies discovered, and noncompliances. This knowledge enables the project to better manage the budget and schedule as well as estimate the risks of noncompliance against the possibly of eliminating some of the planned verification actions to meet budget and schedule constraints.

2.3.5.10   Transition Process

Overview

Purpose As stated in ISO/IEC/IEEE 15288,

[6.4.10.1] The purpose of the Transition process is to establish a capability for a system to provide services specified by stakeholder requirements in the operational environment.

Description

The Transition process installs a SoI into its operational and maintenance environment. This process makes the SoI an integral part of the acquiring organization systems, business processes, and capabilities so the organization starts to benefit from using and sustaining the system's services.

The Transition process coordinates with verification and validation performed in the target environment, with the activities of operation and maintenance of new systems and services, and with the disposal of systems, system elements, materials, and services no longer needed for operation.

Transition may identify system requirements and design gaps. It may also drive changes, augmenting the initial stakeholder and system requirements.

Inputs/Outputs   Inputs and outputs for the Transition process are listed in Figure 2.51. Descriptions of each input and output are provided in Appendix E.

Process Activities The Transition process includes the following activities:

- *Prepare for the Transition.*
    - Analyze the intended environment for the system deployment, including the physical sites, information technology infrastructure, organizational structure, and processes of the receiving organization.
    - Identify the changes to the existing environment to accommodate the system.
    - Identify and obtain (e.g., procure, develop, reuse, rent, schedule, subcontract) the requisite enabling systems, controls, products, or services required for the transition, including the changes in the environment.
    - Plan for coordinating the development of the SoI with the modifications of its intended environment.

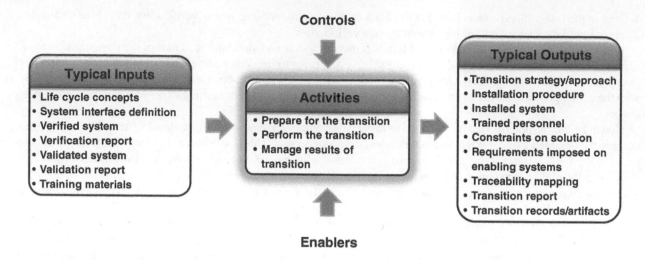

FIGURE 2.51   IPO diagram for Transition process. INCOSE SEH original figure created by Shortell, Walden, and Yip. Usage per the INCOSE Notices page. All other rights reserved.

- Determine the transition team structure, composition, and responsibilities for the transition activities.
- Plan for the system's transition, including allocating time and budget for all parts of the transition.
- Plan for mitigation strategies if the transition if the system encounters difficulties. Ensure that stakeholders understand the process/risk of possible downtime of the system and even actions to restore the predecessor system (roll-back) and the "point-of-no-return."
- Plan the configuration management of the system's adaptation to the local operation and support context.
- Develop procedures for system deployment and service activation, incremental and staged if appropriate.
- Develop procedures to validate the system and services at all relevant sites, either physical or virtual.
- Staff, organize, and train collaborative transition teams.
- *Perform the Transition.*
  - Deploy the system to operation, support, and maintenance sites.
  - Invoke integration and verification processes to realize operable local system configurations.
  - Establish systems, processes, and organizational capabilities for ongoing adaptation of the system to evolving context, including capabilities for integration with other systems, deployment to other sites, performance monitoring, and problem detection, investigation, and correction.
  - Train the operation, maintenance, and other personnel. As applicable, perform complete review and hand-off of the operator, maintenance, and support manuals. Affirm that the personnel have the knowledge and skill levels necessary to operate, maintain, and support the system.
  - Provide as-built information for configuration management.
  - Activate/commission the system's services at each site. Ensure that the system delivers its intended services as expected, including collaboration with other systems and personnel.

- Receive final confirmation that the installed system can provide its required functions and be sustained. Assure that the system has been properly installed and verified and all issues and action items have been resolved. Assure that all agreements about developing and delivering a fully supportable system have been fully satisfied or adjudicated.
- Perform or support contractual acceptance of the system by the acquirer, followed by transfer of control, responsibility, ownership, and custody.
- *Manage results of Transition.*
  - Capture incidents, problems, and anomalies. Investigate and document issues. Perform corrective actions as needed. Use the Quality Assurance process for managing incidents and problem resolution. If the transition is to multiple sites using a phased approach, ensure that any corrective actions are incorporated into the transition approach.
  - Use the experience gained in the current transition instances for improving future instances.
  - Maintain bidirectional traceability of the transitioned system elements, system services, and operational capabilities with the architecture, design, and system requirements. Initiate changes as needed.

Common approaches and tips.

- Identify the enabling systems and materials needed for transition early in the life cycle to allow for the necessary lead time to obtain or access them.

Elaboration

Transition Concepts. The Transition process is not limited to the SoI going into service as a part of the operating organization. Each system element undergoes transition during its integration into a larger element, and the element's transition must be formalized in the agreements between key stakeholders, such as prime contractors and its subcontractors.

The Transition process coordinates the system or system element deployment and activation with the modification of its environment. It pays particular attention to integrating the SoI and other systems in its environment. The Transition process should be fully integrated with an organizational change process led by the receiving organization, usually incremental and staged.

The Transition process comprises all activities required to establish the capability for a system to provide services for the benefit of the organization acquiring the system. The transition transfers the system from the development context ("system-in-the-lab") to the utilization context ("system-in-the field"). Successful transition typically marks the beginning of the SoI or system element's utilization stage.

Transition Considerations. The transition of new systems to a newly created organization (or a new element into a new system) differs from transitioning a new system or element into an existing organization or system. The former is sometimes referred to as "greenfield" or "clean sheet" transitioning, and the latter as "brownfield" or "legacy systems" (see Sections 4.3.1 and 4.3.2). The introduction of the new element disrupts the existing environment, so considerable effort must be invested to transition to the "new norm."

A phase of provisional operation (also referred to as "burn-in") is sometimes included in the transition activities, allowing operations to get used to the new system before acceptance, resulting in concurrent and iterative application of the Transition and Operation processes. Burn-in involves activities taken to operate a system element in the operational or simulated environment to detect failures and improve reliability. Usually, the operation of the system is done at levels that would cover or exceed the range of expected environmental values (heat, vibration, power, etc.). The warranty period may delay the transfer of responsibility for the system maintenance, resulting in concurrency and iteration between the Transition and Maintenance Processes.

2.3.5.11 *Validation Process*

Overview

Purpose As stated in ISO/IEC/IEEE 15288,

> [6.4.11.1] The purpose of the Validation process is to provide objective evidence that the system, when in use, fulfills its business or mission objectives and stakeholder needs and requirements, achieving its intended use in its intended operational environment.

Description The Validation process can be applied to any engineering artifact, entity, or information item that has contributed to the definition and realization of the SoI (e.g., validation of stakeholder needs, stakeholder requirements, system requirements, models, simulations, the system architecture, design characteristics, validation procedures, or a realized system or system element). The Validation process provides objective evidence with an acceptable degree of confidence to confirm:

1. The "right" artifact or entity has been made according to the stakeholder needs and stakeholder requirements.
2. Whether or not these artifacts, entities, or information items, will result in the right SoI, when realized, that can be validated to accomplish its intended use in its operational environment when operated by its intended users.
3. The system does not enable unintended users to negatively impact the intended use of the system or use the system in an unintended way.

As is often stated, validation is intended to ensure that "the right artifact or entity will be or was built," while verification is intended to ensure that "the artifact or entity has been built right."

Inputs/Outputs Inputs and outputs for the Validation process are listed in Figure 2.52. Descriptions of each input and output are provided in Appendix E.

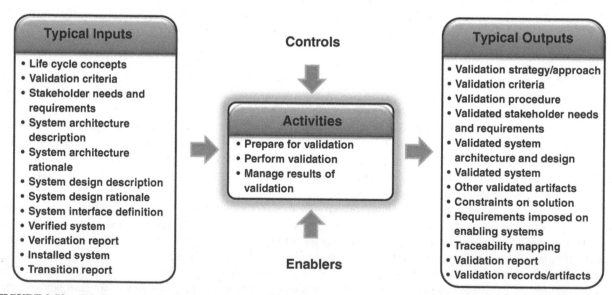

FIGURE 2.52   IPO diagram for Validation process. INCOSE SEH original figure created by Shortell, Walden, and Yip. Usage per the INCOSE Notices page. All other rights reserved.

Process Activities The Validation process includes the following activities:

- *Prepare for validation.*
  - Define the scope (what will be validated) and the validation actions (strategy, method, and success criteria). Validation activities consume resources: time, labor, facilities, and funds. The scope of the organization's validation strategy/approach should be documented within the project's SEMP and system integration, verification, and validation plans.
    - ° Establish a list of artifacts, entities, or information items to be validated.
    - ° Identify the stakeholder needs and stakeholder requirements against which each entity will be validated.
  - Consider and capture constraints that could impact the feasibility or effectiveness of validation actions. The constraints could impact the implementation of the validation actions and include contractual constraints, limitations due to regulatory requirements, cost, schedule, feasibility to exercise a function, safety and security considerations, the laws of physics, physical configurations, accessibility, etc.
  - For each validation action, select one or more validation methods and associated success criteria. Validation methods are similar to the methods defined for verification (inspection, analysis, demonstration, or test) (see Section 2.3.5.9). The success criteria define what the validation actions must do that will result in sufficient objective evidence to show that the entity has fulfilled the need(s) or requirement(s) against which it was validated against, achieving its intended use in its intended operational environment by its intended users.
  - Establish the strategy/approach for validation, including trade-offs between scope and constraints. The validation strategy includes the method that will result in objective evidence that the validation success criteria have been met with an acceptable degree of confidence. Significant collaboration is necessary with the stakeholders and approval authority to ensure there is agreement on what is necessary to accept the validation results.
    - ° Define validation activities. For each validation instance, define a specific validation action that will result in objective evidence needed to validate the SoI meets one or more stakeholder needs or stakeholder requirements per the defined validation strategy.
    - ° Define validation procedure requirements for each validation action. The validation procedure requirements are requirements that will drive the formulation of steps and actions for a given validation procedure.
  - Identify constraints and objectives from the validation strategy to be incorporated within the sets of stakeholder needs and requirements and the system requirements transformed from them.
  - Plan for the necessary enabling systems or services needed through the life cycle for validation. Enabling systems include organizational support, validation equipment, simulators, emulators, test beds, test automation tools, facilities, etc.
  - Ensure enabling system or service access needed to support validation. This includes confirming everything required for the validation activities will be available, when needed. The acquisition of the enablers can be done through several ways such as rental, procurement, development, reuse, and subcontracting.
- *Perform validation.*
  - Define the procedures for the validation actions. A procedure can support one action or a set of actions.
  - Execute the validation procedures for planned validation actions.
    - ° Schedule the execution of validation procedures. Each scheduled validation event represents a commitment of personnel, time, resources, and equipment that would ideally show up on a project's schedule. At the integrated SoI level, this should be done against the actual SoI in the operational environment or one as close to it as possible, by the intended users or equivalent surrogates.
    - ° Ensure readiness to conduct the validation procedure: availability and configuration status of the system/entity, the availability of the validation enablers, qualified personnel or operators, resources, etc. At the integrated SoI level, since it often depends on customer and intended user involvement, this can be particularly important to plan out in advance to be sure the right individuals are present.

- *Manage results of validation.*
  - Record the validation results and any defects identified. Maintain the results in validation reports and records per organizational policy as well as contractual and regulatory requirements.
    ° Analyze the validation results against the validation success criteria to determine whether the entity being validated meets those criteria with an acceptable degree of confidence.
  - Throughout validation, capture operational incidents and problems and track them until final resolution. Problem resolution and any subsequent changes will be handled through the Project Assessment and Control process (see Section 2.3.4.2) and the Configuration Management process (see Section 2.3.4.5). Any changes to the SoI definition (e.g., stakeholder needs, stakeholder requirements, system requirements, system architecture, system design, design characteristics, or interfaces) and associated engineering artifacts are performed within other Technical Processes
  - Obtain agreement from the approval authority that the validation criteria have been met to their satisfaction. Combine the individual validation records into a validation approval package for the entity being validated and submit to the validation approval authority. The validation approval authority is the party authorized to determine whether sufficient evidence has been provided to show that the entity has passed validation with an acceptable degree of confidence.
  - At the integrated SoI level, validation may be performed with or by the acquirer as defined in the supplier agreement. However, at lower levels in the architecture, validation may be performed by the supplier without acquirer direct involvement.
  - Establish and sustain traceability (validation). Establish and maintain bidirectional traceability of the validated entity and validation artifacts with the system architecture, system design, models, and the stakeholder needs and stakeholder requirements against which the entity is being validated.
  - Give CM the information items, work products, or other artifacts needed for baselines. The Configuration Management process (see Section 2.3.4.5) is used to establish and maintain configuration items and baselines. The validation process identifies candidates for baseline, and then provides the items to the Configuration Management process.

Common approaches and tips.

- Identify the enabling systems and materials needed for validation early in the life cycle to allow for the necessary lead time to obtain or access them.
- Validation also reveals the effects the SoI may have on enabling, interfacing, and interoperating systems. Validation actions and analysis should include these system interactions in the scope.
- Involve the broadest range of stakeholders that is practical, including end users and operators,
- Validation should include actions that provide insight as early as possible, such as analysis, modeling, and simulation of anticipated operational characteristics and system behavior.
- Start to develop the validation planning as the OpsCon, operational scenarios, stakeholder needs, and stakeholder requirements are defined. Early consideration of the potential validation actions and methods helps to anticipate constraints, costs, and necessary enablers, as well as start the acquisition of those enablers.
- Validation actions during the Business or Mission Analysis process (see Section 2.3.5.1) include assessment of the OpsCon through operational scenarios that exercise all system operational modes and demonstrating system-level performance.

Elaboration

This section elaborates and provides "how-to" information on the Validation process. Additional guidance on validation can be found in the INCOSE NRM (2022) and INCOSE GtVV (2022).

General Considerations. The stakeholder needs and stakeholder requirements the SoI is being validated against are derived from the mission statement, goals, objectives, critical measures, constraints, risks, and set of life cycle concepts for the SoI defined by the organization or acquirer during the Stakeholder Needs and Requirements Definition and System Requirements Definition processes (see Sections 2.3.5.2 and 2.3.5.3). The life cycle concepts include scenarios and use cases that are performed in a specific operational environment by the intended users for not only operation, but during other life cycle stages including production, operation, support, and retirement. It is common for these scenarios and use cases to be exercised during the conduct of the validation procedures within the operational environment with the intended users. The common saying "test as you fly, fly as you test" applies. When using scenarios and use cases, in addition to nominal operations, it is important to also address off-nominal, alternate cases, misuse cases, and loss scenarios. A positive validation result obtained in a given environment by specific users can turn noncompliant if the environment or class of users change. These changes may not be immediately known by the developer; however, changing stakeholder needs and stakeholder requirements should be accommodated by the acquirer and developer's SE processes.

During validation, especially for walkthroughs and similar activities, it is highly recommended to involve intended users/operators. Validation will often involve going back directly to the users to have them perform an acceptance test under their own local operational conditions in the intended operational environment. When the system is validated at a supplier facility or organization, the acquirer will often want to conduct additional validation activities in their own facility, in the intended operational environment, and with the intended users. The stakeholders who were involved in defining the life cycle concepts and needs must be presented with the results of the validation activities to ensure their needs and requirements have been met.

Validation Planning. Planning for validation should begin when the stakeholder needs and stakeholder requirements are being defined. As they are defined, it is recommended to define the validation success criteria, method, and strategy and obtain acquirer and approval authority approval. Early planning helps drive cost and schedule estimates of the system validation plan earlier in the project—maximizing the chance the full system validation plan will be resourced.

Reduction of Validation Activities and Risk. If validation activities must be reduced due to cost and schedule concerns, this should be done using a risk-based approach. The SE practitioner is urged to resist the temptation to blindly reduce the number of, or the costliest, validation activities due to budget or schedule concerns. Gaps and misses are more costly and time consuming to correct later in the life cycle—especially when these gaps show up at final system acceptance by the acquirer or regulatory agency. If additional resources become available that allow an opportunity to validate lower-risk, non-critical stakeholder needs and stakeholder requirements, the project should do so to reduce risk and increase the degree of confidence.

Notion of a Validation Action.   Validation actions are similar to verification actions, and the reader is referred to the Verification process (see Section 2.3.5.9) for background. Figure 2.53 illustrates several common validation actions.

Examples of validation actions include:

- *Validation of a stakeholder requirement (requirement validation)*—Validate that the stakeholder requirement is the right requirement and clearly and accurately communicates the need of the stakeholder, is in the stakeholder's language, and is actionable (i.e., can be transformed into one or more system requirements). For stakeholder requirements and sets of stakeholder requirements ask, "If a SoI were built to these requirements, would the SoI meet the needs from which these requirements were transformed?"

- *Validation of a system requirement (requirement validation)*—Validate that the system requirement is the right requirement and clearly and accurately communicates the need and requirement of the stakeholder, is expressed in technical terms, and is actionable (i.e., can be transformed into a system architecture and design). For system requirements and sets of system requirements ask, "If a SoI were built to the system architecture and design transformed from these requirements, would the SoI meet the intent of the requirements from which the architecture and design were transformed?"

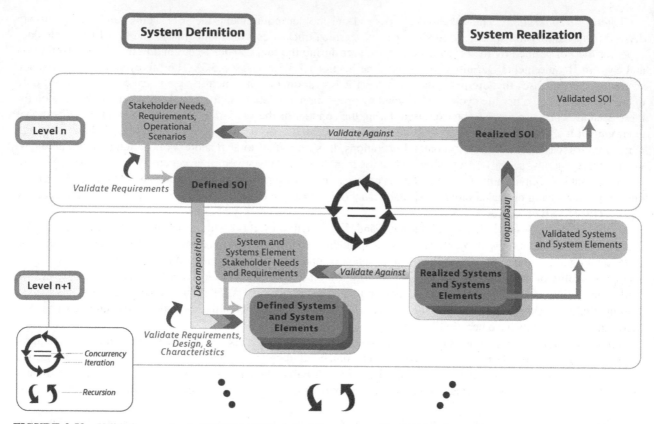

FIGURE 2.53   Validation per level. INCOSE SEH original figure created by Walden from Faisandier. Usage per the INCOSE Notices page. All other rights reserved.

- *Validation of a model or simulation (model or simulation validation)*—(1) Validate that the model/simulation accurately reflects the intended behavior of the entity it represents in its operational environment when operated by the intended users and (2) validate that the model/simulation meets the intended purpose for which it was developed.

- *Validation of the system architecture (architecture validation)*—Validate that the architecture is the right architecture that will result in a design for the SoI that will meet the stakeholder needs and stakeholder requirements.

- *Validation of the system design (design validation)*—Validate that the design, as communicated by the design characteristics, will result in a SoI that meets its intended purpose in its operational environment when operated by the intended users as defined by the stakeholder needs and stakeholder requirements.

- *Validation of a realized SoI (product, service, or enterprise) (system validation)*—Validate that realized SoI meets its intended purpose in its operational environment when operated by the intended users and does not enable unintended users to negatively impact the intended use of the system or use the system in an unintended way with an acceptable degree of confidence as defined by the stakeholder needs and stakeholder requirements.

Validation Outcomes. Typical validation outcomes include:

- **Acceptance**. Acceptance is an activity conducted prior to transition to the acquirer such that the acquirer can decide if this transition is appropriate. A set of operational validation actions is often exercised, or a review of validation results performed by the supplier is systematically performed as part of acceptance.

- **Certification**. Certification is a written assurance that the system has been developed per a defined procedure and can perform its intended functions in accordance with identified legal or industrial standards (e.g., airworthiness standards for aircraft, information assurance). A host of information can be part of the certification package, including development reviews, verification results, and validation results. However, certification is typically performed by outside authorities, without direction as to how the needs are to be validated. For example, this method is used for electronics devices via Conformité Européenne (CE) certification in Europe and via Underwriters Laboratories (UL) certification in the United States and Canada.

- **Readiness for Use**. As part of the analysis of the validation results, the project team and validation authority may need to make a readiness for use assessment. This may occur several times in the life cycle, including upon first article delivery, upon completion of production (if more than a single system is produced), following maintenance actions, or successful completion of field trials with a predefined user population. In the field, particularly after maintenance, it may be necessary to establish whether the system is ready for reintroduction to service.

- **Qualification**. System qualification requires that all verification and validation actions have been successfully performed, documented, and that the SoI is "qualified" for use as intended by the supplier organization. These verification and validation actions cover not only the SoI itself but also all the interfaces with its environment (e.g., for a space system, the validation of the interface between space segment and ground segment). The qualification process must demonstrate that the characteristics or properties of the realized system, including margins, meet the applicable system requirements and/or stakeholder requirements. The qualification is concluded by an acceptance review and/or an operational readiness review.

Validation per Level. The SoI may have a number of hierarchical layers of system elements within its architecture. The planning of the validation is done recursively for each level as the definition of the system or a system element evolves. The execution of the validation actions occurs recursively for each layer as the elements are integrated as shown in Figure 4.53. For example, the stakeholder needs and stakeholder requirements are validated against the stakeholder real world expectations to ensure they are the right stakeholder needs and stakeholder requirements, the systems requirements are validated against the stakeholder needs and requirements to ensure they are right system requirements, and the system architecture and design are validated against the stakeholder needs and requirements to ensure they are the right system architecture and design. Additionally, every layer of realized systems and system elements are validated to ensure they meet their stakeholder needs and stakeholder requirements in their operational environment before being integrated into the next higher level of the SoI architecture. Having passed system verification and system validation at a given level, that system element is integrated into the next higher-level system as defined in the Integration process (see Section 2.3.5.8). System integration, system verification, and system validation continue until the integrated SoI has passed system validation.

Early System Validation and MBSE. With the increased use of models and simulations as part of the design process, validation activities can be conducted earlier in the life cycle prior to implementation. Doing so will reduce the risk of issues and anomalies being discovered during system integration, system verification and system validation activities with the actual physical hardware, mechanisms, and software and reduce the resulting expensive and time-consuming rework.

In addition, modeling and simulations early in the project allows not only expectation management but also early feedback from the acquirer and other stakeholders on the final system architecture and design before implementation. It will be much less expensive and time consuming to resolve issues before the realization of the actual physical hardware and software and before system integration, system verification, and system validation activities.

Because the behavior of a system is a function of the interaction of its elements, a major goal of systems validation is assessing the behavior of the integrated physical system and identifying emergent properties not specifically addressed in the stakeholder needs or stakeholder requirements nor identified during modeling and simulations. Emergent properties may be positive or negative. For example, cascading failures across multiple interface boundaries between the system elements that are part of the SoI's architecture. Relying on models and simulations of the SoI and operational environment may not uncover all the emerging properties and issues that occur in the physical realm.

While validation using models and simulations allows a theoretical determination that the modeled system will meet its needs in the operational environment by the intended users once realized, the assessment of the actual system behavior (system validation) must be done, whenever possible, in the physical realm with the actual hardware and software integrated into the higher-level system which it is a part in the actual operational environment by the intended users.

There are cases when it may not be practical in terms of the intended use and actual operational environment to do all system validation activities. However, the SE practitioner is cautioned to not substitute validation of the realized system with the validation results obtained using models and simulations, unless absolutely necessary. Doing so adds risk to the project and reduces the confidence level (as compared to validation against the actual realized system in its actual operational environment when operated by the intended users) and adds risk of the realized system failing system validation when delivered to the acquirer or submitted to a regulatory agency. As long as the realized system is not completely integrated and/or has not been validated to operate in the actual operational environment by the intended users, no result must be regarded as definitive until the acceptable degree of confidence is realized.

Managing the project's validation program. In the progress of the project, it is important to know, at any time, the status of the validation activities, anomalies discovered, and noncompliances. This knowledge enables the project to better manage the budget and schedule as well as estimate the risks of noncompliance against the possibly of eliminating some of the planned validation actions to meet budget and schedule constraints.

2.3.5.12 Operation Process

Overview

Purpose As stated in ISO/IEC/IEEE 15288,

[6.4.12.1] The purpose of the Operation process is to use the system to deliver its services.

Description
The Operation process focuses on delivering services provided by the system for the benefit of the operating organization. This process is often concurrent with the Maintenance process of sustaining the system's services. During Operation, the SoI functions as an integral part of the operating organization. The SoI contributes to the Business or Mission Analysis process by cooperating with human operators and diverse interfacing systems.

Operation may identify the system requirements and design gaps. It may also drive changes, augmenting the initial stakeholder and system requirements.

Inputs/Outputs Inputs and outputs for the operation process are listed in Figure 2.54. Descriptions of each input and output are provided in Appendix E.

Process Activities The Operation process includes the following activities:

- *Prepare for operation.*
  - Influence the Concept of Operations (ConOps) of the receiving organization, the Operational Concept (OpsCon) of the SoI, the stakeholder needs and requirements, and the system requirements impacting the operation of the SoI.
  - Identify relevant regulations, legal requirements, environmental and ethical constraints.
  - Define business rules related to modifications that sustain existing or enhanced services.
  - Plan for operational capability build-up, including confirmation of site deployment schedules, personnel availability, training, and logistic support availability.
  - Identify and obtain (procure, develop, reuse, rent, schedule, subcontract) the requisite enabling systems, controls, products, or services required for the operation.

Controls

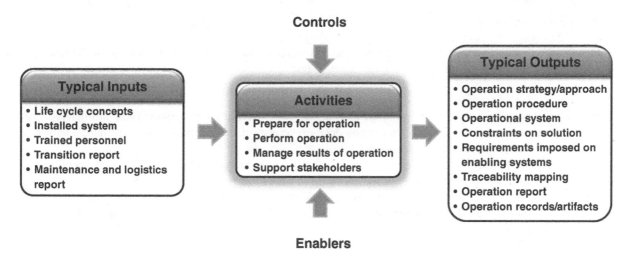

Typical Inputs

- **Life cycle concepts**
- **Installed system**
- **Trained personnel**
- **Transition report**
- **Maintenance and logistics report**

Activities

- **Prepare for operation**
- **Perform operation**
- **Manage results of operation**
- **Support stakeholders**

Typical Outputs

- **Operation strategy/approach**
- **Operation procedure**
- **Operational system**
- **Constraints on solution**
- **Requirements imposed on enabling systems**
- **Traceability mapping**
- **Operation report**
- **Operation records/artifacts**

Enablers

FIGURE 2.54   IPO diagram for Operation process. INCOSE SEH original figure created by Shortell, Walden, and Yip. Usage per the INCOSE Notices page. All other rights reserved.

- Verify that the SoI is accompanied by all relevant information products, such as documentation, manuals, and procedures. Identify gaps and initiate changes as necessary.
- Review the transition, validation, and maintenance strategies for compatibility with the OpsCon and their completeness concerning the expected operational capabilities.

- *Perform operation.*
  - Confirm completion of the system transition at the operational sites.
  - Prepare and verify the system's configurations for delivering specific services or missions.
  - Operate the system according to the established procedures. Update the procedures as experience accumulates.
  - Ensure the flow of materials, energy, and information into and from the SoI. Monitor the functioning of the systems providing inputs for the SoI and utilizing its outputs.
  - Track system performance, including operational availability. Identify, investigate, and correct problems and anomalies.
  - When abnormal operational conditions warrant, conduct planned contingency actions. Perform system contingency operations, if necessary.

- *Manage results of operation.*
  - Capture incidents, problems, and anomalies. Investigate and document the issues. Perform corrective actions as needed. Use the Quality Assurance process for managing incidents and problem resolution.
  - Use the experience gained during the operation for improvement.
  - Maintain bidirectional traceability of the system's assets, services, and operational capabilities with system architecture, design, and system requirements. Initiate changes as needed.

- *Support stakeholders*
  - While the customer is responsible for the Operation process, the supplier should support the customer throughout the system life cycle leveraging the knowledge generated by the customer and the supplier.

Common approaches and tips.

- Identify the enabling systems, products, services, and materials needed for operation early in the life cycle to allow for the necessary lead time to obtain or access them.

Elaboration

Operation Concepts. Successful operation of the SoI as a part of the operating organization is the ultimate goal of SE. The stakeholders' needs and requirements regarding operation constitute a significant source of the system requirements and a significant input to the Validation and Transition processes.

During operation, the SoI interfaces with other systems in its environment (see Section 1.3.3). These systems are SoIs in their own right, and their life cycles must be coordinated with the life cycle of your SoI.

The operational environment may change and evolve while the system is being developed. Considerable effort must be invested in recognizing these changes and updating the life cycle concepts (especially ConOps and OpsCon) and all derived requirements.

2.3.5.13 Maintenance Process

Overview

Purpose As stated in ISO/IEC/IEEE 15288,

[6.4.13] The purpose of the Maintenance process is to sustain the capability of the system to provide a service.

Description

The Maintenance process focuses on sustaining the system's ability to provide services for the operating organization's benefit. This process is often concurrent with the Operation process of delivering the system's services. Maintenance includes the activities to provide operations support, logistics, and material management to sustain satisfactory quality, performance, and availability of the system's services.

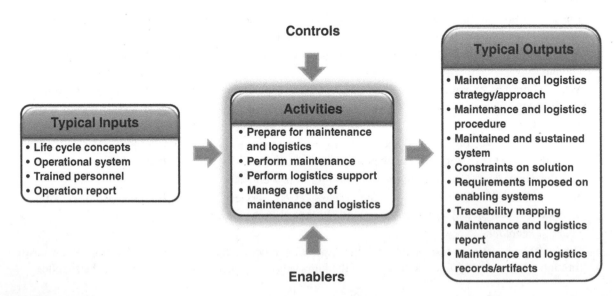

FIGURE 2.55   IPO diagram for Maintenance process. INCOSE SEH original figure created by Shortell, Walden, and Yip. Usage per the INCOSE Notices page. All other rights reserved.

Maintenance may identify requirements and design gaps. It may also drive changes in the SoI, augmenting the initial stakeholder and system requirements.

Inputs/Outputs Inputs and outputs for the Maintenance process are listed in Figure 2.55. Descriptions of each input and output are provided in Appendix E.

Process Activities The Maintenance process includes the following activities:

- *Prepare for maintenance and logistics.*
  - Define and maintain the maintenance and logistics strategies of the SoI and its elements and update the system requirements and attribute specifications impacting the maintenance and logistics support.
  - Define business rules related to modifications that sustain existing or enhanced services.
  - Identify relevant regulations, legal requirements, and ethical constraints and generate corresponding requirements.
  - Plan for maintenance and logistics support capability build-up, including site deployment schedules, personnel availability, and training, including the logistic support availability.
  - Establish appropriate warranty and licenses (e.g., software, legal) and the lines of communication to activate more support when needed.
  - Identify and obtain (procure, develop, reuse, rent, schedule, subcontract) the requisite enabling systems, controls, products, or services required for maintenance and logistics support.
  - Review the transition, validation, and operation strategies for compatibility with the support concept and their completeness concerning the expected maintenance and logistics support capabilities.
- *Perform maintenance.*
  - Confirm completion of the system transition at the maintenance sites.
  - Maintain the system according to the established procedures. Update the procedures when experience accumulates.
  - Detect, identify, and repair physical and logical damage to the system, including data corruption or inaccessibility. Identify and replace faulty or obsolete parts, including software updates.
  - Monitor the SoI and its environment to detect or predict system failures or performance degradation, identifying and resolving operational problems minimizing operational interruptions.
  - Prevent operation disruptions by scheduling repairs and replacements before failures occur, based on operations history or failure prediction.
  - Ensure availability of materials and parts for replacement and repairs by production, acquisition, or repairs, including operations and maintenance of logistics processes and systems. Conduct logistics operations according to the established procedures. Update the procedures when experience accumulates.
  - Track all maintenance repairs for analysis, which may lead to performance trends that can trigger warranty claims or new project needs.
- *Perform logistic support.*
  - Conduct acquisition logistics actions
  - Conduct operational logistics actions
- *Manage results of maintenance and logistics.*
  - Capture incidents, problems, and anomalies. Investigate and document the issues. Perform corrective actions as needed. Use the Quality Assurance process for managing incidents and problem resolution.
  - Use the experience gained while performing maintenance for improvement.
  - Maintain bidirectional traceability of the maintenance and logistics assets, services, and capabilities with system architecture, design, and system requirements. Initiate changes as needed.
  - Manage the configuration data items.

Common approaches and tips.

- Identify the enabling systems, products, services, and materials needed for maintenance and logistics support early in the life cycle to allow for the necessary lead time to obtain or access them.
- The maintenance of the SoI must be coordinated with the maintenance of other systems in its environment (the interoperating and enabling systems). The failure or malfunction of any system can trigger maintenance actions in other systems due to technical, organizational, economic, or political concerns.

Elaboration

The Maintenance process supports the operation of the SoI and its elements throughout its life cycle. The maintenance and logistics activities regarding the SoI must be integrated into the operating organization's existing support and logistics networks. This includes provisions for sustaining the skills and competencies of personnel performing operation and maintenance.

Different modes of maintenance should be considered:

- Corrective maintenance restores system services to normal operations (e.g., remove and replace hardware, reload software, apply a software patch).
- Preventive maintenance prevents failures and malfunctions by scheduling routine maintenance actions to sustain optimal system operational performance.
- Predictive maintenance is a more advanced preventive maintenance that utilizes data collected during the system operations to predict failures and malfunctions and schedule the maintenance actions in advance.
- System modification is a form of maintenance that extends the system's useful life by changing the system to sustain existing capabilities in the changing environment. Adding new capabilities (system upgrades) is sometimes considered part of the maintenance.

2.3.5.14   Disposal Process

Overview

Purpose As stated in ISO/IEC/IEEE 15288,

[6.4.14.1] The purpose of the Disposal process is to end the existence of a system element or system for a specified intended use, appropriately handle replaced or retired elements, appropriately handle any waste products, and to properly attend to identified critical disposal needs.

The Disposal process is conducted in accordance with applicable guidance, policy, regulations, and statutes throughout the system life cycle.

Description The Disposal process generates requirements and constraints that must be balanced with defined stakeholders' needs and requirements and other design considerations. Further, environmental concerns drive the designer to consider reclaiming the materials or recycling them into new systems. Incremental disposal can be applied at any point in the life cycle (e.g., prototypes that are not to be reused or evolved, waste materials during manufacturing, parts that are replaced during maintenance). The Disposal process may also be used to manage the transition of system elements from a current SoI to a different system.

The Disposal process also includes any steps necessary to return the environment to an acceptable condition; handle all system elements and waste products in an environmentally sound manner in accordance with applicable legislation, organizational constraints, and stakeholder agreements; and document and retain records of disposal activities, as required for monitoring by external oversight or regulatory agencies.

Inputs/Outputs Inputs and outputs for the Disposal process are listed in Figure 2.56. Descriptions of each input and output are provided in Appendix E.

Process Activities The Disposal process includes the following activities:

- *Prepare for disposal.*
  - Review the retirement concept (may be called a disposal concept), including any hazardous materials and other environmental impacts to be encountered during disposal.
  - Plan for disposal, including the development of the strategy.
  - Impose associated constraints on the system requirements.
  - Ensure that the necessary enabling systems, products, or services required for disposal are available, when needed. The planning includes the identification of requirements and interfaces for the enablers. The acquisition of the enablers can be done through various ways such as rental, procurement, development, reuse, and subcontracting. An enabler may be a complete enabling system developed as a separate project from the project of the SoI.
  - Identify elements that can be reused and that cannot be reused. Special methods may need to be implemented for hazardous materials.
  - Specify containment facilities, storage locations, inspection criteria, and storage periods, if the system is to be stored.
- *Perform disposal.*
  - Decommission the system or system elements to be disposed.
  - Disassemble the elements for ease of handling. Include identification and processing of reusable elements.
  - Extract all elements and waste materials that are no longer needed—this includes removing materials from storage sites, consigning the elements and waste products for destruction or permanent storage, and ensuring that the waste products or elements not intended for reuse cannot get back into the supply chain.

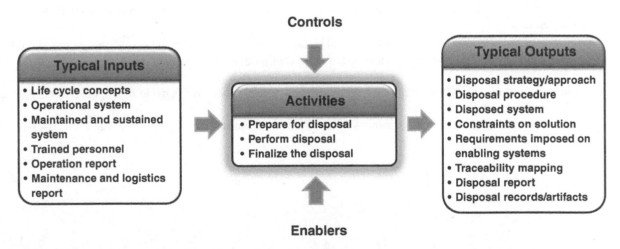

FIGURE 2.56 IPO diagram for Disposal process. INCOSE SEH original figure created by Shortell, Walden, and Yip. Usage per the INCOSE Notices page. All other rights reserved.

- Dispose of deactivated system elements per the disposal procedure.
- Ensure the disposal staff adheres to safety, security, privacy and environment regulations or policies and capture their tacit knowledge for future needs.
- *Finalize the disposal.*
  - Confirm no adverse effects from the disposal activities and return the environment to its original state.
  - Maintain documentation of all disposal activities and residual hazards.

Common approaches and tips:

- Consider donating an obsolete system—Many items, both systems and information, of cultural and historical value have been lost to posterity because museums and conservatories were not considered as an option during the retirement stage.
- Concepts such as zero footprint and zero emissions drive current trends toward corporate social responsibility that influence decision making regarding cleaner production and operational environments and eventual disposal of depleted materials and systems.
- Design the SoI to support the circular economy (see Section 3.1.10). Maintaining materials in closed loops maximizes material value without damaging ecosystems (McDonough, 2013).

Elaboration

The project team conducts analyses to develop solutions for disposition of the system, system elements, and waste products based on evaluation of alternative disposal methods. Methods addressed should include storing, dismantling, reusing, recycling, reprocessing, and destroying systems, system elements, materials, consumables, waste, and enabling systems,.

Disposal analyses are essential to ensure the planning and feasibility of disposal throughout the life cycle. The following are key points with respect to the analyses.

- Analyses include consideration of costs (including LCC), disposal sites, environmental impacts, health and safety issues, responsible agencies, handling and shipping, supporting items, and applicable international, national, and local regulations.
- Analyses support selection of system elements and materials that will be used in the system design and should be readdressed to consider design and project impacts from changing laws and regulations throughout the life cycle.

Disposal strategy and design considerations are updated throughout the system life cycle in response to changes in applicable laws, regulations, and policy.

The ISO 14000 (2015) series includes standards for environmental management systems and life cycle assessment.

3

LIFE CYCLE ANALYSES AND METHODS

3.1 QUALITY CHARACTERISTICS AND APPROACHES

3.1.1 Introduction to Quality Characteristics

ISO/IEC/IEEE 15288 (2023), Section 3.36 defines *Quality Characteristic* (QC) as: *inherent characteristic of a product, process, or system related to a requirement.* QCs are how the stakeholders will judge the quality of the system. Approaches exist that help ensure these characteristics are present in the SoI and its broader context or environment.

The objective of the following sections is to give enough information to a Systems Engineering (SE) practitioner to appreciate the significance of various QC approaches, even if they are not an expert in the subject. In previous editions of the handbook, the QC approaches were known as Specialty Engineering or the Engineering Specialties. These approaches are also known as Design for X (DFX) and Through-Life Considerations. The QCs are informally known as the -ilities since many, but not all, end in "ility" in the English language.

QC approaches, as used in this handbook, are life cycle perspectives that need to be considered to ensure the system is developed and its ecosystem cultivated so that QCs are present when the system is produced, utilized, supported, and ultimately retired. QC approaches often generate non-functional requirements. Some QC approaches, such as safety, security, and resilience may also generate functional requirements. These QC approaches are applied throughout the system's life cycle, as notionally shown in Figure 3.1. Consideration beyond the engineered system, including the system, SoS, or enterprise that it is a part of, and its interoperating and enabling systems, is also necessary.

The QC approaches in this section are covered in alphabetical order by name to avoid giving more weight to one over another. Table 3.1 summarizes the QC approaches included in the handbook. Not every QC approach will be applicable to every system or every application domain. It is recommended that subject matter experts are consulted and assigned as appropriate to conduct QC approaches. More information about the QC approaches can be found in references to external sources.

INCOSE Systems Engineering Handbook: A Guide for System Life Cycle Processes and Activities, Fifth Edition.
Edited by David D. Walden, Thomas M. Shortell, Garry J. Roedler, Bernardo A. Delicado, Odile Mornas, Yip Yew-Seng, and David Endler.
© 2023 John Wiley & Sons Ltd. Published 2023 by John Wiley & Sons Ltd.

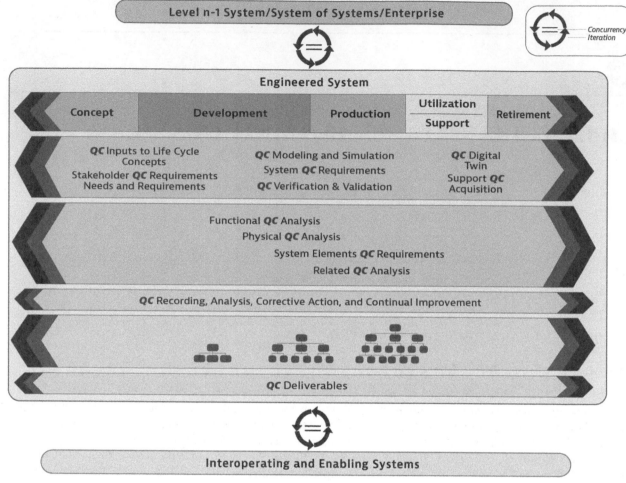

FIGURE 3.1 Quality characteristic approaches across the life cycle. INCOSE SEH original figure created by Taljaard, Kemp, and Walden. Usage per the INCOSE Notices page. All other rights reserved.

This handbook includes a set of QC approaches that are generally applicable in various applications and domains. However, the SE practitioner should ensure that any additional applicable QC approaches are also addressed.

3.1.2 Affordability Analysis

Definition   Affordability Analysis is an approach that maximizes value, providing cost-effective capability over the entire life cycle.

INCOSE has defined system affordability as follows:

> Affordability is the balance of system performance, cost, and schedule constraints over the system life while satisfying mission needs in concert with strategic investment and organizational needs.

Key Concepts   As stated in Blanchard and Fabrycky (2011),

Many systems are planned, designed, produced, and operated with little *initial* concern for affordability and the total cost of the system over its intended lifecycle... The technical [*aspects* are] usually considered first, with the economic [aspects] deferred until later.

TABLE 3.1   Quality Characteristic approaches

QC approach	An approach that …	Representative QCs
Affordability Analysis	maximizes value, providing cost effective capability over the entire life cycle	Affordability, Cost-Effectiveness, Life Cycle Cost (LCC), Value Robustness
Agility Engineering	enables change in a timely and cost-effective manner	Adaptability, Agility, Changeability, Evolvability, Extensibility, Flexibility, Modularity, Reconfigurability, Scalability
Human Systems Integration	integrates technology, organizations, and people effectively	Desirability, Ergonomics, Habitability, Human Factors, Human-Computer Interaction (HCI), Human-Machine Interface (HMI), Usability, User Interface (UI). User eXperience (UX)
Interoperability Analysis	ensures the system interacts effectively with other systems	Compatibility, Connectivity, Interoperability
Logistics Engineering	enables support for the entire life cycle	Supportability
Manufacturability/ Producibility Analysis	enables production in a responsible and cost effective manner	Manufacturability, Producibility
Reliability, Availability, Maintainability Engineering	enables the system to perform without failure, to be operational when needed, and to be retained in or restored to a required functional state	Accessibility, Availability, Interchangeability, Maintainability, Reliability, Repairability, Testability
Resilience Engineering	provides required capability when facing adversity	Resilience, Robustness, Survivability
Sustainability Engineering	supports the circular economy over its life	Disposability, Environmental Impact, Sustainability
System Safety Engineering	reduces the likelihood of harm to people, assets, and the wider environment	Safety
System Security Engineering	identifies, protects from, detects, responds to, and recovers from anomalous and disruptive events, including those in a cyber contested environment	Cybersecurity, Information Assurance (IA), Physical Security, Trustworthiness

This section addresses economic and cost factors under the general topics of affordability and cost-effectiveness. The concept of life cycle cost (LCC) is also discussed. Improving design methods for affordability is critical for all application domains (Bobinis, et al., 2013; Tuttle and Bobinis, 2013). Case 4 (Design for Maintainability-Incubators) from Section 6.4 provides an illustration of its importance.

A system is "affordable" if it can be developed to meet its requirements within cost and schedule constraints. The concept can seem straightforward. The difficulty arises when an attempt is made to specify and quantify the affordability of a system. This is significant when writing requirements or when comparing two solutions to conduct an affordability trade study. Affordability analysis is contextually sensitive, often leading to a misunderstanding and incompatible perspectives on what an "affordable system *is*."

Key affordability concepts include:

- Affordability context, system(s), and portfolios (of systems capabilities) need to be consistently defined and included in any understanding of what an affordable system is.

- An affordability process/framework needs to be established and documented.

- Accountability (system governance) for affordability needs to be assigned across the life cycle, which includes stakeholders from the various contextual domains.

Affordability costs include acquisition, operating, and support costs. It may be expanded to encompass additional elements required for the Life Cycle Cost (LCC) of a system, as an outcome of various contexts in which any system is embedded. In the SE domain, affordability as an attribute must be determined both inside the boundaries of the system of interest (SoI) and outside. The concept of affordability must encompass everything from a portfolio (e.g., family of automobiles) to an individual project (specific car model).

An affordability design model must be able to provide the ability to effectively manage and evolve systems over long life cycles. One of the major assumptions for measuring the affordability of competing systems is that given two systems, which produce similar output capabilities, it will be the *nonfunctional* attributes of those systems that differentiate system value to its stakeholders. As shown in Figure 3.2, the affordability model is concerned with operational attributes of systems that determine their value and effectiveness over time, typically expressed as the system's quality characteristics as they are called in this handbook. These attributes are properties of the system as a whole and as such represent the salient features of the system and are measures of the ability of the system to deliver the capabilities it was designed for over time.

Managing a system within an affordability trade space means that we are concerned with the actual performance of the fielded system, defined in one or more appropriate metrics, bounded by cost over time. The time dimension extends a specific "point analysis" (static) to a continuous life cycle perspective (dynamic). Quantifying a relationship between cost, performance, and time defines a functional space that can be graphed and analyzed mathematically. Then it becomes possible to examine how the output (e.g., performance, availability, capability) changes due to changes in the input (e.g., cost constraints, budget availability). This functional relationship between cost and outcome defines an affordability trade space to analyze the relationship between money spent and system performance and possibly determine the point of diminishing returns. This is illustrated in Figure 3.3. The capabilities and schedule have been fixed leaving either the cost or the performance to be the evaluation criteria, while the other becomes the constraint. This results in a relatively simple relationship between performance and cost. The maximum budget and the minimum performance are identified.

Below the maximum budget line in Figure 3.3 lie solutions that meet the definition of "conducting a project at a cost constrained by the maximum resources." The solutions to the right of the minimum performance line satisfy the threshold requirement. Thus, in the shaded rectangle lie the solutions to be considered since they meet the minimum performance and are less than the maximum budget. On the curve lay the solutions that are the "best value," in the sense that for a given cost the corresponding point on the curve is the maximum performance that can be achieved. *In actuality, the curve is rarely smooth or continuous and multiple curves need to be considered simultaneously.* Similarly, for a given performance, the corresponding point on the curve is the minimum cost for which that performance can be

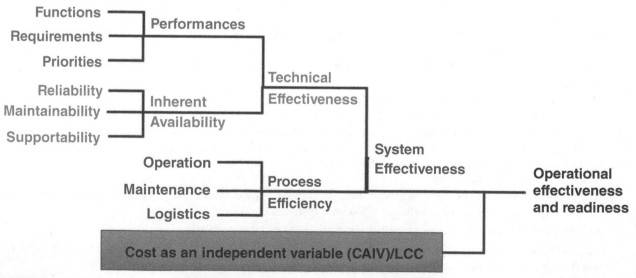

FIGURE 3.2   System operational effectiveness. From Bobinis et al. (2013). Used with permission. All other rights reserved.

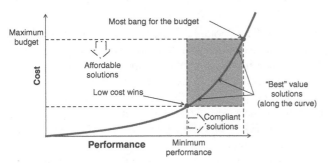

FIGURE 3.3 Cost versus performance. INCOSE SEH original figure created by Bobinis on behalf of the INCOSE Affordability Working Group. Usage per the INCOSE Notices page. All other rights reserved.

achieved. Selecting the decision criterion as cost will result in achieving the threshold performance. If the decision criterion is performance, all of the budget would be expended. Consequently, to specify affordability for a system or project requires determining which affordability element is the basis for the decision criteria and which elements are being specified as constraints.

Affordability is the result of a disciplined decision-making process requiring systematic methodologies that support selection of the most affordable technologies and systems.

Elaboration

Cost-Effectiveness Analysis

Cost-effectiveness (CE) is a measure relating cost to system effectiveness. It is defined below with the achieved systems effectiveness as the numerator and cost as the denominator (Blanchard, 1967):

$$CE = SE/(IC + SC)$$

Where SE = System Effectiveness, IC = initial cost and SC = sustainment cost.

Reliability and maintainability are major factors in determining the cost effectiveness of a system since they impact sustainment costs.

System effectiveness is a term used in a broad context to reflect the technical characteristics of a system (e.g., performance, availability, supportability, dependability) such as examples mentioned in the preceding section. It may be expressed differently depending on the specific application. Sometimes a single-figure of merit is used to express system effectiveness and sometimes multiple figures-of-merit are employed (Blanchard and Fabrycky, 2011). The IC and SC can also be expressed in different ways depending on the application or system parameters under evaluation. It may include costs for concept, development, production, utilization, support, and retirement.

Cost-Effectiveness Analysis (CEA) is distinct from cost–benefit analysis (CBA). The approach to measuring costs is similar for both techniques, but in contrast to CEA where the results are measured in performance terms, CBA uses monetary measures of outcomes. This approach has the advantage of being able to compare the costs and benefits in monetary values for each alternative to see if the benefits exceed the costs. It also enables a comparison among projects with very different goals if both costs and benefits can be placed in monetary terms. Other closely related, but slightly different, formal techniques include cost–utility analysis, economic impact analysis, fiscal impact analysis, and social return on investment (SROI) analysis.

The concept of cost effectiveness is applied to the planning and management of many types of organized activity. It is widely used in many system aspects. Some examples are:

- Studies of the desirable performance characteristics of commercial aircraft to increase an airline's market share at lowest overall cost over its route structure (e.g., more passengers, better fuel consumption)
- Urban studies of the most cost-effective improvements to a city's transportation infrastructure (e.g., buses, trains, motorways, and mass transit routes and departure schedules)
- In health services, where it may be inappropriate to monetize health effect (e.g., years of life, premature births averted, sight years gained)
- In the acquisition of military hardware when competing designs are compared not only for purchase price but also for such factors as their operating radius, top speed, rate of fire, armor protection, and caliber and armor penetration of their guns

LCC Analysis

LCC refers to the total cost incurred by a system throughout its life. This "total" cost varies by circumstances, the stakeholders' points of view, and the system. For example, when purchasing an automobile, the major cost factors are the cost of acquisition, operation, maintenance, and disposal (or trade-in value). A more expensive car (acquisition cost) may have lower LCC because of lower operation and maintenance costs. But the car manufacturer has other costs such as development and production costs, including setting up the production line, to be considered. The SE Practitioner needs to look at costs from several aspects and be aware of the stakeholders' perspectives. LCC should not be equated to Total Cost of Ownership (TCO), Total Ownership Cost (TOC), or Whole Life Cost (WLC). These measures may only include costs once the system has been purchased or acquired.

LCC estimates are sometimes used to support internal project trade-off decisions and need only be accurate enough to support the relative trade-offs. The analyst should always attempt to prepare as accurate cost estimates as possible and assign risk to them. These estimates should be reviewed by upper management and potential stakeholders. Future costs, while unknown, can be predicted based on assumptions and risk assigned. All assumptions when doing LCC analysis should be documented.

LCC analysis can be used in affordability and system cost-effectiveness assessments. The LCC is *not* the definitive cost proposal for a project since LCC "estimates" (based on future assumptions) are often prepared early in a project's life cycle when there is insufficient detailed design information. Later, LCC estimates should be updated with actual costs from early project stages and will be more definitive and accurate due to hands-on experience with the system. A major purpose of LCC studies is to help identify cost drivers and areas in which emphasis can be placed during the subsequent life cycle stages to obtain the best decisions. Accuracy in the estimates will improve as the system evolves and the data used in the calculation is less uncertain.

LCC analysis helps the project team understand the total cost impact of a decision, compare between project alternatives, and support trade studies for decisions made *throughout* the system life cycle. LCC normally includes the following costs, represented in Figure 3.4:

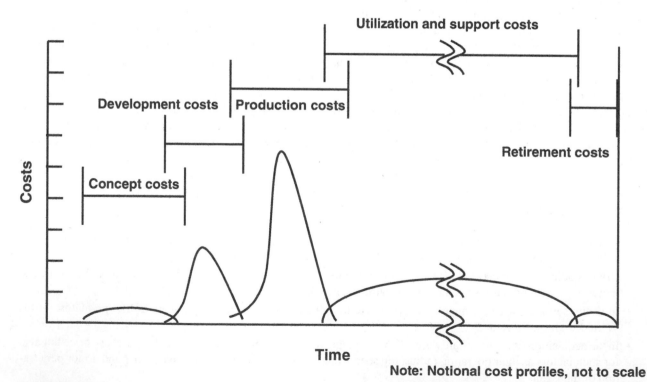

FIGURE 3.4 Life cycle cost elements. INCOSE SEH original figure from INCOSE SEH v2 Figure 4-83. Usage per the INCOSE Notices page. All other rights reserved.

- *Concept costs*—Costs for the initial concept development efforts. These could be estimated based on average staffing and schedule spans and may include overhead, general and administrative (G&A) costs, and fees, as necessary.
- *Development costs*—Costs for the system development efforts. Similar to concept costs, these can be estimated based on average staffing and schedule spans and may include overhead, G&A costs, and fees, as necessary. Parametric cost models may also be used.
- *Production costs*—Usually driven by tooling and material costs for large-volume systems. Labor cost estimates are prepared by estimating the cost of the first production unit and then applying learning curve formula to determine the reduced costs of subsequent production units.
- *Utilization and support costs*—Typically based on future assumptions for ongoing operation and maintenance of the system, for example, fuel costs, personnel levels, and spare parts.
- *Retirement costs*—The costs for removing the system from operation and includes an estimate of trade-in or salvage costs. Could be positive or negative and should be mindful of the environmental impacts to dispose.

For global products, other sources of cost may include compliance costs (government regulations, import/export requirements, etc.) or other incidental costs of international business.

Common methods/techniques for conducting LCC analysis that may be suitable for different situations and/or used in combinations follow:

- *Analogy*—Reasoning by comparing the proposed project with one or more completed projects that are judged to be similar, with corrections added for known differences. May be acceptable for early estimations.
- *Bottom up*—Identifies and estimates costs for each lower-level element separately and rolls them up for the total cost.
- *Delphi technique*—A structured approach to build estimates iteratively from multiple domain experts. Surveys are used, and in each round feedback on the group statistics is provided for experts to help revise their estimates.
- *Design-to-Cost (DTC)*—Using a predetermined cost (e.g., the SoI material cost) as a constraint on the design solution.
- *Expert judgment*—Estimate performed by one or more experts using their experience and judgment. It can be used for comparison and sanity check against other methods.
- *Parametric (algorithmic)*—Uses mathematical algorithms to compute cost estimates as a function of cost factors based on historical data. This technique is supported by public domain and commercial tools and models. Examples include the Constructive Systems Engineering Cost Model (COSYSMO) for SE effort and the Constructive Cost Model (COCOMO) for software engineering effort.
- *Parkinsonian technique*—Work estimates based on the available resources or schedules (Parkinson's Law states that work expands to fill the available volume).
- *Price to win*—Focuses on providing an estimate, and associated solution, at or below the price judged necessary to win the contract.
- *Taxonomy method*—Using a hierarchical structure or classification scheme as a basis of the estimates.
- *Top down*—Developing costs based on overall project characteristics at the top level of the architecture.

3.1.3 Agility Engineering

Definition   Agility Engineering is an approach that enables change in a timely and cost-effective manner.

Key Concepts   Agility is the ability to thrive and survive in uncertain, unpredictable operational environments; and manifests as effective response to situations presented by the environment (Dove and LaBarge, 2014). Effective response has four metrics:

- timely (fast enough to deliver value),
- affordable (at a cost that can be repeated as often as necessary),
- predictable (can be counted on to meet the need), and
- comprehensive (anything and everything within mission boundary).

Agile systems-engineering and *agile-systems engineering* are two different things (Haberfellner and de Weck, 2005) that share the word agile. In the first case the SoI is an engineering process (e.g., using an agile SE process). This is addressed in Section 4.2.2. In the second case, the SoI is what is produced by an engineering process (e.g., engineering an agile system). This is the subject of this section. Sustained agility is enabled by an architectural pattern and a set of design principles that are fundamental and common to both agile SE processes and engineered agile systems.

Elaboration
Agility Architectural Framework
The architecture that enables agility will be recognized in a simple sense as a drag-and-drop plug-and-play loosely coupled modularity, with some critical aspects not often called to mind with the general thoughts of a modular architecture. The architectural objective is to enable rapid and effective composability of processes and systems from available resources, appropriate for the needs at hand (Dove and LaBarge, 2014). Construction toys, like Lego or Meccano sets, are iconic architectural examples.

There are three critical elements in the architecture: a roster of drag-and-drop *encapsulated modules*, a *passive infrastructure* of minimal but sufficient rules and standards that enable and constrain plug-and-play operation, and an *active infrastructure* that designates specific responsibilities that sustain agile operational capability:

Encapsulated modules—Modules are self-contained encapsulated units complete with well-defined interfaces that conform to the plug-and-play passive infrastructure. They can be dragged and dropped into a system of response capability with relationship to other modules determined by the passive infrastructure. Modules are encapsulated so that their interfaces conform to the passive infrastructure, but their methods of functionality are not dependent on the functional methods of other modules except as the passive infrastructure dictates.

Passive infrastructure—The passive infrastructure provides drag-and-drop connectivity between modules. Its value is in isolating the encapsulated modules so that unexpected side effects are minimized and new operational functionality is rapid. Selecting passive infrastructure elements is a critical balance between requisite variety and parsimony—just enough in standards and rules to facilitate module connectivity but not so much to overly constrain innovative system configurations.

Active infrastructure—An agile system is not something designed and deployed in a fixed event and then left alone. Agility is most active as new system configurations are assembled in response to new requirements—something which may happen very frequently, even daily in some cases. In order for new configurations to be enabled when needed, five responsibilities are required:

- Module mix evolution—Who (or what process) is responsible for ensuring that new modules are added to the roster and existing modules are upgraded in time to satisfy response needs?
- Module readiness—Who (or what process) is responsible for ensuring that sufficient modules are ready for deployment at unpredictable times?
- Situational awareness—Who (or what process) is responsible for monitoring, evaluating, and anticipating the operational environment?
- System assembly—Who (or what process) assembles new system configurations when new situations require something different in capability?
- Infrastructure evolution—Who (or what process) is responsible for evolving the passive and active infrastructures as new rules and standards are anticipated and become appropriate?

Responsibilities for these five activities must be designated and embedded within the system to ensure that effective response capability is possible at unpredictable times

Agility Architectural Design Principles
Ten reusable, reconfigurable, scalable design principles are briefly itemized in this section:
Reusable principles are as follows:

- *Encapsulated modules*—Modules are distinct, separable, loosely coupled, independent units cooperating toward a shared common purpose.
- *Facilitated interfacing (plug compatibility)*—Modules share well-defined interaction and interface standards and are easily inserted or removed in system configurations.
- *Facilitated reuse*—Modules are reusable and replicable, with supporting facilitation for finding and employing appropriate modules.

Reconfigurable principles are as follows:

- *Peer–peer interaction*—Modules communicate directly on a peer-to-peer relationship; and parallel (rather than sequential) relationships are favored.
- *Distributed control and information*—Modules are directed by objective (rather than method); decisions are made at point of maximum knowledge, and information is associated locally and accessible globally.
- *Deferred commitment*—Requirements can change rapidly and continue to evolve. Work activity, response assembly, and response deployment that are deferred to the last responsible moment avoid costly wasted effort that may also preclude a subsequent effective response.
- *Self-organization*—Module relationships are self-determined where possible, and module interaction is self-adjusting or self-negotiated.

Scalable principles are as follows:

- *Evolving infrastructure standards*—Passive infrastructure standardizes intermodular communication and interaction, defines module compatibility, and is evolved by designated responsibility for maintaining current and emerging relevance.
- *Redundancy and diversity*—Duplicate modules provide capacity right-sizing options and fail-soft tolerance, and diversity among similar modules employing different methods is exploitable.
- *Elastic capacity*—Modules may be combined in responsive assemblies to increase or decrease functional capacity within the current architecture.

Agility Metrics
Agility measures are enabled and constrained principally by architecture—in both the process and the product of development:

- *Time to respond*, measured in both the time to understand a response is necessary and the time to accomplish the response.
- *Cost to respond*, measured in both the cost of accomplishing the response and the cost incurred elsewhere as a result of the response.

- *Predictability of response*, measured before the fact in architectural preparedness for response and confirmed after the fact in repeatable accuracy of response time and cost estimates.
- *Scope of response*, measured before the fact in architectural preparedness for comprehensive response capability within mission and confirmed after the fact in repeatable evidence of broad response accommodation.

3.1.4   Human Systems Integration

Definition   Human Systems Integration (HSI) is an approach that integrates technology, organizations, and people effectively.

HSI is an essential, transdisciplinary, sociotechnical, and management approach of SE used to ensure that the system's technical, organizational, and human elements are appropriately addressed across the whole system life cycle, service, or enterprise system. HSI considers systems in their operational context together with the necessary interactions between and among their human and technological elements to make them work in harmony and cost effectively, from concept to retirement.

Key Concepts
Human
The "human" in HSI includes all individuals and groups interacting within the SoI. Within HSI, these are typically referred to as "stakeholders." Stakeholders can include system acquirers, owners, users, operators, maintainers, trainers, support personnel, and the general public. While most people who interact within the SoI will be cooperative or have a vested interest in its performance, consideration may also need to be given to non-cooperative people or those with malign intent such as competitors, adversaries, criminals (physical and cyber), and those seeking to use the system outside of its design intent. Life, human, and social sciences have different representations of the human element and can all bring different perspectives to HSI activities.

Systems
HSI adopts a sociotechnical system perspective that considers a system as a representation of natural and artificial elements that are organizations of humans and machines (where machines include both hardware and software). Therefore, HSI considers that all systems include both humans and machines, and to optimize the system, all of these elements must be considered within SE activities.

Integration
HSI considers integration from two key viewpoints. The first is the effective integration of the human and technological elements in a system. The second is the efficient integration of the different perspectives of both human and machine elements within the system. An example of these different HSI perspectives can be seen in Figure 3.5. The specific perspectives relevant to a project will vary depending on the nature of the system and the organization's activities.

All systems involve or affect people and exist within a wider sociotechnical and organizational context. Therefore, HSI is an essential enabler to SE practice. The sociotechnical approach provided by HSI supports analysis, design, and evaluation activities in holistically understanding and effectively integrating the technological, organizational (including processes), and human elements of a system. As shown in Figure 3.5, HSI emerges from the overlapping of three main circles: (1) technology, organization, and people (the TOP Model) within an environment at the heart; (2) HSI perspectives; and (3) contributing disciplines associated with the operational domain shown in the periphery. It is particularly important that systems are designed to meet human capabilities, limitations, and goals.

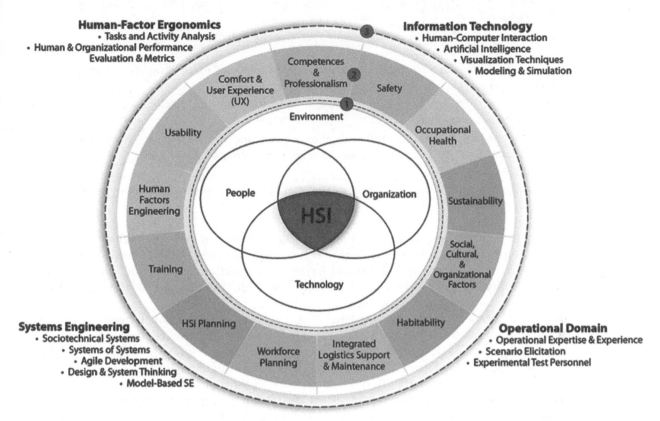

FIGURE 3.5   HSI technology, organization, people within an environment. INCOSE SEH original figure created by Boy. Usage per the INCOSE Notices page. All other rights reserved.

Elaboration

Purpose and Value of HSI

The purpose of HSI is to optimize total system performance and stakeholder satisfaction through the mutual integration of technology, organizations (including processes), people, and environment.

The benefits which can be realized by HSI vary from domain to domain, depending on their priorities and purpose (e.g., safety, cost, efficiency, performance, acceptability) and the nature of the system. They can be broken down into the following areas:

- *holistic optimization of system performance and efficiency*: participatory design, and human-in-the-loop (HITL) activities
- *improved safety*: hazard, risk, performance limitations and emergent properties analysis
- *reduced development costs*: consider the TOP Model
- *reduced system LCC costs*: HSI from the beginning of the SE life cycle
- *improved sales*: resulting from product or service usability
- *user experience (UX) and desirability*: focus on Human-Centered Design (HCD) and user needs
- *improved adoption of new systems by the workforce or user groups*: considering sociotechnical factors
- *HSI value to a project*: from intuition to expertise in HSI

Scope and Breadth of HSI

HSI is based on the convergence of four key communities of practice (third circle in Figure 3.5):

- *human factors and ergonomics (HF/E)* that provides human-centered and organization-centered analysis, performance evaluation techniques, and metrics (Boehm-Davis, et al., 2015);
- *information technology (IT)* that includes human-computer interaction, artificial intelligence, visualization techniques, and modeling and simulation;
- *systems engineering* that includes socio-technical systems, systems of systems (see Section 4.3.6), agile development (see Section 4.2.2), design and system thinking (see Sections 3.2.7 and 1.5), and model-based SE (MBSE) (see Section 4.2.1); and
- *the operational domain* that includes operational expertise and experience, scenario elicitation, and experimental test personnel (see Section 4.4).

These communities enable support of HSI through HCD as a major process that involves development and use of domain ontology, prototypes and digital modeling, scenario-based design, modeling and HITL activities (simulations and physical tests), formative evaluations, agile design and development, as well as human performance and organizational metrics (e.g., maturity and flexibility) (Boy, 2013) (Boy, 2020). HCD validation both requires certification approval and contributes to certification rules evolution.

HSI considers systems complexity analysis as a baseline. It seeks simplification (where possible) and familiarity with complex systems (where necessary). HITL activities enable discovery and elicitation of complex systems' emergent behaviors, properties, functions, and structures, which are incrementally integrated into the SoI through its whole life cycle. HITL activities provide SE and HCD teams with improved understanding of the SoI early in the life cycle, contributing to design flexibility and better resource management. HSI is a foundational enabler for industrial endeavors, such as Industry 4.0, where digital engineering, enabling virtual HCD, requires increased physical and cognitive tangibility testing across the life cycle of a system (see Section 5.4). Case 5 (Artificial Intelligence in Systems Engineering - Autonomous Vehicles) from Section 6.5 illustrates the importance of all these aspects.

HSI can be considered as both an *enabling process*, associating HCD and SE during the life cycle of a system, and a *product* resulting from this process. HSI is the result of this HCD-based convergence, which requires optimizing the TOP Model. User eXperience (UX) and User Interface (UI) development are integral parts of the HSI process from the early stages and throughout the system life cycle. HSI processes are iterative and supported by two main types of assets, methods, and tools: expertise elicitation and creativity. The former enables effective elicitation from subject matter experts through knowledge and know-how, supporting design teams during system formative evaluations, agile development, and certification. The latter enables out-of-the-box projections that are validated using prototyping and HITL activities.

HSI Perspectives

HSI encompasses several important perspectives displayed in Figure 3.5 (second circle) and described in more detail in Table 3.2.

A wide variety of HSI methods, models, knowledge, and approaches can be used to support decisions made across the whole system life cycle. This can include support to requirements analysis, trade-studies, life cost benefit analysis, options or tender down select, risk management, safety case development, design decisions, acceptance testing, and workforce planning. Human-related trade studies are critical to determining holistic of operational concept (OpsCon) and thereby informing the design team in terms of effectivity, efficiency, suitability, usability, safety, and affordability. See the INCOSE HSI Primer (2023) for more detail.

TABLE 3.2   HSI perspective descriptions

Human Factors Engineering (*HFE*) is the scientific discipline concerned with the understanding of interactions among humans and other elements of a system, and the profession that applies theory, principles, data, and other methods to design in order to optimize human well-being and overall system performance.

Social, Cultural, and Organizational Factors consider the organizational aspects of socio-technical systems and includes the organizations who will be using and supporting the operational system, as well as the organizations who are involved throughout the entire life cycle of the system.

HSI Planning addresses the implementation of HSI through the SE process to ensure the human element is effectively integrated with the system. HSI strategies and priorities need to be set up-front, can be formalized in the HSI Plan, and potentially adjusted during the life cycle, upon mission definition, and carried throughout the allocation of resources and project personnel.

Integrated Logistics Support (ILS) & Maintenance covers human performance during the whole life cycle of a system based on an ILS plan supported by an HSI plan. ILS includes training, operations, maintenance, potential redesign, and dismantling.

Workforce Planning addresses the number and type of personnel and the various occupational specialties required and potentially available to develop, train, operate, maintain, and support the system.

Competences and Professionalism consider the type of knowledge, skills, experience levels, and aptitudes (cognitive, physical, and sensory) required to operate, maintain, and support a critical system and the means to provide such people (through selection, recruitment, training, etc.).

Training encompasses designing to account for ease and reduction of operation time needed to provide training through trade studies evaluated to assess their impact on training, as well as the instructions and resources required to provide personnel with requisite competence, knowledge, skills, and attitudes to properly operate, maintain, and support systems.

Safety promotes system characteristics and procedures to minimize the risk of accidents or mishaps that cause death or injury to operators, maintainers, support personnel, or others who could come into intentional or unintentional contact with the system; threaten systems operations; or cause cascading failures in other systems. It includes survivability.

Occupational Health promotes system design features and procedures that serve to minimize physiological mental and social health hazards which might result in injury, acute or chronic illness, and disability; and to enhance job performance and wellbeing of personnel who operate, maintain, or support the system.

Sustainability covers the environmental considerations that can affect operations and particularly human performance and considers wider ranging concerns and long-term goals of how the humans within the system can affect the environment, society, and economy without compromising future generations' needs.

Habitability involves characteristics of system living and working conditions.

Usability involves objective evaluation methods to address aspects such as efficiency, conformity to human expectations, tolerance/resistance toward human errors, and learnability to improve the degree to which humans can reach their objectives when interacting with a system.

Comfort and UX are personal internal human aspects such as joy, guilt, opinions, and unconscious aspects which are to be considered, not only in regard to the primary users of the final product, but in regard to all humans involved in the systems engineering process.

3.1.5   Interoperability Analysis

Definition   Interoperability Analysis is an approach that ensures the system interacts effectively with other systems. In the domains of data/information exchange and communications, there are four definitions of interoperability:

- The capability of systems to communicate with one another and to exchange and use information including content, format, and semantics (NIST SP 500-230, 1996).
- The ability of two or more systems or system elements to exchange data and use information (IEEE 610.12, 1990).
- The ability of two or more systems to exchange information and to mutually use the information that is exchanged (US Army, 1997).
- The condition achieved among communications-electronics systems or items of communications—electronics equipment when information or services can be exchanged directly and satisfactorily between them and/or their users (US DoD, 2021).

Key Concepts   Interoperability reflects the ability of a system to work in conjunction with other system(s) to achieve an outcome. For example, a mobile phone can operate on different networks across the world, agricultural implements from different companies can work on each other's tractors, or a system provides an interface allowing remote control of its capabilities. Originally described in terms of computer/software systems, the concept of interoperability applies more widely, such as human interactions. A broad definition of interoperability also takes into account social, political, and organizational factors that impact system-to-system performance. Interoperability is a key enabler for a System of Systems (SoS), because it allows the elements of a large and complex system to work together as a single entity, toward a shared purpose (see Section 4.3.6).

Interoperability may be achieved in two principal ways, which can also be combined:

- Agreeing on one or more *published standards* as the definition of the interface. This exposure of interfaces complying with open interfaces is increasingly common in the consumer product area where "plug and play" is expected.
- Defining and implementing a *custom interface*. When a standard interface does not exist, or is not suitable, a custom interface can be defined as the agreed way in which two or more systems will connect, communicate, interact, or cooperate to achieve their shared purpose.

Elaboration   Interoperability will increase in importance as the world grows smaller due to expanding communications networks (e.g., the internet of things (IoT)), as nations continue to perceive the need to communicate seamlessly across international coalitions of commercial organizations or national defense forces, and as individuals increasingly expect that products and services will "work together."

The Øresund Bridge (see Section 6.2) demonstrates the interoperability challenges faced when just two nations collaborate on a project. For example, the meshing of regulations on health and safety, interfacing a left-handed (Sweden) and right-handed (Denmark) railway, and the resolution of two power supply systems for the railway. Hence careful choices were necessary for the standards selected for the bridge itself, and for its interfaces at the Swedish and Danish ends.

3.1.6   Logistics Engineering

Definition   Logistics Engineering is an approach that enables support for the entire life cycle.

Key Concepts   Logistics engineering, which may also be referred to as product support engineering, is the engineering discipline concerned with the identification, acquisition, procurement, and provisioning of all support resources required to sustain operation and maintenance of a system (Blanchard and Fabrycky, 2011). Logistics engineering is also concerned with engineering the inherent supportability of the design. Logistics should be addressed from a life cycle perspective and be considered in all stages and especially as an inherent part of system concept and development. Furthermore, logistics should be approached from a system perspective to include all activities associated with design for supportability, the acquisition and procurement of the elements of support, the supply and distribution of required support material, and the maintenance and support of systems throughout their planned period of utilization.

The scope of logistics engineering is thus
- to determine logistics support requirements,
- to design the system for supportability,
- to acquire or procure the support, and
- to provide cost-effective logistics support for a system during utilization and support stages.

Logistics engineering has evolved into several related elements such as supply chain management (SCM) in the commercial sector and integrated logistics support (ILS) in the defense sector.

Elaboration
Support Elements
Support planning starts with the definition of the support (including maintenance) concept in the concept stage and continues through supportability analysis in the development stage, to the ultimate development of a support plan. The support concept describes the support environment in which the system will operate and which inherent supportability and support system elements are required for establishing the system operational capability.

The following elements of support are to be fully integrated with the system at the lowest possible LCC:
- *Product support integration and management*—Plan and manage cost and performance across the product support value chain, from concept to retirement.
- *Design interface*—Participate in the SE process to impact the design from inception throughout the life cycle. Facilitate supportability to maximize availability, effectiveness, and capability at the lowest LCC. Early application of the support concept drives the design inherent supportability objectives and trade-offs. It is an important mechanism for aligning design Reliability, Maintainability, and Supportability (RMS), maintenance planning, and establishment of support capabilities for the operational environment. It guides design modularity, reliability, maintainability, testability, and overall repair policies.
- *Sustained logistics engineering of the fielded system*—This effort spans those technical tasks (engineering investigations and analyses) to ensure continued dependable operation, including maintenance, for the life cycle. It characterizes the system and support capabilities' RMS performance as an input to dependable planning of operational use. It involves applying improved confidence level RMS characteristics data, gained from the operational experience, to enhance maintenance strategy and the support system, and to propose design RMS improvements.
- *Maintenance planning*—Identifying the system maintenance requirements, determining the maintenance strategy, and implementing the maintenance capabilities required to deliver the system operational capability. The support concept guides overall repair policies, such as "repair vs. replace" criteria.
- *Operation and maintenance personnel*—Identify, plan, and acquire personnel, with the training, experience, and skills required to operate, maintain, and support the system.
- *Training and training support*—Establish and maintain the required operator and maintainer skill levels across the system life cycle. Identify, develop, and acquire Training Aids, Devices, Simulators, and Simulations (TADSS) to maximize the effectiveness of the personnel to operate and sustain the system equipment.
- *Supply support*—determine requirements for supply, and acquire, catalog, receive, store, transfer, issue, and dispose of spares, repair parts, and supplies. This means having the right spares, repair parts, and all classes of supplies available, in the right quantities, at the right place, at the right time, at the right price.
- *Computer resources (hardware and software)*—Computers, associated software, networks, and interfaces necessary to enable long-term logistics engineering, maintenance management, system technical and associated support operations data management, and storage.

- *Technical data, reports, and documentation*—Represents recorded information of scientific or technical nature (e.g., equipment technical manuals, engineering drawings), engineering data, specifications, and standards.
- *Facilities and infrastructure*—This includes facilities (e.g., buildings, warehouses, hangars, waterways, associated facilities equipment) and infrastructure (e.g., IT services, fuel, water, electrical service, machine shops, dry docks, test ranges).
- *Packaging, handling, storage, and transportation (PHS&T)*—Ensure that all system equipment and support items are preserved, packaged, handled, and transported properly, including environmental considerations, equipment reservation for short and long storage, and transportability. Some items may require special environmentally controlled, shock-isolated containers for transport to and from storage, operational, and repair facilities via all modes of transportation (e.g., land, rail, sea, air, space).
- *Support equipment*—All equipment (mobile and fixed) required to sustain the operation and maintenance of a system, including, but not limited to, handling and maintenance equipment, trucks, air conditioners, generators, tools, metrology and calibration equipment, and manual and automatic test equipment.

Supportability Analysis

As shown in the Figure 3.1, supportability analysis addresses all elements of design supportability and of the support system required during all life cycle stages:

- *Functional failure analysis*—A Functional Breakdown Structure (FBS) is used as reference to perform functional FMECA (Failure Mode Effects and Criticality Analysis), FTA (Fault Tree Analysis) and/or RBD (Reliability Block Diagram) analysis. These analyses can be used to identify functional failure modes and to classify them according to criticality (e.g., severity of failure effects and probability of occurrence). The functional failure analysis can also provide valuable system design input (e.g., redundancy requirements). In describing functional failure compensation means, including compensation by support, the functional failure analysis provides early means of illustrating the system supportability interface and criticality of support.
- *Physical failure analysis*—A Product Breakdown Structure (PBS) is used as reference to perform hardware FMECA, FTA, and/or RBD analysis with the objective of optimizing the design and to identify all maintenance tasks for potential failure modes. An objective of logistic engineering is to minimize operational maintenance tasks and resource requirements. The FMECA (in assessing the design inherent reliability, protection, and testability versus reliance on preventive or corrective maintenance) allows in context trade-offs of the operational value of improving the design versus defaulting to reliance on operational maintenance. The FMECA findings are used to balance the level of repair allocation. Failure probability, criticality, detection means, the design modularity, and the complexity of failure restoration need to be in balance with the level of repair capabilities framed by the system support concept.
- *Task identification and optimization*—Corrective maintenance tasks are primarily identified using FMECA, while preventive maintenance tasks are identified using RCM (Reliability-Centered Maintenance). Trade-off studies may be required to achieve an optimized maintenance strategy. Associated support tasks, such as operational transportation, are identified from analysis of the operational concept and support workflows.
- *Detail task analysis*—Detail procedures for support tasks should be developed, and support resources identified and allocated to each task. The system Level of Repair Analysis (LORA), in conjunction with the support concept, may be used to determine the most appropriate location for executing these tasks.
- *Support element specifications*—Support element specifications should be developed for all support deliverables. Depending on the system, specifications may be required for training aids, facilities, support equipment, publications, and packaging material. Establishment of support elements, such as facilities, may involve extended lead times requiring identification of requirements and initiation of acquisition from as early as the system concept stage. The support element requirements analysis is therefore iterated from the system concept stage to highlight long-lead time support element acquisition requirements.
- *Support deliverables, test, and evaluation*—All support deliverables should be acquired based on the individual specifications. The support deliverables should be tested and evaluated against support element specifications and the overall system requirements.

- *Support modeling and simulation*—Modeling and simulation are integral parts of supportability analysis that should be initiated during the early stages to frame and develop a compliant and optimized system design, maintenance strategy, and support system. Modeling and simulation during acquisition are progressed to become decision and planning optimization tools for the operational and support stages. The predictive modeling information during acquisition is progressively matured as experience is gained with the operational system and operational support capabilities (e.g., digital twin for operation and support).

- *Recording and corrective action*—Failure recording and corrective action during the utilization and support stages form the basis for continuous improvement. System operational value delivery metrics should be applied to continuously monitor the system to improve support where deficiencies are identified, and to highlight focus areas for operational enhancements to system inherent reliability, maintainability, and supportability.

3.1.7 Manufacturability/Producibility Analysis

Definition   Manufacturability/Producibility Analysis is an approach that enables production in a responsible and cost-effective manner.

Key Concepts   Production involves the repeated manufacture of the developed system. The capability to manufacture or produce a system or its elements is as essential as the ability to properly develop it. A system that cannot be effectively produced causes unnecessary costs and may lead to rework and project delays with associated cost overruns. For this reason, manufacturability/producibility analysis is an integral part of the SE process.

Producibility considerations differ depending upon the type and number of systems being produced. For example, the manufacture of satellites (limited production runs), military tanks (medium production runs), and mobile phones (high production runs) would be vastly different. A unique aspect of infrastructure systems is that production typically takes place on-site, rather than in a factory (see Section 4.4.5). Multiple production cycles require the consideration of production maintenance and downtime.

One objective is to determine if existing production enabling systems are satisfactory (see Section 1.3.3), since this could be the lowest risk and most cost-effective approach. If not, the requirements for the production enabling systems and processes need to be determined, and the production enabling systems developed so they are ready when needed. A SE approach to manufacturing and production is necessary because the production enabling systems can sometimes cost more than the system being produced (Maier and Rechtin, 2009).

Elaboration   Producibility analysis is a key task in developing cost-effective, quality products. Multidisciplinary teams work to simplify the design and stabilize the manufacturing process to reduce risks, manufacturing costs, lead times, and cycle times and to minimize strategic or critical material use. Producibility analysis draws upon the production and support life cycle concepts. Producibility requirements are identified in the Business or Mission Analysis and Stakeholder Needs and Requirements Definition processes (see Sections 2.3.5.1 and 2.3.5.2) and included in the project risk analysis, if necessary. Similarly, long-lead-time items, sole source items (where only one supplier for the required item is available), material limitations, special processes, and manufacturing constraints are evaluated. Design simplification also considers ready assembly and disassembly for ease of maintenance and preservation of material for recycling. When production requirements create a constraint on the system, they are communicated and documented. The selection of manufacturing methods and processes is included in early decisions. Manufacturing test considerations are captured and are taken into account in built-in test and automated test equipment.

IKEA® is often used as an example of supply chain excellence. IKEA has orchestrated a value creating chain that begins with motivating customers to perform the final stages of furniture assembly in exchange for lower prices and a fun shopping experience. They achieve this through designs that support low-cost production and transportability (e.g., the bookcase that comes in a flat package and goes home on the roof of a car).

3.1.8 Reliability, Availability, Maintainability Engineering

Definition   Reliability, Availability, Maintainability Engineering is an approach that enables the system to perform without failure, to be operational when needed, and to be retained in or restored to a required functional state.

RAM (sometimes expressed as RMA) is a well-known acronym for Reliability, Availability, and Maintainability. These QCs are completely interrelated with each other and have a strong relationship with logistics and supportability.

Key Concepts   From a SE perspective, RAM should not only be viewed as quality characteristics, but as nonfunctional requirements. RAM activities are often neglected during system development, resulting in a substantial increase in risk of project failure or stakeholder dissatisfaction. Since RAM often drives other system requirements, it is essential that these activities be selected, tailored, planned, and executed in an integrated manner with other SE processes. A practical way to achieve this is to develop detailed reliability and maintainability plans early in the system development process and to integrate these plans with the SE management plan (SEMP).

RAM, being important inputs to the system maintenance concept, support other SE processes in two ways. First, they should be used to influence both system and system support definitions (e.g., the system architecture depends on RAM requirements). Second, they should be used as part of system verification (e.g., system analysis or system test).

Depending on the particular industry, availability is often seen as the most important of these three quality characteristics, especially from the viewpoint of a user or acquirer. Any availability loss can usually easily be translated to mission or production loss and increased costs.

Elaboration
Reliability
The IEEE Reliability Society defines:

> Reliability is a design engineering discipline which applies scientific knowledge to assure a product will perform its intended function for the required duration within a given environment. This includes designing in the ability to maintain, test, and support the product throughout its total life cycle. Reliability is best described as product performance over time. This is accomplished concurrently with other design disciplines by contributing to the selection of the system architecture, materials, processes, and system elements—both software and hardware; followed by verifying the selections made by thorough analysis and test.

"To be reliable, a system must be robust—it must avoid failure modes even in the presence of a broad range of conditions including harsh environments, changing operational demands, and internal deterioration" (Clausing and Frey, 2005). An in-depth understanding of the interaction between the system, the environment where it will be used, the operating conditions it will be subjected to, and potential failure modes and failure mechanisms is thus essential to design and manufacture reliable systems. Figure 3.6 shows the interaction between these aspects.

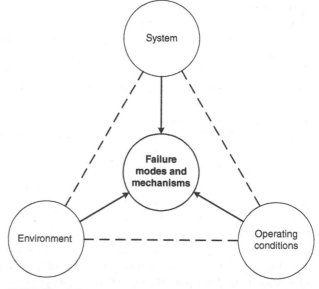

FIGURE 3.6 Interaction between system, environment, operating conditions, and failure modes and failure mechanisms. INCOSE SEH original figure created by Barnard. Usage per the INCOSE Notices page. All other rights reserved.

Reliability can be formally defined as "the ability of a system to perform as designed, without failure, in an operational environment, for a stated period of time" (Tortorella, 2015). Since "ability" is an abstract concept, many reliability metrics are available which can be used to measure and manage the reliability of a system during the development, utilization, and support stages (e.g., number of failures per time period, failure-free period, expected lifetime of non-repairable parts, Mean Time Between Failure [MTBF]).

O'Connor and Kleyner (2012) state the objectives of reliability engineering, in the order of priority, are:

1. To apply engineering knowledge and specialist techniques to prevent or to reduce the likelihood or frequency of failures.
2. To identify and correct the causes of failures that do occur, despite the efforts to prevent them.
3. To determine ways of coping with failures that do occur, if their causes have not been corrected.
4. To apply methods for estimating the likely reliability of new designs and for analyzing reliability data.

The priority emphasis is important, since proactive prevention of failure is always more cost-effective than reactive correction of failure. Timely execution of appropriate reliability engineering activities is of utmost importance in achieving the required system reliability.

Modern approaches to reliability place strong emphasis on the engineering processes required to prevent failure during the expected life of a system. The concept of "design for reliability" has recently shifted the focus from a reactive "test-analyze-fix" approach to a proactive approach of designing reliability into the system. This requires that design attention be given to the early identification of potential failure modes, with subsequent mitigation actions implemented during development (i.e., reliability objective 1). Understanding of "how" (i.e., failure modes) and "why" (i.e., failure mechanisms) a system can fail is key to the achievement of reliability. In practice, this proactive approach to reliability is always complemented by a reactive approach where observed failure modes are managed and corrected (i.e., reliability objectives 2 and 3). Finally, reliability prediction, test, and demonstration play an important role during development stages (i.e., reliability objective 4).

"Design for reliability" implies that reliability should receive adequate attention during requirements analysis. Reliability requirements may be specified either in qualitative or quantitative terms, depending on the specific industry. Care should be taken with quantitative requirements, since verification by test of reliability is often not practical (especially for high reliability requirements). Also, the misuse of some reliability metrics (e.g., MTBF) frequently results in "playing the numbers game" during system development, instead of focusing on the engineering effort necessary to achieve reliability (Barnard, 2008). For example, MTBF is often used as an indicator of "expected life" of an item, which is incorrect. It is therefore recommended that other reliability metrics be used for quantitative requirements (e.g., reliability (as success probability) at a specific time, or failure-free period).

Appropriate reliability engineering activities should be selected and tailored according to the objectives of the specific project. These activities should be captured in the reliability program plan. The plan should indicate which activities will be performed, the planned timing of the activities, the level of detail required for the activities, and the parties responsible for execution of the activities. ANSI/GEIA-STD-0009 Reliability Program Standard for Systems Design, Development, and Manufacturing which supports a system life cycle approach to reliability engineering, can be referenced for this purpose. This standard addresses not only hardware and software failures but also other common failure causes (e.g., manufacturing, operator error, operator maintenance, training, quality). "At the heart of the standard is a systematic 'design-reliability-in' process, which includes three elements:

- Progressive understanding of system-level operational and environmental loads and the resulting loads and stresses that occur throughout the structure of the system.
- Progressive identification of the resulting failure modes and mechanisms.
- Aggressive mitigation of surfaced failure modes."

ANSI/GEIA-STD-0009 (2008) consists of the following objectives:

- Understand acquirer / user requirements and constraints.
- Design and redesign for reliability.
- Produce reliable systems / products.
- Monitor and assess user reliability.

The reliability program plan is often used to capture a forward-looking view on how to achieve reliability objectives. Complementary to the reliability program plan is the reliability case which provides a retrospective (and documented) view on achieved objectives during the system life cycle.

"Failure mode avoidance" approaches attempt to improve reliability of a system primarily early during development stages. It is performed by evaluating system functions, technology maturity, system architecture, redundancy, design options, etc., in terms of potential failure modes. The most significant improvements in system reliability can be achieved by avoiding physical failure modes in the first place and not by minor improvements after the system has been conceived, designed, and produced.

Reliability engineering activities can be divided into two groups: engineering analyses and tests and failure analyses. These activities are supported by various reliability management activities (e.g., design procedures, design checklists, design reviews, electronic part derating guidelines, preferred parts lists, preferred supplier lists).

Engineering analyses and tests refer to traditional design analyses and test methods performed during system development. Included in this group are Finite Element Analysis (FEA), Computational Fluid Dynamics (CFD), vibration and shock analysis, load-strength analysis, thermal analysis and measurement, electrical and mechanical stress analysis, wear-out life prediction, Accelerated Life Testing (ALT), and Highly Accelerated Life Testing (HALT).

Failure analyses refer to traditional reliability engineering analyses to improve understanding of cause-and-effect relationships. Included in this group are Failure Mode and Effects Analysis (FMEA), Fault Tree Analysis (FTA), Reliability Block Diagram (RBD) analysis, systems modeling, Monte Carlo simulation, failure data analysis, root cause analysis, and reliability growth analysis.

Availability

As part of system effectiveness, availability requirements should be carefully derived from user needs and specified during system definition. These requirements play a key role in influencing a multitude of design decisions and availability should be monitored during the utilization and support stages. The simplest definition of availability is the ratio between uptime and total time of a system, usually expressed as a percentage. Since total time consists of uptime and downtime, availability is therefore dependent on the reliability (influencing uptime) and maintainability (influencing downtime) of the system. Furthermore, downtime is obviously highly dependent on the system support environment during the support stage (influencing delay times). Due to these direct relationships, availability is governed by reliability, maintainability, and various logistics engineering aspects. Since availability is a function of both reliability and maintainability (including logistics aspects), achievement of a required availability usually requires trade-offs between reliability and maintainability, and other requirements and constraints (e.g., performance, cost).

Availability can be formally defined as "the probability that a system, when used under stated conditions, will operate satisfactorily at any point in time as required" (Blanchard, 2004). It may be expressed and defined as inherent, achieved, or operational availability:

- *Inherent availability (Ai)* is based only on the inherent reliability and maintainability of the system. It assumes an ideal support environment (e.g., readily available tools, spares, maintenance personnel) and excludes preventive maintenance, logistics delay time, and administrative delay time.
- *Achieved availability (Aa)* is similar to inherent availability, except that preventive (i.e., scheduled) maintenance is included. It excludes logistics delay time and administrative delay time.

- *Operational availability (Ao)* assumes an actual operational environment and therefore also includes logistics delay time and administrative delay time.

Inherent availability thus focusses primarily on "design for reliability and maintainability" activities. Achieved availability takes a broader view to include preventive maintenance, and operational availability includes possible logistics and administrative delays.

A service-level agreement (SLA) between a service provider and an acquirer typically includes availability performance, usually measured for a certain period (e.g., one year) and is then translated into the maximum duration of downtime allowed for that period.

Maintainability

An objective in SE is to design and develop a system that can be maintained effectively, safely, in the least amount of time, in a cost-effective manner, and with a minimum expenditure of support resources without adversely affecting the mission of that system. Maintainability refers to all measures and activities implemented during the design, production, and use of a system that reduces the required maintenance (as measured in maintenance frequency, repair hours, tools, cost, skills, and facilities). Maintainability is thus the ability of a system to be maintained, whereas maintenance constitutes a series of actions to be taken to restore or retain a system in an effective operational state. Maintainability must be inherent or "built into" the design, while maintenance is the result of design. Maintainability can formally be defined as "the ability of a system to be repaired and restored to service when maintenance is conducted by personnel using specified skill levels and prescribed procedures and resources" (Tortorella, 2015). Case 4 (Design for Maintainability-Incubators) from Section 6.4 illustrates the importance of maintainability.

Maintenance can be broken down into the following groups:
- *Corrective maintenance*: unscheduled maintenance accomplished, as a result of failure, to restore a system to a specified level of performance.
- *Preventive maintenance*: scheduled maintenance accomplished to retain a system at a specified level of performance by providing systematic inspection and servicing or preventing impending failures through periodic item replacements.
- *Predictive maintenance*: scheduled maintenance based on the in-service condition of a system to estimate when maintenance should be performed.
- *System upgrades*: periodic maintenance to support system life extension and performance upgrades.

A maintainability engineering plan is often used to capture activities such as quantitative maintainability modeling and simulation, development of the system maintenance concept, level of repair analysis (LORA), diagnostic capabilities, identification of preventive maintenance activities, etc. It is thus closely related to logistics engineering (see Section 3.1.6). The maintainability engineering plan should consider various aspects such as interchangeability of parts, accessibility to parts for removal, and testability of equipment. Testability includes aspects such as built-in test (BIT) capability, diagnostic test equipment, and support software. Service providers such as telecommunication operators that serve the mass market may use OTA (Over-the-Air) technology to remotely provide maintenance (e.g., data transfer to update software or firmware). Like reliability, maintainability requirements should be derived from system availability requirements.

Various maintainability metrics can be used to specify or measure maintainability. The most widely used metric, Mean Time to Repair (MTTR), measures the elapsed time to perform a certain maintenance activity. It typically includes time for activities such as failure detection/failure isolation (FD/FI), disassembly, active repair, reassembly, and finally system testing. It is important to note that MTTR refers to the mean time of the underlying probability distribution. Maintenance times tend to be lognormally distributed, especially for electronic systems without a built-in test capability and for many other electromechanical systems.

Relationship with Other Engineering Disciplines
As discussed in this section, RAM engineering is closely related to several other engineering disciplines. The primary objective of reliability engineering is prevention of failure. The primary objective of safety engineering is prevention and mitigation of harm under both normal and abnormal conditions (see Section 3.1.11). The primary objective of logistics engineering is the development of efficient logistics support (see Section 3.1.6). Furthermore, RAM is also related to engineering disciplines such as affordability (see Section 3.1.2), resilience engineering (see Section 3.1.9), and reusability of products in a product line (see Section 4.2.4). The life cycle cost (LCC) of a system is highly dependent on reliability and maintainability, which are considered major drivers in support resources and related in-service costs (see Section 3.1.2).

Many of these not only have "failure" as common theme, but they may also use similar activities, albeit from different viewpoints. For example, an FMEA may be applicable to reliability, safety, and logistics engineering. However, a reliability FMEA will be different to a safety or logistics FMEA, due to the different objectives. Common to all disciplines is the necessity of early implementation during the system life cycle.

More information on RAM can be found in ANSI/GEIA-STD-0009 (2008), Barnard (2008), Blanchard (2004), Clausing and Frey (2005), O'Connor and Kleyner (2012), and Tortorella (2015).

3.1.9   Resilience Engineering

Definition   Resilience Engineering is an approach that provides required capability when facing adversity.

Resilience is a relatively new term in SE, appearing in the 2006 timeframe and becoming popularized around 2010. Resilience typically subsumes survivability. The recent application of "resilience" to engineered systems has led to a proliferation of alternative definitions. While the details of definitions will continue to be discussed and debated, there is general agreement that resilience of engineered systems is the ability to provide required capability when facing adversity.

Key Concepts   System development often focuses on system capability under nominal conditions. Resilience directs the SE focus to the system's ability to deliver capability when faced with adverse conditions. This perspective can be important to stakeholders but is sometimes overlooked. Resilience in the realm of SE involves identifying:

- the capabilities that are required of the system,
- the adverse conditions under which the system is required to deliver those capabilities, and
- the architecture and design that will ensure the system can provide the required capabilities.

It is important to emphasize that resilience focuses on providing the required capability—not necessarily with maintaining the architecture or composition of the system. While system continuity is one approach to achieving resilience, so is adaptability.

Elaboration
Scope of Resilience
The fundamental objectives of resilience are avoiding, withstanding, and recovering from adversity. In non-engineering contexts, resilience is often limited to the ability to recover after degradation. In the context of engineered systems, it is recommended that "avoiding" and "withstanding" adversity be considered in scope (Jackson and Ferris, 2016). Resilience, as does SE, applies to cyber-physical, organizational, and conceptual systems.

Scope of the Adversity
For the purpose of resilience, adversity is anything that might degrade the capability provided by a system. Achieving resilience requires consideration of all sources (e.g., environmental sources, human sources, system failure) and types of adversity (e.g., from adversarial, friendly, or neutral parties; adversities that are malicious or accidental; adversities

that are expected or not). Adversities may be issues, risks, or unknown-unknowns. Adversities may arise from inside or outside the system. Adversity may be a single event or may take the form of complex causal chain of conditions and events that stress the system over multiple periods of time.

Taxonomy of Resilience Objectives
Resilience, and engineering its achievement, can be facilitated by considering a taxonomy of its objectives. A two-layer objectives-based taxonomy includes:

- First layer, the *fundamental objectives* of resilience and
- Second layer, the *means objectives* of resilience.

The layers relate by many-to-many relationships (Brtis, 2016) (Jackson and Ferris, 2013).
   Taxonomy Layer 1: Resilience can be said to equate to achieving its three *fundamental objectives*. These are:

- **Avoid**: eliminate or reduce exposure to stress.
- **Withstand**: resist capability degradation when stressed.
- **Recover**: replenish lost capability after degradation.

Taxonomy Layer 2: These fundamental objectives can be achieved through the pursuit of *means objectives*. Means objectives are not values or ends in themselves. Their value resides in their ability to help achieve resilience and its three fundamental objectives. The means objectives include:

- **Adaptive Response:** reacting appropriately and dynamically to the specific situation to limit consequences and avoid degradation of system capability.
- **Agility:** ability of a system to adapt to deliver required capability in unpredictably evolving conditions.
- **Anticipation:** establishing awareness of the nature of potential adversities, their likely consequences, and appropriate responses, prior to the adversity stressing the system.
- **Constrain:** limit the propagation of damage within the system.
- **Continuity:** ensuring the endurance of the delivery of required capability, while and after being stressed.
- **Disaggregation:** dispersing missions, functions, or system elements across multiple systems or system elements.
- **Evolution:** restructuring the system to address changes to the adversity or needs over time.
- **Graceful Degradation:** ability of the system to transition to a state that has acceptable, potentially limited capabilities.
- **Integrity:** the quality of being complete and unaltered (ISO 13008 (2022)).
- **Prepare:** developing and maintaining courses of action that address predicted or anticipated adversity.
- **Prevent:** deterring or precluding the realization of adversity.
- **Re-architect:** modifying the system architecture for improved resilience.
- **Redeploy:** restructuring resources to provide capabilities after stress.
- **Robustness:** the ability of a structure to withstand adverse and unforeseen events or consequences of human errors without being damaged (damage insensitivity) (ISO 8930 (2021)).
- **Situational Awareness**: perception of elements in the environment, and a comprehension of their meaning, and could include a projection of the future status of perceived elements and the risk associated with that status (ISO 17757 (2019)).
- **Tolerance:** the ability of a material/structure to resist failure due to the presence of flaws for a specified period of unrepaired usage (damage tolerance) (ISO 21347 (2005)).

- **Transform:** changing aspects of system behavior.
- **Understand:** developing and maintaining useful representations of required system capabilities, how those capabilities are generated, the system environment, and the potential for degradation due to adversity.

The SEBOK section on resilience provides a more extensive taxonomy of design, architecture, and operational techniques for achieving resilience.

Key Activities, Methods, and Tools

While resilience must be considered throughout the SE life cycle, it is critical that resilience be considered in the early life cycle stages: those that lead to the development of resilience requirements. Once resilience requirements are established, they can, and should, be managed along with all other requirements in the trade space through the system life cycle. As shown in Table 3.3, Brtis and McEvilley (2019) identify specific considerations that need to be included in the early life cycle activities.

TABLE 3.3   Resilience considerations

Business or Mission Analysis Process

- Defining the problem space includes identification of adversities and expectations for performance under those adversities.
- ConOps, OpsCon, and solution classes consider the ability to avoid, withstand, and recover from the adversities
- Evaluation of alterative solution classes consider the ability to deliver required capabilities under adversity

Stakeholder Needs and Requirements Definition Process

- The stakeholder set includes persons who understand potential adversities and stakeholder resilience needs.
- Identifying stakeholder needs identifies expectations for capability under adverse conditions, and degraded/alternate, but useful, modes of operation.
- Operational concept scenarios include resilience scenarios.
- Transforming stakeholder needs to stakeholder requirements includes stakeholder resilience requirements.
- Analysis of stakeholder requirements includes resilience scenarios in the adverse operational environment.

System Requirements Definition Process

- Resilience is considered in the identification of requirements.
- Achieving resilience and other adversity-driven considerations is addressed holistically.

System Architecture Definition Process

- Viewpoints selected support the representation of resilience.
- Resilience requirements significantly limit and guide the range of acceptable architectures. It is critical that resilience requirements are mature when used for architecture selection.
- Individuals developing candidate architectures are familiar with architectural techniques for achieving resilience.
- Achieving resilience and other adversity-driven considerations are addressed holistically.

Design Definition Process

- Individuals developing candidate designs are familiar with design techniques for achieving resilience.
- Achieving resilience and the other adversity-driven considerations are addressed holistically.

Risk Management Process

- Risk management is planned to handle risks and opportunities identified by resilience activities.

Content, Structure, and Development of Resilience Requirements

Brtis and McEvilley (2019) investigated the content and structure needed to specify resilience requirements. Resilience requirements often take the form of a resilience scenario. There can be many such scenario threads in the ConOps or OpsCon. The following information is often part of a resilience scenario:

- Operational concept/scenario name
- System or system element of interest
- Capability(s) of interest their metric(s) and units
- Target value(s) (required amount) of the capability(s)
- System modes of operation during the scenario (e.g., operational, training, exercise, maintenance, update)
- System states expected during the scenario
- Adversity(s) being considered, their source, and type
- Potential stresses on the system, their metrics, units, and values (Note: Stresses are a type of adversity. They are proximate forces or influences, directly affecting the system that can cause degradation of the system's ability to deliver required capability.)
- Resilience related scenario constraints (e.g., cost, schedule, policies, regulations)
- Timeframe and sub-timeframes of interest
- Resilience metric(s), units, determination method(s), and resilience metric target(s) (e.g., expected availability of required capability, maximum allowed degradation, maximum length of degradation, total delivered capability). Note: There may be multiple resilience targets (e.g., threshold, objective, As Resilient as Practicable (ARAP)).

Importantly, many of these parameters may vary over the timeframe of the scenario. Figure 3.7 notionally shows the required capability, the stress on the system, and the delivered capability as they vary as a function of time. A single resilience scenario may involve multiple stresses, which may be involved at multiple times throughout the scenario.

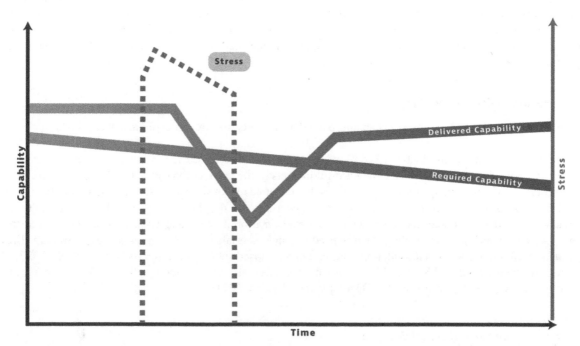

FIGURE 3.7   Timewise values of notional resilience scenario parameters. INCOSE SEH original figure created by Brtis and Cureton. Usage per the INCOSE Notices page. All other rights reserved.

Requirement Patterns for Resilience
An example of a natural language pattern for representing this information would be:

The`<system, mode(t), state(t)>`*encountering*`<adversity(t), source, type>`,*which imposes*`<stress(t),metric,units,value(t)>`*thus affecting delivery of*`<capability(t), metric, units>`*during*`<scenario timeframe, start time, end time, units>`*and under* `<scenario constraints>`,*shall achieve*`<resilience target(t) (include excluded effects)>`*for*`<resilience metric, units, determination method>`.

Here, "(t)" is used to indicate that the item may vary as a function of time.

System State Considerations
When a system encounters an adversity, the system may pass through a number of states, from a fully capable state to a state minimally acceptable to the system stakeholders. Intermediate states include damaged or partially capable states. The transitions between these states fall into three categories. The *robustness* category defines the transitions for which the system maintains its level of capability. These transitions include maintaining fully capable or partially capable states. The *tolerance* category includes passing from any higher level of capability to a lower level of capability (e.g., a degraded capability). The *recovery* category includes passing from any lower-level capability to a higher-level capability including the original fully capable state. The system should be designed applying design principles that will manage the transitions between states to result in context appropriate behavior. Guidance on techniques is provided in Jackson and Ferris (2013) and Brtis (2016).

Related Quality Characteristics
Resilience has commonality and synergy with a number of other QCs. Examples include availability, maintainability, reliability, safety, security, and sustainability. This group of quality areas are referred to as loss-driven areas because they all focus on potential losses involved in the development and use of systems (see Section 3.1.13). These areas frequently share: the assets considered, losses considered, adversities considered, requirements, and architectural, design, and process techniques. It is imperative that these areas work closely with one another and share information and decision-making in order to achieve a holistic approach that avoids unbalanced emphasis in any one area.

Further information and references on resilience (including the state-of-the-art) can be found in the resilience section of the SEBoK.

3.1.10   Sustainability Engineering

Definition   Sustainability Engineering is an approach that supports the circular economy over its life.

Key Concepts   Design for sustainability is defined as the process that considers environmental and social aspects as key elements in product design to reduce the harmful impacts of the product throughout its life cycle (Sharma, et al., 2020). It entails environmentally conscious decisions that promote responsible disposability via product recycling and materials reuse as options for the preservation of scarce material resources. Sustainability and disposability are critical components toward the circular economy, which is based on a production and consumption model that involves sharing, reusing, repairing, and recycling existing products and materials as much as possible, expanding the life cycle of products, minimizing waste and pollution, and creating a closed-loop system (Geissdoerfer, et al., 2020). These goals are consistent with the 17 Sustainable Development Goals that were adopted by all UN Member States in 2015, as part of the 2030 Agenda for Sustainable Development (Haskins, 2021).

Elaboration
Role of Sustainability and Disposability in SE
Addressing sustainability effectively in context is highly complex, requiring the integration of multiple disciplines in balancing a wide range of interdependent issues (Pearce, et al., 2012). Sustainability is essential given the significant

impacts that the SE processes, and the resulting systems, have had, and continue to have, on the environment. Achieving sustainability must include a holistic adoption of environmental stewardship in engineering activities (Alwi, et al., 2014) (Rosen, 2012).

The focus of environmental impact analysis is on potential harmful effects of a proposed system's development, production, utilization, support, and retirement stages. Concern extends over the full life cycle of the system, from the materials used and scrap waste from the production process, operation of the system, replacement parts, consumables and their packaging, to final disposal of the system.

Disposal analysis is a significant analysis area within environmental impact analysis. During System Architecture Definition and Design Definition (see Sections 2.3.5.4 and 2.3.5.5), one goal is to maximize the economic value of the residual system elements after useful life and minimize the amount of waste materials. Design for disassembly has become an important consideration in the design process so that the products are created in a way that minimizes destructive separation of system elements such that the material can be reused in future generations of products, remanufacturing, or recycling processes (Abuzied, et al., 2020). This may include designing for transformation (e.g., decomposition, biodegradation).

Key Activities, Tools, and Methods of Sustainability and Disposability
The ISO 14000 (2015) series of environmental management standards are an excellent resource for methods to analyze and assess industrial operations and their impacts on the environment. Attention to environmental regulations should be addressed in the earliest activities of requirements analysis. The Øresund Bridge (see Section 6.2.) is an example of how early analysis of potential environmental impacts ensures that measures are taken in concept, development, and production to protect the environment with positive results. Two key elements of the success of this initiative were the continual monitoring of the environmental status and the integration of environmental concerns into the requirements of the two countries.

Another effort in the ISO community is the development of a standard for Environmental Product Declarations (EPD), based on carbon footprints, as an indicator of the global environmental impact of a product expressed in carbon emission equivalents (He, et al., 2018). EPD and labeling, such as the Nordic Swan and Blue Angel, offer consumers assistance in their purchasing decisions. Methods associated with life cycle assessment (LCA), life cycle impact assessment (LCIA), life cycle optimization (LCO), and life cycle management (LCM) are increasingly sophisticated and supported by software (Avraamidou, et al., 2020).

Related QCs
Achieving a circular closed-loop system relies on integrating additional quality characteristics. Useful life extensions rely on reliability and maintainability (see Section 3.1.8) alongside efficient logistical support (see Section 3.1.6) and products designed to be resilient (see Section 3.1.9). Recovery of valuable resources after useful life is highly dependent on decisions made when considering manufacturability (see Section 3.1.7).

More information on Sustainability/Disposability can be found in the Journal of Cleaner Production (2023), the Journal of Environmental Management (2023), Wood et al. (2023), ICE (2023), and MDPI (2023).

3.1.11   System Safety Engineering

Definition   System safety engineering is an approach that reduces the likelihood of harm to people, assets, and the wider environment.

Key Concepts   The goal of system safety engineering is to reduce and mitigate hazards of systems to an acceptable level of risk. Engineered systems have safety risks; they are not 100% safe. The definition of what is acceptably safe, safety regulations, processes, and culture vary across different industries and countries.

System safety engineering is not limited to ensuring that the engineered system is acceptably safe. It includes minimizing the risks to everyone involved in the production, utilization, support, and retirement of the system, as well as third parties who could also be affected by these activities. System safety engineering is about engineering the SoI, the wider socio-technical operational system, and the socio-technical (or even purely social) management system.

Safety is an emergent property of the engineered system in its real operational environment. How the system is used, maintained, and managed can have as big an impact on system safety as its inherent design. Understanding, and aligning, the mental models of designers, operators, and managers is critical.

Safety is managed by minimizing the hazards that can lead to an accident. This is either through reducing the likelihood the hazard will occur or minimizing the impact if it does. This is either through designing the hazard out, technical mitigations in the system, or procedural controls. This requires a mixture of suitably qualified people, effective processes, appropriate governance, and culture.

In-service systems need careful monitoring to ensure that design assumptions remain valid, no new hazards have been identified, and that operations/maintenance is as expected. Slow feedback loops and misaligned mental models can be particularly problematic, as issues can grow unseen for years before they appear with catastrophic impact.

Good systems safety engineering seeks to ensure operators do not misuse systems, leaders set the right tone and culture, and maintainers don't take shortcuts. System safety engineers and senior leaders are often accountable for predictable misuse of systems, failure to address poor behavior, and ineffective oversight of maintenance and operations.

Elaboration
Acceptably Safe
The safety regulations, processes, and culture vary across different industries and countries. What is acceptable in one industry and country may not be acceptable in another. There is a wide diversity in regulators, definitions, evidence, and perceived benefits that adds further complexity. Even the definition of what is "acceptably safe" varies. Typical perceptions of "acceptably safe" include:

- *"We have complied with the necessary [product] regulations."* This is generally accepted for simple, well understood, standardized system elements (e.g., electrical cable, bolts).
- *"We have evaluated all identified hazards and have mitigated each to be 'as low as reasonably practicable' (ALARP)."* This would be the typical approach adopted for complicated, safety critical civilian systems (e.g., a railway signaling system, passenger aircraft).
- *"We have evaluated all the identified hazards of the system, and they are either ALARP, or the level of hazard of the new system is less than the alterative (of not having it)."* This would be the typical approach adopted for military, medical, or emergency response systems (e.g., artillery, pacemaker).

Eliminating all safety risk is not possible; therefore, no system can be described as 100% safe. There may be unknown hazards and hazards that cannot be eliminated but are determined to be acceptable given the perceived benefit. Similarly, assuming that the system must be safe because there haven't been any accidents yet, or reported, is equally incorrect. There may have been near misses (see below), the hazardous functionality/performance may not have been used, or the effect/damage has not yet been recognized.

Emergence, Accidents, and Hazards
Safety is an emergent property of a system. It is not the sum of system element level safety or reliability. Rather, it is impacted by interactions between system elements and affected by the environment in which it is used. Many factors affect the level of safety risk (e.g., who uses and maintains a system, how they do it, in what environment). Safety is not a static property as systems (and the surrounding ecosystems) are dynamic and evolve. The designers' expectations of risks often differ from the system under test and evaluation and from the system in operation (which continues evolves over time). The mental models of the humans interacting with the system and the related processes are also dynamic and subject to change.

Most accidents result from more than a single causal factor. When an unmitigated hazardous situation does occur but does not result in harm, it is referred to as a *near miss*. Near misses serve as critical feedback on the system safety level in operation. Safety culture often determines how near misses are treated and responded to. However:

- Accident investigations often reveal dependencies between hazards exist, despite earlier beliefs that causal factors were independent. A failure to recognize dependencies when applying statistical methods can result in flawed safety decisions.
- A hazardous situation can occur without a system element failure. This could be because of a design error, an implementation error, or a misalignment between mental models of designer and operators or maintainers.

Regulators typically assess safety risk in terms of both:

- the *likelihood* of hazards occurring and leading to harm and
- the *severity* of the resulting harm.

Safety is managed by eliminating hazards where possible; and when not possible, by reducing risks to an acceptable level. When possible, design changes to eliminate potential hazards are the preferred options. The next preferred options are design mitigations to reduce the likelihood of hazards occurring. When designs changes or mitigations are not possible, other means are typically employed such as operational controls and limitations, maintenance inspections or activities, warnings via labeling, and training.

Hazards may result from a range of sources such as intrinsic, functional, socio-technical, or management/wider culture. Intrinsic hazards are typically caused by the material, or other design factors of the system elements used in the system. Functional hazards result from incorrect, unexpected, or undesirable functions or performance of the system. Socio-technical hazards result from interactions between the physical system and its operators and the wider environment. Finally, management/cultural hazards relate to the system and the wider management controls needed to realize and sustain the system.

Examples of safety hazards include:

- Interactions between system elements, the operating environment, and operators: A car on an icy road and an inexperienced driver is likely to result in an accident. Traction control, trained drivers, or not driving in icy conditions mitigates this hazard.
- Mistakes in system/system element requirements, design, manufacturing, or installation: A failure to specify the Maneuvering Characteristics Augmentation System (MCAS) system element as safety critical resulted in loss of two 737 Max aircraft (Cantwell, 2021).
- The system creates hazards in the wider SoI: Bull-bars / kangaroo bars reduce the risk of injury to a driver in vehicle to vehicle or vehicle to large animal collisions. However, they significantly increase the risks to pedestrians in vehicle to pedestrian collisions in urban areas (Desapriya, et al., 2012).
- The inherent material used in system elements: Asbestos or flammable substances present inherent hazards.
- Incorrect operation or maintenance: A failure to properly remove old wiring led to the Clapham junction rail accident (Hidden, 1989).
- Misaligned mental models between operators, maintainers, and designers: Prior to the Smiler accident, false alarms were a common occurrence. This led to operators believing all alarms were false positives. This resulted in alarms being overridden without investigation (Kemp and O'Neil, 2018).
- The activities undertaken to design, manufacture, test and maintain the system: The Piper-Alpha oil platform accident was caused by a failure to properly manage design changes, poor maintenance management, and poor contingency planning (Cullen, 1990).

Managing and Controlling Hazards
System safety engineering seeks to control hazards by:

- Understanding the system environment, wider SoI, proposed system, and how it will be used.

- Specifying the safety requirements for the system. System requirements flow from stakeholder needs and requirements and are derived into system element requirements (see Section 2.3.5.3). Maintaining traceability is key, as it may not be immediately obvious to a system element designer that a specific requirement is safety critical (see Section 3.2.3).
- Analyzing the potential safety hazards. There is a range of hazard analysis techniques looking at the system, and its functions, physical, process, and human interactions. An effective hazard analysis will use multiple techniques.
- Mitigating/controlling the known hazards, either by reducing the likelihood or severity of a hazard occurring. Approaches include removing the hazard entirely, designing-in passive or active controls to mitigate the hazard, or including operational or maintenance controls. A key element of including controls is ensuring effective feedback loops and recognizing the full control model beyond the engineered system (including the operating environment and management systems) (Leveson, 2011).
- Establishing a safety management system to ensure the system remains safe throughout its life cycle.

Establish an Appropriate Safety Management System
Each organization's safety management system needs to be tailored based on country, region, and industry/application considerations. Organizations need to understand the regulatory, operational, and physical environments that the systems they develop will be used. For example, the safety management system needed for a safety critical industry such as rail or aerospace would be inappropriate for consumer electronics.

The safety management system needs to: define the organizations approach to developing safe systems; manage the infrastructure, processes and information required to support system delivery; oversee the operations and maintenance of in-service systems; and ensure an effective safety culture.

The safety management system needs to be fully integrated into the wider organizations' business management systems. It needs to ensure that:

- Projects to deliver new systems are given clear safety objectives, measures, and targets.
- In-service systems are operated and maintained safely, with appropriate monitoring of changes to use, environment, and material state of the system.
- Incidents and near misses are reported without fear of retribution, and on-going monitoring and implementation of mitigation actions create an ongoing learning/improvement cycle (Dekker, 2014).

A key facet of the safety management system is the organization's safety and ethical culture, or "how we behave when no-one is looking" (see Section 5.1.4). Regular measurement of the safety is useful to track organization's safety culture (Hudson, 2001). Regular, frequent communications and stories of the behavior needed are necessary by senior and influential people to reinforce the safety culture (Kemp and O'Neil, 2018).

Establish and Run Projects with Safety at Their Core
Key issues to be aware of include:

- Ensuring that the tailoring of the SE approach is appropriate for the system under development.
- Ensuring that the safety engineering processes integrated as tightly as possible to the wider engineering and business processes, driving both operational efficiencies and reducing the risk of incorrect assumptions between the safety team and the wider project.
- Accepting the need to make assumptions, but recognizing that when the assumptions turn out to be incorrect and appropriate rework will be required (and accepting the cost and time impact of the rework).
- Ensuring the selection of an appropriate life cycle model. As Akroyd-Wallis (2018) notes, direct adoption of Agile software techniques to safety critical systems may be problematic. SE Practitioners should exercise caution when employing agile for safety critical systems.

- Terminating projects when they can no longer meet their safety objectives at a reasonable cost, timescales, or performance.

Embed Safety in the Core of the SE Process
Specific safety engineering considerations include:

- Ensuring Business or Mission Analysis (see Section 2.3.5.1) includes a high-level assessment of hazards and that analysis of alternatives includes their inherent safety potential
- Including a comprehensive hazard analysis during Stakeholder Needs and Requirements Definition (see Section 2.3.5.2) to identify potential hazards with the system being developed and the wider operational capability the SoI will be deployed into
- Ensuring that key safety functions and safety performance is captured during System Architecture Definition (see Section 2.3.5.4). A safety viewpoint will help ensure traceability from high level needs down to system element requirements.
- Where the system is to be part of a wider SoS, ensuring that system functions/performance necessary to mitigate SoS hazards are captured as safety requirements (see Section 4.3.6).
- Ensuring that verification and validation of safety requirements is sufficiently rigorous to meet the agreed levels of "acceptably safe."
- Ensuring that hazards that are managed by Operational or Maintenance (see Sections 2.3.5.12 and 2.3.5.13) activities are clearly embedded in relevant processes and are clearly communicated, precisely defined, and reasonable.
- Ensuring that the assumptions in the safety case remain valid through life. (Is the environment as expected? Have the operators (and their mental models) changed? Has the use of the system changed? Is maintenance being done as required and is the maintenance as effective as planned? Are there any new potential new hazards being seen? Is the frequency and severity of hazards as expected?)

Ensure the System Is Delivered Safely
System safety engineering needs to ensure that systems can be designed, built, and verified safely. The systems that manufacture, test, and maintain complicated and complex systems can be as hazardous as the SoI itself. The traditional split between product safety and occupational safety is becoming less clear as:

- Development becomes more agile.
- Organizations shift from product to service delivery.
- Increased use of non-expert operators and maintainers.

Suitably Qualified and Experienced Personnel Drive Safety Performance
The effectiveness of safety management system is driven by the people who work within it. Good people may build safe systems despite poor processes and tools. Good processes and tools cannot make up for poorly qualified people.

Effective safety practitioners need to be numerate, critical thinkers, and system thinkers who understand the technologies being used, the environment the system will be deployed into, and the mental models of operators and maintainers. In addition, they need the influencing and persuasion skills to convince others of the right approach to take and the moral courage to "say no" when necessary (see Section 5.1.2).

Safety practitioners need to be in the room when key decisions are made. They need to lead the safety decision making, capturing key information in an audit trail. If they are involved too late, organizations can be left with systems that cannot be deployed. This happened to the UK Air Force when acquiring eight Chinook Mk3 helicopters (NAO, 2008).

For more illustrations on the importance of system safety, refer to Case 1 (Radiation Therapy - The Therac-25) from Section 6.1 and Case 5 (Artificial Intelligence in Systems Engineering – Autonomous Vehicles) from Section 6.5.

3.1.12   System Security Engineering

Definition   System Security Engineering is an approach that identifies, protects from, detects, responds to, and recovers from anomalous and disruptive events, including those in a cyber contested environment.

Key Concepts   System Security Engineering (SSE) is focused on ensuring a system can function under anomalous and disruptive events associated with misuse and malicious behavior. SSE involves a disciplined application of SE principles in analyzing security threats and vulnerabilities to the system and assessing and mitigating security risks to assets of the system during the life cycle. It blends technology, management principles and practices, and operational rules to ensure sufficient protections are available at all times.

Sources of potential anomalous and disruptive events (threats) are many and varied. They may emanate from external sources (e.g., theft, denial of service attacks, power interruptions) or may be caused by internal forces (e.g., user actions, supporting systems). A disruption may be unintentional (misuse) or intentional (malicious) in nature

Physical security protects a system from unauthorized access, misuse, or damage caused by physical actions and events such as theft, vandalism, and intrusion. Protecting physical facilities, equipment, resources, and personnel can involve the use of multiple layers of interdependent systems such as surveillance and intrusion detection systems, deterrent systems, security guards, protective barriers, locks, and access control. Hardware devices can employ anti-tampering features to detect unauthorized opening or altering of the packaging, either to ensure the content is authentic or to trigger actions to protect sensitive information in the devices.

As our world becomes increasingly digital, both hardware and software systems are increasingly at risk for disruption or damage caused by threats taking advantage of digital technologies. Integrating and implementing systems security using SSE approaches is the most efficient and effective way to ensure that security is addressed at each stage of the life cycle and becomes part of the overall SE solution instead of being done separately and isolated from other SE activities (NIST SP 800-160 Vol. 1, 2022 and NIST 800-160 Vol. 2, 2021). SSE provides the needed complementary engineering capability that extends the notion of trustworthiness to deliver trustworthy secure systems, which are less susceptible to the effects of modern adversity such as attacks orchestrated by an intelligent adversary (NIST SP 800-160 Vol. 1, 2022).

Cybersecurity generally refers to the confidentiality, integrity, and availability of information assets. Security management includes controls (e.g., policies, practices, procedures, organization structures, and software). Trustworthiness is a concept that includes privacy, reliability, resilience, safety, and security, therefore worthy of being trusted to fulfill whatever critical requirements may be needed for a particular system element, system, network, application, mission, business function, enterprise, or other entity (NIST 800-160 Vol. 2, 2021).

Elaboration   SSE practitioners should have skills, expertise, and experience in multiple areas. Examples include security requirements, security architecture views, threat assessment, networking, security technologies, hardware and software security, security test and evaluation, vulnerability assessment, penetration testing, and supply chain security risk assessment. A major challenge in managing engineering projects is unclear security roles, responsibilities, and accountability. To assist in the security role development and understanding responsibilities, an SE/SSE roles and responsibilities framework can be used to break down tasks into a matrix format that enables the SE practitioner to understand the role contributions and identify the types of artifacts created by the execution of the SE life cycle processes. (Nejib, et al., 2017)

Through NIST 800–160 Vol. 1 (2022) and Vol. 2 (2021), it has been determined that the best way to integrate cybersecurity into systems is through an SE process. NIST 800–160 Vol. 1 (2022) and Vol. 2 (2021) are based on ISO/IEC/IEEE 15288 (2023) and this handbook. They use the same terminology so that both SE and SSE practitioners can understand the key relationship that exists between the two disciplines. There is a direct correlation between SE and SSE, and SE practitioners need to understand and incorporate security components into each SE life cycle process. Table 3.4 shows an example of how the SSE technical processes in NIST SP 800–160 Vol. 1 (2022) can be reused and referenced by SE, SSE, and other disciplines practitioners. Specifically, Table 3.4 is an example of the Implementation process (see Section 2.3.5.7) breakout defined with extensions for SSE to include the purpose, outcomes, activities and

TABLE 3.4   Implementation process breakout

Implementation Process Breakout	
Purpose	• Realize the security aspects of the system element • Results in a system element that satisfies specified system security requirements, architecture, and design
Outcomes	• Security aspects of the implementation strategy are developed • Security aspects of implementation that constrain the requirements, architecture, or design are identified • Security system element • System elements securely packaged and stored • Enabling systems or services needed for security aspects of implementation • Traceability of security aspects of implemented system elements
Activities and Tasks	1. Prepare for the security aspects of implementation 2. Perform the security aspects of implementation 3. Manage results of the security aspects of implementation
Inputs	Security strategy, plan, traceability, requirements, design, architecture, secure system elements, assurance evidence, assurance results, and anomalies report
Responsible and Supporting Roles	Responsible: Systems Security Engineer (SSE) Supporting: Program Manager (PM), Chief Engineer (CE), Systems Engineer (SE), Systems Architect (SA), and Test Engineer (TE)

tasks, inputs, and responsible and supporting roles. This same format was used to breakout each of the technical SSE processes in NIST 800–160 Vol. 1 Rev. 1 (2022) and the SE Technical Processes (see Section 2.3.5) to build an understanding of the relationships between the SE and SSE processes.

Case Study 3, Cybersecurity Considerations in Systems Engineering-The Stuxnet Attack on a Cyber Physical System (see Section 6.3) provides an example of the importance of system security.

3.1.13   Loss-Driven Systems Engineering

Loss-Driven Systems Engineering (LDSE) is the value adding unification of the QCs that address the potential losses associated with developing and using systems (Brtis, 2020). SE methodologies often focus on the delivery of desired capability. As a result, SE methodologies are largely capability-driven, and may not provide integrated attention to the potential losses associated with developing and using systems. Loss and loss-driven QCs are often considered in isolation—if at all. Examples of loss-driven QCs include resilience, safety, security, sustainability/disposability, and availability.

There is significant commonality and synergy among the loss-driven QCs, which needs to be leveraged. To do this, work on the loss-driven QCs should be collaborative on:
- the adversities considered,
- the weakness, defects, flaws, exposures, hazards, and vulnerabilities considered,
- the assets and losses considered, and
- the coping mechanism considered.

Further, SE practitioners should:
- elicit, analyze, and capture loss-driven requirements as an integrated part of the overall stakeholder and system requirements development,
- make architectural and design decisions holistically across the loss-driven QC areas, and
- integrate the management of risks associated with all loss-driven areas into the project's risk management activities.

3.2   SYSTEMS ENGINEERING ANALYSES AND METHODS

Part II of this handbook provided a set of SE life cycle processes used across the system life cycle. Each process contains a set of process activities and elaborations in the context of that specific life cycle process. This section provides insight into topics, techniques, and methods that cut across the SE life cycle processes, reflecting various aspects of the concurrent, iterative, and recursive nature of SE.

3.2.1   Modeling, Analysis, and Simulation

Overview and Purpose   The INCOSE Systems Engineering Vision 2035 (2022) predicts that "The future of systems engineering is model-based, leveraging next generation modeling, simulation, and visualization environments powered by the global digital transformation, to specify, analyze, design, and verify systems. High fidelity models, advanced visualization, and highly integrated, multidisciplinary simulations will allow SE Practitioners to evaluate and assess an order of magnitude more alternative designs more quickly and thoroughly than can be done on a single design today."

The essential artifact of modeling, analysis, and simulation (MA&S) is an explicit model, an idealized representation of one or more aspects of the as-is or to-be SoI. Systems Modeling and Simulation has been defined (NAFEMS and INCOSE, 2019) as the use of interdisciplinary functional, architectural, and behavioral models (with physical, mathematical, and logical representations) for all life stages.

The terms "modeling," "analysis," and "simulation" are sometimes used interchangeably. However, they clearly refer to distinct activities. *Modeling* is the conception, creation, and refinement of models. *Analysis* is the process of systematic, reproducible examination to gain insight. *Simulation* is the process of using a model to predict and study the behavior or performance of the SoI—for aspects represented in the model. Simulation is often performed to support a particular kind of analysis, but not all analysis is performed through simulation. In the classification of models presented below, two major types of models are distinguished: *physical models*, and *digital models*. The SE discipline primarily makes use of digital models, since they provide many benefits to the SE processes in a timely and affordable manner, in particular during the early life cycle stages. Other engineering disciplines make use of both physical and digital models throughout the product life cycle. Although sometimes the term "simulation" is used in conjunction with a *physical model*, in this section *simulation* always involves a *digital model*, and any examination involving a *physical model* is always a *test*.

For effective MA&S, digital models are often parameterized to enable analysis or simulation of multiple configurations or situations with one model. Each configuration is typically defined by grouping selected parameter values in an *experiment*, see also Minsky (1965). Alternatively, the terms "analysis case," "load case," or "verification case" are used, depending on the application domain or convention. An *experiment* specifies the purpose of the analysis or simulation as well as the combination of target scenario, environment, initial and boundary conditions, and any other user-defined parameters. Figure 3.8 shows a typical workflow.

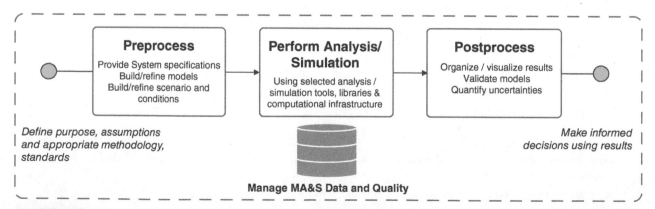

FIGURE 3.8   Schematic view of a generic MA&S process. INCOSE SEH original figure created by the NAFEMS-INCOSE Systems Modeling and Simulation Working Group (SMSWG)). Usage per the INCOSE Notices page. All other rights reserved.

MA&S links directly to the Core Competency "Systems Modeling and Analysis" in the INCOSE SE Competency Framework (INCOSE SECF, 2018). MA&S supports all Technical, Management and Integrating Competencies. Appropriate MA&S practices can clearly support SE professionals (individuals and teams) to perform better SE, more efficiently.

MA&S Related to Life Cycle Processes   Creation and refinement of descriptive models in the Business or Mission Analysis process (see Section 2.3.5.1) and the Stakeholder Needs and Requirements Definition process (see Section 2.3.5.2) can be used to ensure that the business or mission proposition is understood correctly, the problem is understood correctly, is specified at the appropriate level of detail, and is fully shared with the stakeholders. MA&S can be used in the System Requirements Definition process (see Section 2.3.5.3) to flow-down the system requirements to system elements. This may include models that specify functional, interface, performance, and physical requirements, as well as other nonfunctional requirements (e.g., reliability, maintainability, safety, and security). In addition to bounding the system design parameters, MA&S can also be used to validate that the system requirements reflect stakeholder needs and requirements before proceeding with subsequent life cycle processes.

MA&S can be used in the System Architecture Definition process (see Section 2.3.5.4) concurrently with the Design Definition process (see Section 2.3.5.5) to synthesize and define alternative system concepts, compare and evaluate candidate options, and enable discovery of the best architecture and design, including the integration with other systems and unambiguously defining the system's capabilities and the value it is expected to deliver to its stakeholders (e.g., in the form of MOEs and MOPs). MA&S is often used extensively to realize an iterative model-based or model-driven design workflow. In many application domains, it is much more cost—and schedule-effective to perform analysis and simulation with digital models than to prototype with physical models. Digital models also allow for full and continuous access to all model parameters and properties, which is often infeasible with physical models. MA&S lends itself to fast iterations between problem specification, architectural design, detailed design, and V&V, as well as between system elements at different levels of decomposition. Using MA&S in the System Analysis process (Section 2.3.5.6), related system analyses can be used to explore a trade space by modeling alternative system solutions, or even generate many candidate solutions, and assessing the impact of critical properties such as mass, speed, energy consumption, accuracy, reliability, and cost on the overall adequacy and performance.

MA&S can be used in the Implementation process (see Section 2.3.5.7) to support definition, understanding and prediction of behavior for various aspects of the enabling production (manufacturing) and supply chain processes for the envisaged SoI. Models that reflect the "as produced" state of the SoI can be used to develop production facilities (factory) and "digital twins." MA&S can be used in the Integration process (see Section 2.3.5.8) to support integration of the elements into a system, as well as in the Verification process (see Section 2.3.5.9) to support verification that the system satisfies its requirements. This often involves integrating lower-level hardware and software design models with system-level design models, thereby allowing verification that the system requirements are satisfied. Systems integration and verification may also include replacing selected hardware and design models with actual hardware and software products to incrementally verify that system requirements are satisfied: so-called hardware-in-the-loop and software-in-the-loop testing. In cases where testing is impossible or carries prohibitive cost, verification of the SoI can be done by analysis or simulation using high fidelity digital models. Models can be used to simulate relevant operational environments where actual environments are unattainable, too costly, or not reproducible. Simulation can use observed data as inputs for computation of critical parameters that are not directly observable.

In the Operation process (see Section 2.3.5.12), MA&S can support definition, understanding, and prediction of behavior for various aspects of the envisaged or actual operation of the SoI, to help train (future) users to interact with the system, and to develop training material. Models may form a basis for developing a simulator of the system with varying degrees of fidelity to represent user interaction in different usage scenarios. Models and simulators can also be used to perform dry runs to prepare for complex or risky operations with the real, deployed system. In the Maintenance process (see Section 2.3.5.13), models that reflect the "as maintained" state of deployed systems can be used to develop "digital twins," possibly for individual deployed systems. Such models can be connected to data acquisition in the field and provide valuable insight and support for health monitoring and preventive maintenance. They can also be used to plan system upgrades and evolutions. MA&S in the Disposal process (see Section 2.3.5.14) can be

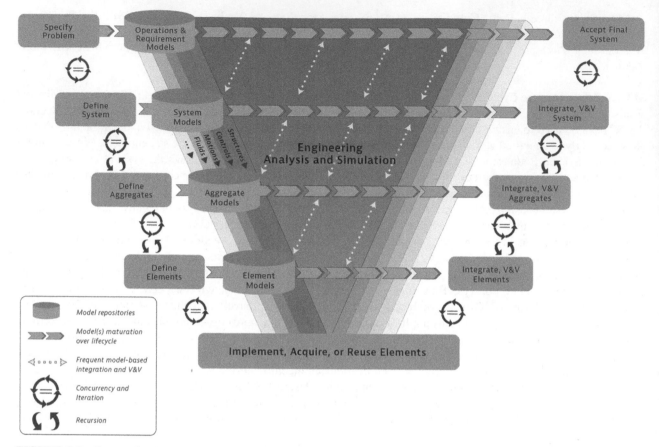

FIGURE 3.9 System development with early, iterative V&V and integration, via modeling, analysis, and simulation. Derived from NAFEMS and INCOSE (2019), based on NDIA, et al. (2011). Used with permission. All other rights reserved.

used to predict and monitor system disposal. MA&S also enables model-based iterations between the development processes mentioned above, as depicted in Figure 3.9.

Cross-Cutting MA&S   There are several cross-cutting uses of MA&S that are not tied to a particular life cycle process, including:

Characterizing an existing system—Existing systems may be poorly documented (in whole or in part). Modeling such a system can provide a concise way to capture its architecture and design. This model can then be used to facilitate maintenance of the system or its further evolution.

Knowledge transfer within teams—MA&S enables the creation and maintenance of more precise, elaborate, and consistent specifications, including the rationale behind many requirements or design choices. Capturing specifications in rich models helps to mitigate the risk of loss of knowledge in case of team changes in long-duration projects.

Automated mapping and transformation—Digital models can be transformed by declarative or procedural algorithms (i.e., automated generation of a new or modified digital model from an existing one). Model transformations are a very powerful means to increase the value of model-based engineering (e.g., convert model formulations from one modeling language to another, move from a systems architecture to a (partial) software architecture, package a model for use outside the owning organization by encapsulating and protecting its intellectual property, creation of a surrogate model from a much more detailed discipline-specific engineering model). The value of transformation increases even more when it can be made bi-directional.

Knowledge capture and system design evolution—MA&S can be an effective means for capturing knowledge about a system as part of the Knowledge Management process (see Section 2.3.3.6). Established modeling can help transform tacit into explicit organizational knowledge. MA&S in projects enables identification and capture of reusable patterns or modules from problem and solution models that have proven their worth. Catalogs of such reusable model patterns and modules can become important assets for organizational knowledge management.

Benefits   MA&S have many advantages including:

Separation of representation and presentation—Models capture representations of an SoI. MA&S tooling can then be used to maintain an "authoritative source of truth" and produce many different presentations, or views, of the model(s) that are *correct-by-construction* and enable effective communication with all engineering and non-engineering actors (humans and machines). The model-based approach can thus overcome the main problem of the document-centric approach in which representation and presentation are often combined in single information containers, which leads to a lot of information duplication and therefore cumbersome maintenance and potential errors. This is one of the most important advantages of the model-centric over the traditional document-centric approach.

More explicit SE—SE Practitioners can use MA&S to systematically check their own thinking, assumptions, and decision making with quantitative analyses. They can capture rationales and decisions in an accessible, traceable, consistent way.

Better problem specification—The stakeholder needs and requirements can be refined and formalized as an integral and structured set of goals, assumptions, requirements, constraints, actors, typical usage scenarios, and critical capabilities, with full traceability between all model elements. Anticipated system behaviors and performances can be explored and vetted with the stakeholders before proceeding with the development of an actual solution and committing significant resources.

Rigorous, well-documented design—MA&S facilitates development of solutions in a more rigorous and consistent way which leads to higher quality specifications. Systematic design space exploration and structured trade-studies become possible. Interfaces can be defined rigorously. Simulation with digital models enables design experimentation and optimization that is impossible, not affordable, or not on time with physical models. Technical and business decision-making can also be integrated into the model repositories for future consultation.

Early V&V to reduce risk—Early validation and verification of solutions with respect to the problem specification can be performed. This enables stakeholders to be informed of the implications of their preferences, provides perspective for evaluating alternatives, and builds confidence in the solution as it develops. Systematic, regular checking of interfaces and actual interconnections is feasible. It also allows catching issues early in the life cycle, when mitigation is affordable and change of scope is still feasible. The ability to detect limitations and incompatibilities early in a project helps avoid cost and schedule overruns in later life cycle stages.

Multi-user collaboration—Use of modern MA&S tooling allows for multi-user and multi-discipline collaboration with integrated configuration and version control, including splitting work into distinct parallel branches and merging results back into a main branch, using well-established workflows. Once deployed, this workflow is much more sophisticated and effective than what could be achieved with a document-centric approach.

Better change impact assessment—Modeled specifications with traceability (see Section 3.2.3) and MA&S tooling allow for change impact assessments by highlighting the consequences of a considered change.

Improved mastering of complexity—The value of MA&S increases with the complexity of the SoI, be it functional or physical. MA&S is a means to master greater complexity by structuring, refining, evaluating, and sharing all information within integrated project teams. On-demand multiple views with dynamic filtering to a suitable level of detail are possible. Quick views to share and capture information for effective communication with other actors, such as non-engineering disciplines and stakeholders, become more feasible and affordable.

Better team planning and handover—Development, deployment, and operational staff can more easily comprehend the design specifications, appreciate imposed limits from technology and management, and ensure an adequate degree of sustainability. Adequate, accurate, and timely MA&S helps an organization and its suppliers to plan and put in place the necessary and sufficient personnel, methods, tools, and infrastructure for system realization.

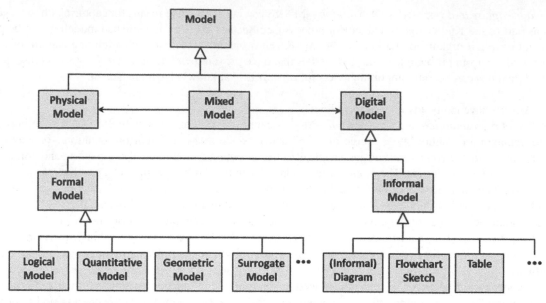

FIGURE 3.10   Illustrative model taxonomy (non-exhaustive). INCOSE SEH original figure created by the NAFEMS-INCOSE Systems Modeling and Simulation Working Group (SMSWG) derived from Friedenthal. Usage per the INCOSE Notices page. All other rights reserved.

Enable efficient maintenance and traceability—Digital models (supported with appropriate tools) enable book-keeping of SE artifacts as they evolve through the life cycle stages in a reliable, consistent, traceable, and timely manner, while also providing hyperlinked navigation.

Flexible and repeatable querying—A rigorous MA&S approach enables gaining insight into many different aspects by querying the models with pertinent questions (what-if, impact of change, etc.) and getting answers efficiently, in support of decision making. Queries can generate many different views from the modeled information that address concerns of selected stakeholders. Query formulations can also be persisted for reuse.

Classifying and Characterizing Models   There are many different kinds of models to address different system aspects and different kinds of systems. Generally, a specific type of model focuses on some subset of the system characteristics, such as timing, process behavior, measures of performance, interfaces, and connections. It is useful to classify the types of models to assist in selecting the appropriate one. Figure 3.10 shows one possible (non-exhaustive) taxonomy as an example.

Physical model—A physical model represents (aspects of) a system with real parts. Examples are a physical mockup, a scaled model airplane, a wind tunnel model, and a 3D-printed scale model from a digital model specification (the latter could be considered a physical view of a digital model). If simulation is performed with a physical model, it is typically called a *test*.

Digital model—Digital models can have many different expressions to represent (e.g., a system, entity, phenomenon, or process), each of which may vary in degrees of formalism. Therefore, the next level of classification is between informal and formal models.

Formal models—A formal model is expressed in a machine-readable language with explicitly defined semantics. The language may be textual and/or graphical, but with only one way of interpretation. Formal models can be further classified as logical, quantitative (i.e., mathematical), geometric, or surrogate models. A logical model, also referred to as a descriptive model or a conceptual model, represents logical relationships about the system such as whole–part relationships, interconnection relationships between elements, or precedence relationships between activities, to name a few. Logical models are often depicted using network graphs (with nodes and edges) or tables. A quantitative model represents quantitative relationships (e.g., mathematical equations) about the system or its elements that yield numerical results. A geometric model represents the geometry, geometric shapes, and spatial relationships of the system or any of its (physical) elements. A surrogate model is a reduced model that is derived from a higher fidelity, more detailed model using a data-driven, typically automated, transformation. The goal (and challenge) is to create a surrogate that adequately represents essential aspects of the modeled system while requiring substantially less computational resources. Surrogate models then enable running large numbers of (parameterized) experiments in order to facilitate design exploration, optimization, or validation.

Informal models—An informal model is expressed using some convention understood by humans, where the convention is defined casually without formal semantics. The model does not need to be machine-readable. An informal model can be created by hand or with simple tools (e.g., word-processing, spreadsheet, diagramming, mind-mapping). While such informal representations can be useful, they often lack the rigor to be considered a type of model that is truly usable for MA&S for non-trivial systems. Informal model presentations may be used as views that are generated from or ingested into formal models in order to communicate with people not familiar with the notation.

Mixed models—A mixed model is a combination of physical and digital models.

In addition to a selected type of model, any model can be further characterized for its intended purpose through the following three characteristics:

- The *model breadth* reflects what aspects of the SoI—and possibly its (actual or intended) environment(s)—are represented, and to what extent.
- The *model granularity* characterizes the amount of visible detail captured in the model, in terms of the represented depth of system decomposition as well as the represented level of details of individual system elements.
- The *model fidelity* indicates how accurately the model represents the real-world system. Where applicable, this includes the computational precision to be achieved and the discretization scheme to be used.

The type of model and the model characteristics must be balanced against project needs and resources. Another important aspect of modeling is to explicitly state the assumptions and limitations that almost inevitably apply to any model.

Model Interoperability   Since the development of complex systems requires collaboration between all project members and disciplines, it is very important to have the ability to exchange and share models as well as analysis and simulation results across disciplines, projects, organizations, and life cycle stages. This is also referred to as *digital interchange*. In most projects and (extended) enterprises it is not possible to standardize on a single set of tools. The alternative is to develop and utilize open, tool-independent standards that enable information exchange and sharing. There is an increasing awareness and consensus between user communities and tool developers on the merit of international royalty-free standards. Standards can be categorized in terms of how MA&S is supported. The main categories are:

- Standardized data exchange file,
- Application programming interface (API),
- Modeling language, and
- Process.

Data exchange files are used for on-demand transfer of complete models or results. APIs usually support more fine-grained data access and sharing, often implementing a service-oriented software architecture. Modeling languages can be graphical, textual or both, and are used to standardize the way of expressing a model. Process standards specify (aspects of) the MA&S processes. Most modeling languages do not prescribe a particular methodology to be followed. This flexibility is a feature of a general-purpose modeling language that enables economies of scale for implementations which are in the interest of the SE community as a whole. However, in order to align how SE practitioners in a team, organization, or application domain approach MA&S, a methodology is needed. A methodology provides guidance and examples on how to organize MA&S over a typical system life cycle, how to structure model artifacts, as well as what stages and milestones to respect. A methodology can also capture proven modeling patterns and checklists, as well as good practices in general. For further details se Section 4.2.1 or consult the OMG MBSE Wiki (2023).

Tools   For physical models, the tools are generally the same as those used for production of the final SoI, plus general or dedicated test facilities. For digital models, the tools are typically MA&S software applications running on general-purpose digital computers. For some computationally intensive applications, HPC (High Performance Computing) facilities may be needed. A MA&S software application may consist of one integrated tool or a set of tools that each implement part of the needed capabilities. The typical features of such tools are a graphical user interface with a hierarchical model structure browser, palettes of model constructs, a graphical and/or textual editor for creation and modification of the model, and multiple views for visualization, reporting, diagnostics, etc. The tool typically checks on-the-fly for adherence to the supported modeling formalism or language. If analysis or simulation support is included, there are also model execution views. For further details consult INCOSE SETDB (2021) or NAFEMS (2021).

Modeling Quality and Metrics   The quality of a model should not be confused with the quality of the design that the model represents. A perfect model can represent a bad design. On the other hand, a low-quality model can in principle represent a good design, although that is not very useful.

A completed model or simulation can be considered a system or a product in its own right. Therefore, the general steps in the development and application of a model are closely aligned to the SE processes described within this handbook. MA&S needs to be planned and tracked, just like any other developmental effort. An essential good practice is to define clearly the purpose and intended life cycle of any (type of) model upfront. In particular, verification and validation of the MA&S methods, procedures, and infra-structure themselves are essential to ensure that the resulting models, analyses, and simulations possess the required quality and credibility that make them "fit for purpose" in an application domain or project. The required rigor of the approach depends on the criticality of the SoI. As an example, the US DoD Modeling and Simulation Enterprise has developed comprehensive guidance on Verification, Validation and Accreditation (VV&A, 2021).

A valuable feature of digital models is that they are amenable to many other kinds of computation than pure analysis or simulation. This enables the assertion of many modeling metrics such as:

- compliance with design or certification rules, including naming conventions, and associated model quality requirements,
- structural consistency of the system architecture,
- compliance of system element interconnections with interface specifications,
- coverage, consistency, and completeness of traceability, such as requirements satisfaction and verification, as well as function allocation,
- consistency and completeness of logical to physical architecture mapping and allocation,
- statistics that assist in monitoring and establishing specification maturity, uncertainty quantification, and resource planning,

MA&S Industrial Practice   A big driver for the adoption of MA&S via all engineering disciplines is the trend that complex systems are becoming more and more software intensive. Analysis and simulation using digital models is much more economical and scalable than prototyping and testing with physical models, especially in the earlier stages of the life cycle. In the later stages, a mixed approach is often used. An example of the latter is an incremental hardware-in-the-loop approach in performing dynamic simulations. When it can be justified, verification through a purely digital model may be used.

Since a major responsibility of SE is to regard the system as a whole and coordinate between all disciplines in a multidisciplinary team, it follows that SE MA&S must interoperate at some level with the modeling and simulation of each of the other engineering disciplines in a team. As shown in Figure 3.11, the trend is to use an integrated system model to ensure information consistency between all engineering disciplines through a hub-and-spokes pattern, where a system model repository forms the hub.

A different way to look at multidisciplinary MA&S coordination is shown in the life cycle view presented in Figure 3.12. The information needed by two or more disciplines in a project team is shared via the integrated system model repository, which acts as the "authoritative source of truth." Within a project, there is then one authoritative repository. The one authoritative repository is a logical concept that may be implemented as a federation of distributed

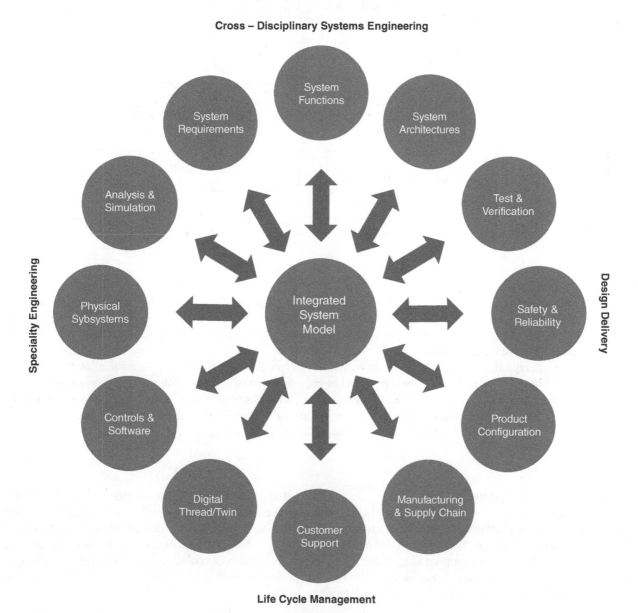

FIGURE 3.11   Model-based integration across multiple disciplines using a hub-and-spokes pattern. Derived from NAFEMS and INCOSE (2019). Used with permission. All other rights reserved.

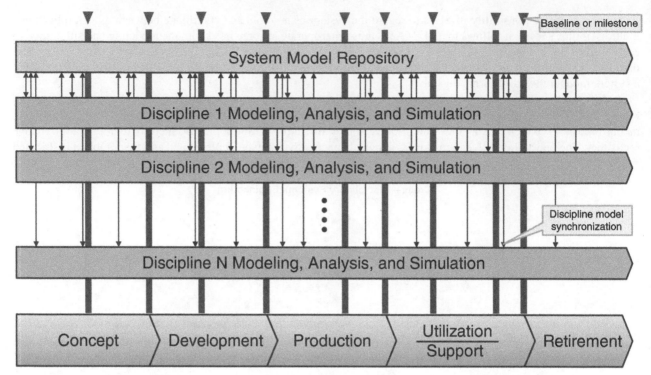

FIGURE 3.12 Multidisciplinary MA&S coordination along the life cycle. INCOSE SEH original table created by the NAFEMS-INCOSE Systems Modeling and Simulation Working Group (SMSWG). Usage per the INCOSE Notice pages. All other rights reserved.

physical repositories. The figure schematically depicts several examples of synchronizations, milestones, and baselines (in a real project there will of course be many more). In addition, MA&S data management supporting versioning, branching, merging, and archiving needs to be implemented for each of the threads, as well as across all organizations in an extended enterprise.

When different organizations collaborate in an extended enterprise, the need may arise to protect intellectual property, which naturally includes the know-how captured in models. To enable such collaboration, often so-called *black box* (also known as *opaque box*) *models* are created and maintained, which hide or obfuscate intellectual property, while still providing a publicly accessible external interface for using them to perform simulations. In contrast, *white box* (also known as *transparent box*) *models* provide full visibility of their internals.

3.2.2 Prototyping

Prototyping is a technique that can significantly enhance the likelihood of providing a system that will meet the stakeholder's needs. In addition, a prototype can facilitate both the awareness and understanding of stakeholder needs and requirements. The original use of a prototype was as the first-of-a-kind product from which all others were replicated. However, prototypes are not "the first draft" of production entities. Prototypes are intended to enhance learning and should be set aside when this purpose is achieved. Once the prototype is functioning, changes will often be made to improve performance or reduce production costs. Thus, the production entity may exhibit different behavior. Two types of prototyping are commonly used: rapid and traditional.

Rapid prototyping is an easy and one of the fastest ways to get system performance data and evaluate alternate concepts (Noorani, 2008). A rapid prototype is a particular type of physical model or simulation quickly assembled from a menu of existing physical, graphical, or mathematical elements. Examples include tools such as laser lithography or computer simulation shells. 3-D printing, or additive manufacturing, has significantly enhanced the physical elements that can be prototyped (Gebhardt and Hotte, 2016). Rapid prototypes are frequently used to investigate form and fit, human–system interface, operations, or producibility considerations. They are widely used and are particularly useful; but except in rare cases, they are not traditional "prototypes."

Traditional prototyping is a tool that can reduce risk or uncertainty and has two primary variants. A *partial prototype* is used to verify critical elements of the system. A *full prototype* is a complete representation of the system. They must be complete and accurate in the aspects of concern. Objective and quantitative data on performance times and error rates can be obtained from higher-fidelity interactive prototypes. SE practitioners are in a much better position to evaluate modifications that will be needed to develop the system because of the existence of a traditional prototype.

3.2.3 Traceability

Traceability for products and systems is defined as "the ability to trace the history, application, or location of an object/entity/item" (ISO 9000, 2015). From an SE perspective, traceability is establishing an association or relationship between two or more objects/entities/items such as life cycle concepts, needs, requirements, architectural definition artifacts (e.g., systems, system elements), design definition artifacts, verification artifacts, validation artifacts, information, models, and acquired or supplied systems or system elements.

Bidirectional traceability is the ability to trace an object/entity/item to another object/entity/item while automatically establishing a reverse link back to the initial object/entity/item. Thus, once a given object/entity/item has been linked to its source/destination, the source/destination is automatically linked to that object/entity/item. Bidirectional traceability is facilitated by SE tools which support the establishment of two-way links (bidirectional traceability) between objects/entities/items.

Vertical traceability is most often referred to in context of organization levels or architectural levels of the system or product under development. From a hierarchical architecture view (see Section 1.3.5), there are various system levels. The SoI level (Level n) has lower-level systems elements (Level n+1), some of which are further decomposed into lower-level system elements (Levels n+2, n+3, etc.) until the elements are defined to the level at which they can be made, bought, or reused. Entities at each level have objects/entities/items defined at various levels of abstraction. As the objects/entities/items are refined level-by-level, bidirectional traceability is established. Many times, these vertical traceability relations are referred to as "parent" and "child" relationships, depending upon the perspective (parent being the relationship to the higher level, child being the relationship to the lower level).

Horizontal traceability involves traceability across the elements of a given level of the architectural or system structure and across the life cycle. From a hierarchical architecture view, as relationships between objects/entities/items at the same level (i.e., Level n) are identified, bidirectional traceability is established. Many times, these horizontal traceability relations are referred to as "peer" relationships. Horizontal traceability also links objects/entities/items generated in one life cycle stage or process to data, information, and artifacts generated in other life cycle stage or process, resulting in connecting these objects/entities/items across the life cycle. For example, from a life cycle stage perspective, concept objects/entities/items are traced to development; which are traced to production; which are traced to utilization and support; which are ultimately traced to retirement. From a life cycle process perspective, a stakeholder requirement can be traced to its system requirements; which can be traced to architecture and design artifacts; which can be traced to the realized product; which can be traced to the system verification and system validation artifacts.

Establishing traceability is a critical activity of the Technical Processes (see Section 2.3.5), especially Business or Mission Analysis; Stakeholder Needs and Requirements Definition; System Requirements Definition; System Architecture Definition; Design Definition; System Analysis; Verification; and Validation. Traceability is facilitated through the appropriate application of the Configuration Management (CM) process (see Section 2.3.4.5). The CM

Identification activity enables the SE practitioner to "connect the dots" and understand the identity, location, relationships, pedigree, origin of data, materials and parts of the objects/entities/items. CM also enables the traceability of the history and location of the product after delivery. The management of products/systems, their system elements, and their configuration information requires unique identification so traceability of these items can be accurately determined. For traceability purposes the product/system identifier consists of a unique identifier which, once issued to a specific project/product/system, should never be reused.

Traceability is also a crucial component of the digital thread, enabling the connection between uniquely identified configurations of digital system models, digital twins, and physical assets (see Section 5.4). In an MBSE environment (see Section 4.2.1), the underlying system model enables a stakeholder requirement to be traced through the functional representations, to the physical product, thus enabling the identification of specific physical elements (and their specific configurations) that are impacted by a change in a given requirement. Vice versa, traceability enables the identification of the requirements that will need to be assessed when a given physical element is modified (e.g., due to a change of supplier or manufacturing process). Digitally enabled traceability methods help ensure the stakeholders get what they asked for. Digitally enabled traceability also supports the transparency of information.

More information on traceability can be found in INCOSE GtNR (2022) and the INCOSE NRM (2022).

3.2.4   Interface Management

The purpose of Interface Management is to facilitate and manage the identification, definition, design, and management of interfaces of the system across the system life cycle. It manages interface boundaries and interactions across those boundaries, the definition and agreement for each interaction, and interface requirements for all interactions identified by the various Technical Processes. Interface Management cuts across the Agreement, Technical Management, and Technical Processes. Because of its importance, the project team should focus on Interface Management as a distinct activity across all life cycle process activities.

Given that the behavior of a system is a function of the interaction of its elements and the interaction of the SoI and external systems, it is critical for the project team to identify and define each of the interactions between all system elements that make up the integrated system as well as interactions of the integrated system with external systems and users. Failing to do so will result in costly and time-consuming rework during system integration, system verification, and system validation. Because of the criticality of interfaces, the project team must define how they will manage interfaces in their project planning (e.g., SEMP). For more complex systems, projects often develop a separate Interface Management Plan. It is often useful to have the interfaces managed using an Interface Control Working Group. Additional elaboration concerning interface identification, interface definition, interface requirements, risk assessment, and managing interfaces across the life cycle is included in the INCOSE NRM (2022).

When interface management is applied as a distinct objective and focus of the SE processes, it will help highlight underlying critical issues much earlier in the project than would otherwise be revealed that could impact the project's budget, schedule, and system performance. Identifying interface boundaries and interactions across those boundaries early in the life cycle facilitates definition of the SoI's boundaries and clarifies the dependencies the SoI has on other systems and dependencies other systems have on the SoI (see Sections 1.3.1 and 1.3.3). Identifying interface boundaries and interactions across those boundaries also helps ensure compatibility between the SoI and those external systems in which it interacts. Of particular importance is the Human Machine Interface (HMI), as ultimately it is the interaction between users, operators, and maintainers that will result in acceptance of the SoI for its intended use by its intended users (see Section 3.1.4). Failure to identify all interface boundaries and interactions across those boundaries is a significant risk to the project, especially during system integration, system verification, system validation, operations, and maintenance. Because of this, it is extremely important the project defines life cycle concepts for how it will make sure the system will work safely and securely with all the external systems and personnel with which it must interact in the intended operational environment when operated by its intended users and is protected from outside threats across those interfaces.

A key characteristic of today's increasingly complex, software-intensive systems is the number of internal interactions within systems and between a system and external systems. The increased number of interactions relates directly to the complexity of a system. It greatly increases the complexity of integration of the system elements that are part of the SoI, and integration of the realized SoI within the system it is part of. It also increases the complexity of assessing the behavior of the integrated system when operated as part of a larger system. Another key characteristic of modern software-intensive systems is the form of the interactions. In the past, when many of the systems were mostly mechanical/electrical, the interactions were more visible involving connectors, wires, pipes, mechanical parts, bolts, etc. In software-intensive systems, there can be multiple computer modules, each with software that communicates commands, messages, and data across one or more communication busses. For example, in modern automobiles there can be more than 150 computer modules connected to each other and multiple sensors and actuators.

Interface Management Related to Life Cycle Processes   Major interface boundaries between the SoI and external systems are identified during preliminary life cycle concept definition activities within the Business or Mission Analysis process (see Section 2.3.5.1). Through the application of the Stakeholder Needs and Requirements Definition process (see Section 2.3.5.2) the life cycle concepts are further elaborated, interface boundaries between external systems are further refined to include all interface boundaries, and interactions across each of those boundaries are identified. Risks associated with each interface boundary and associated interactions are assessed as part of the Risk Management process (see Section 2.3.4.4). Using the System Requirements Definition process (see Section 2.3.5.3), the interactions are further refined and the characteristics of what is involved in each interaction are defined. Using this information, interface requirements are defined.

As the system architecture and system elements are defined, the System Architecture Definition process (see Section 2.3.5.4) concurrently with the System Requirements Definition process identify and define interface requirements for external systems, including enabling systems, which are also allocated to the applicable system elements. Internal interface boundaries and interactions across those boundaries are identified, and interface requirements defined for each of the interactions across the interface boundaries internal to the SoI (i.e., between system elements). The focus is on defining and agreeing on the characteristics of what is involved in the interactions, not on how those interactions are realized. The interface identification, definition, and requirements continue to evolve as the system requirements, architecture, design, and models evolve. These definitions are recorded in some form of interface control artifacts (e.g., Interface Control Document (ICD)) that are put under configuration control. For each interaction across an interface boundary, the identified interaction is input to the System Requirements Definition process to define the interface requirements.

How those interactions are realized is addressed by the Design Definition process (see Section 2.3.5.5). Definitions of interactions across interface boundaries are refined to include what each system element involved in the interaction looks like at the interface boundary and the media (e.g., a data bus, a wiring harness, a physical connection, Wi-Fi, Bluetooth) involved in the interaction is determined. Additional interface boundaries and interactions may need to be identified and defined that were not addressed by the System Architecture Definition and System Requirements Definition processes. The definition of these additional interfaces often drives additional iteration between these processes to capture the interface characteristics and requirements definition. Interactions across interface boundaries are primary considerations in both horizontal and vertical integration across the life cycle as part of the Integration Process (see Section 2.3.5.8).

A major issue concerning interface definition is that when a system element is contracted out to a supplier or the SoI interacts with other supplier-developed system elements. Often the contracts are issued prior to design and thus the design definitions of what the SoI and system elements look like at the interface boundary and the media involved in the interaction have not yet been defined. In addition, it is common for the suppliers to have little insight into the workings of other suppler-developed system elements with which they interact and how changes to those system elements could affect the interactions and performance of their system element (or changes to their system element could affect other system elements). In these cases, it is important that the acquirer clearly addresses how each supplier will support, participate in, and comply with the interface management activities during interface definition, design, system integration, system verification, and system validation in the agreements via the Agreement Processes (see Section 2.3.2).

Using the System Analysis process (see Section 2.3.5.6), the level and type of analysis needed to understand the trade space with respect to the interface requirements and definition is determined and performed. This can include mathematical analysis, modeling, simulation, experimentation, and other techniques. The analysis results are input to trade-offs made through the Decision Management process (see Section 2.3.4.3).

The Implementation process (see Section 2.3.5.7) is used to develop the system element interfaces and record evidence of meeting the interface requirements for an implemented system element.

The Integration process (see Section 2.3.5.8) considers the integration of the system and system element interfaces in the integration planning and integrates the implemented system elements together at the interface boundaries. Each system element is verified to have met their interface requirements using the Verification process (see Section 2.3.5.9). The system interfaces are validated using the Validation process (see Section 2.3.5.11) against the stakeholder needs and stakeholder requirements concerning interactions with systems or users external to the SoI in the operational environment. The Transition process (see Section 2.3.5.10) checks the installation and operational state of the interfaces in the operational environment.

Key activities that are part of interface management include facilitating cooperation and agreements with other stakeholders, defining roles and responsibilities, enabling open communication concerning issues, establishing timing for providing interface information, problem resolution, and agreeing on the interaction characteristics across interface boundaries early in the project. These functions are done through the Project Planning process (see Section 2.3.4.1). An important interface analysis activity is assessing and managing risks as part of the Risk Management process (see Section 2.3.4.4), avoiding potential impacts especially during system integration, system verification, system validation, operation, and maintenance. Other processes may also contribute to the management of the interfaces.

After establishing baselines for interface requirements, interface definitions, architecture, and design, the Configuration Management process (see Section 2.3.4.5) provides the ongoing management and control of the interface requirements and definitions, as well as any associated artifacts.

Recording Definitions of Interactions across Interface Boundaries   In a document-centric practice of SE, definitions concerning interface boundaries, the interaction across those boundaries, and the media involved are commonly recorded in some type of interface definition artifact (e.g., Interface Control Document (ICD), Data Dictionary (DD), Interface Definition Document (IDD), Interface Agreement Document (IAD)) or within the project's integrated dataset from which the associated report may be generated. In a data-centric practice of SE, these are often captured in databases and models. A data-centric practice enables effective impact and change analysis, as well as helping ensure consistency of interface requirements and definitions across the architecture.

Interface Analysis   In conjunction with the other Technical Processes, the System Analysis process (see Section 2.3.5.6) applies various analysis methods to identify interface boundaries, and interactions across those boundaries, to better understand how the SoI interacts with the other systems that make up the system of which it is a part and to help ensure there are no missing interface boundaries, definitions, and interface requirements. Some common diagrams, methods, and tools used for analysis include functional flow block diagrams (FFBD), data flow diagrams (DFD), context diagrams, boundary diagrams, external interface diagrams, input, process, output (IPO) diagrams, N^2 diagrams, and internal interface diagrams, Failure Modes and Effects Analysis (FMEA), System-Theoretic Process Analysis (STPA), language-based models (e.g., SysML diagrams), and simulations.

A critical part of interface analysis includes an assessment of each interaction across an interface boundary in terms of maturity, stability, documentation, threats, and risks. The SoI is particularly vulnerable when interfacing with external systems over which they may have little or no control. Because of this, the SoI is vulnerable to undesirable effects at and across the interface boundaries. Therefore, identifying and managing risks associated with interface boundaries and interactions across those boundaries is key to exposing potential risks to the project across applicable life cycle stages. Many of the major issues discovered during system integration, system verification, and system validation involve interfaces.

An example analysis tool is the N^2 *diagram* shown in Figure 3.13, which enables a systematic approach to identifying interface boundaries and interactions across those boundaries. N^2 diagrams enable the SE practitioner to

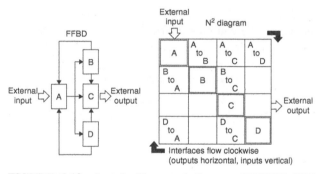

FIGURE 3.13 Sample N-squared diagram. INCOSE SEH original figure created by Krueger and Forsberg. Usage per the INCOSE Notices page. All other rights reserved.

assess and identify interface boundaries and interactions across those boundaries in a structured, bidirectional, fixed framework. The N^2 diagrams can be used at several levels of abstraction of the SoI: a functional view and a physical view.

An N^2 diagram is created using an N × N matrix. The system elements (functional or physical) are placed in squares forming a diagonal from upper left to lower right. The rest of the squares in the matrix represent potential interactions (interfaces) between the elements. In an N^2 diagram, interactions between elements flow in a clockwise direction. For example, the entity being passed from element A to element B, can be defined in the appropriate off-diagonal square. A blank square indicates there is no interaction between the respective elements. Sometimes, characteristics of the entity passing between elements may be included in the *off-diagonal* square where the interacting entity is identified. When all elements have been compared to all other elements, then the matrix is complete. If lower-level elements are identified in the process with corresponding lower-level interactions, then they can be successively described in expanded lower-level N^2 diagrams. The Design Structure Matrix (DSM) is very similar in appearance and usage to the N^2 diagram, but a different input and output convention is typically used (inputs on the horizontal rows and outputs on the vertical columns) resulting in interactions between elements flowing in a counterclockwise direction (Eppinger and Browning, 2012). Figure D-1 illustrates an N^2 diagram for the interactions amongst the system life cycle processes.

One of the main functions of the N^2 diagram, besides the identification of interactions, is to pinpoint areas where conflicts may arise between elements so that systems integration later in the development cycle can proceed efficiently (Becker, et al., 2000) (DSMC, 1983) (Lano, 1977). Alternatively, or in addition, functional and physical diagrams can be used with N^2 diagrams to characterize the flow of information among system elements and between system elements and the external systems. As the system architecture is decomposed to lower levels, it is important to ensure the interface interaction definitions keep pace and that interactions are defined so that decompositions of lower levels are considered.

Coupling matrices (a type of N^2 diagram, shown in Figure 3.14) are a basic method to define the aggregates and the order of integration (Grady, 1994). They can be used during System Architecture Definition (see Section 2.3.5.4), with

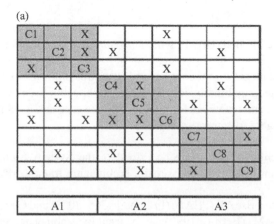

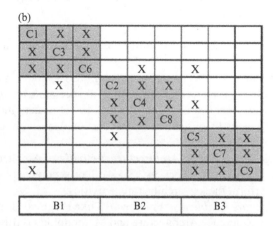

FIGURE 3.14 Sample coupling matrix showing: (a) Initial arrangement of aggregates; (b) final arrangement after reorganization. INCOSE SEH original figure created by Faisandier. Usage per the INCOSE Notices page. All other rights reserved.

the goal of keeping the interfaces as simple as possible. Simplicity of interfaces can be a distinguishing characteristic and a selection criterion between alternate architectural candidates. The coupling matrices are also useful for optimizing the aggregate definition and the verification of interfaces during Integration (see Section 2.3.5.8). Integration can be optimized by reorganizing the coupling matrix in order to group the system elements into aggregates and minimize the number of interfaces to be verified between aggregates. When verifying the interactions between aggregates, the coupling matrix can be an aid for fault detection.

3.2.5   Architecture Frameworks

An *architecture description framework* is defined in ISO/IEC/IEEE 42010 (2022) as:

> A set of "conventions, principles and practices for the description of architectures established within a specific domain of application or community of stakeholders."

The term "description" is used in the definition to avoid confusion between architecture description frameworks and other frameworks (e.g., enterprise architecture framework, architecture evaluation framework). Other definitions for *architecture framework* (AF) can be found in the technical literature, for example The Open Group Architecture Framework (OMG TOGAF, 2023) defines AF as:

> A foundational structure, or set of structures, which can be used for developing a broad range of different architectures.

Architecture frameworks are used in various domains to help ensure harmonization, consistency, and re-use. When a commonly agreed upon architecture framework is adhered to by all project teams involved, better aligned project artifacts typically result. This benefit will be particularly evident in distributed teams and within enterprises when architecture descriptions and architectural artefacts are reused across projects.

Most architecture frameworks are organized to provide one or more *viewpoints* to cover the target domains and their typical stakeholders' concerns (e.g., NATO AF (NAF), Unified AF (UAF), and Department of Defense AF (DoDAF)). Some frameworks also provide one or more of the following:

- A method for describing systems in terms of a set of architecture building blocks, and for showing how the building blocks fit together.
- A set of tools and a common vocabulary.
- Multiple dimensions with coordinates for relating particular groups of concerns or solutions along the dimensional aspects.

Figure 3.15 provides an overview of the Unified Architecture Method (UAM), which provides dimensions for perspectives and aspects.

Others advocate that architecture frameworks should include a list of recommended standards, libraries of patterns, and compliant products that can be used to accelerate architecting. Finally, it is useful for architecture frameworks, or more broadly, architecting environments, to define activities and resources for architecture governance, in addition to the governance of skills and competencies in place, with regard to enterprise objectives.

Framework Support to Architecture Activities   Architecture activities are described in the System Architecture Definition process (see Section 2.3.5.4). This section explains how some of the major architecture frameworks can be used to perform the key architecture activities.

Architecture Enablement Frameworks like Pragmatic Enterprise AF (PEAF) (Pragmatic 365, 2023) and Generalized Enterprise Reference Architecture and Methodology (GERAM) (Bernus, 1999) can be used to establish and maintain a set of capabilities, services, and resources that support the architecture process. The enablement activities include:

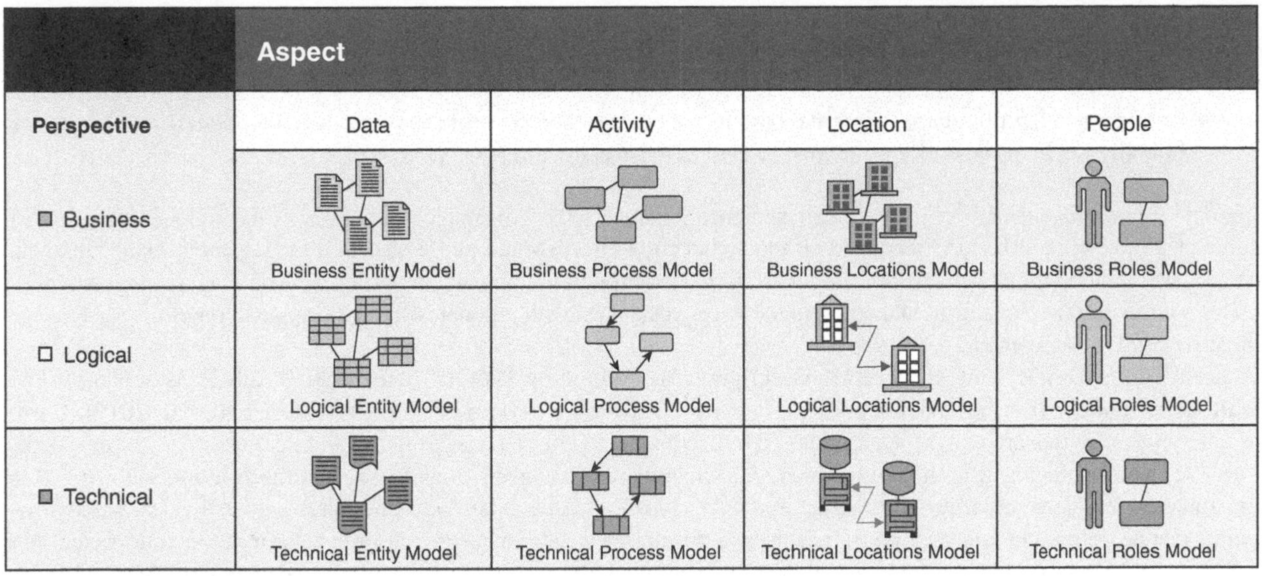

Perspective	Aspect			
	Data	Activity	Location	People
☐ Business	Business Entity Model	Business Process Model	Business Locations Model	Business Roles Model
☐ Logical	Logical Entity Model	Logical Process Model	Logical Locations Model	Logical Roles Model
☐ Technical	Technical Entity Model	Technical Process Model	Technical Locations Model	Technical Roles Model

FIGURE 3.15 Unified Architecture Method. From UAM (2022). Used with permission. All other rights reserved.

- Analysis of context in the organization where the architecture activities can take place.
- Definition of the main principles and the overall organization where the processes, methods, roles, and technologies can be used for architecting.
- Implementation of these high-level principles with methodologies provided by frameworks like TOGAF for IT domain or NAF and DoDAF for defense domains. This implementation comprises development of a metamodel to capture the terminology and a collection of architecture development methods, standards library, architecture repository and registry, and architecture capability. Architecture capability includes skills and governance logic.
- Reference documents like Evans (2014) help assess the architecture context and environment with regard to the architecting styles of the enterprise programs and projects.

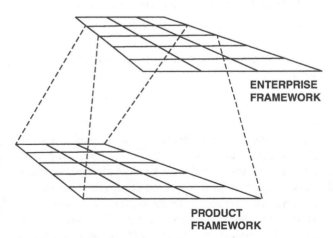

ENTERPRISE FRAMEWORK

PRODUCT FRAMEWORK

FIGURE 3.16 Enterprise and product frameworks. From Sowa and Zachman (1992). Used with permission. All other rights reserved.

Architecture Governance Related concepts are objective, goal, strategy, policy, directive, roadmap, life cycle stage, and statement of work. Per Sowa and Zachman (1992), as illustrated in Figure 3.16, two levels of architecture frameworks should be established to ensure consistency of products and systems regarding the enterprise strategy. Within most contexts, there exists a need to consider the architecture of multiple entities, each with its own life cycle, and correspondingly a framework, that helps describe or model that entity. In addition, architecting must consider that the life cycles of these entities are interrelated, often in a recursive manner (by one entity contributing to some or all activities in the life cycle of another), and that these activities may have to be synchronized (e.g., for complexity reduction purposes, to achieve or to maintain selected system quality characteristics).

Frameworks like NAF and TOGAF consider two levels of architecture governance:

- At *enterprise* level, in accordance with enterprise objectives, goals, strategy, roadmap, policies and directives.
- At *project* level, with regard to internal or external contracts of the projects of produced architecture(s) along the whole life cycle of the entity of interest related to the architecture(s).

Architecture Management Architecture management implements the governance directives in the frame of the project contract. Frameworks like NAF explain that the architecting effort should be defined in a management plan. This plan should also include the management of the work products throughout their whole life cycle. The management activities coordinate the architecting effort and report to the governance level with rationales about the application of the governance directives and the agreement.

Architecture Description Architecture Description, as defined by ISO/IEC/IEEE 42010 (2022), is accomplished with *Architecture Conceptualization* and *Architecture Elaboration*, as defined by ISO/IEC/IEEE 42020 (2019). These architecting activities are performed as planned by the management plan covering the architecture effort. Frameworks like Zachman explain how architecture viewpoints characterize the problem and represent the solution with regard to the stakeholders' concerns, possibly structured as architecture aspects and stakeholders' perspectives. These viewpoints can be defined in architecture frameworks or developed within the project for the benefit of the enterprise. ISO/IEC/IEEE 42010 defines *views* and *viewpoints* as they apply to the architecture description. Annex F of ISO/IEC/IEEE 42010 includes tables of requirements compliance for Architecture Description Frameworks. Development of project-specific viewpoint specifications needs to be justified because they imply additional effort for architecture description, evaluation, management, and usage of the work products.

Frameworks like NAF and TOGAF include a methodology to develop the architecture views governed by the viewpoints. As far as possible, this development should be based on patterns and standards already proven in the business domain where the entity of interest resides. Formalisms, model kind specifications, and modeling languages are typically defined in AFs.

Architecture Evaluation As defined by ISO/IEC/IEEE 42030 (2019), the evaluation activities determine the extent to which one or more architectures meet their objectives, address stakeholder concerns, and meet relevant requirements. These activities are performed as planned by the management plan covering the architecture effort. Frameworks like Architecture Tradeoff Analysis Method (ATAM) (CMU/SEI, 2000) and Method Framework for Engineering System Architectures (MFESA) (Firesmith, et al. 2008) allow performing architecture evaluation in three steps:

- Definition of the objectives and evaluation criteria agreed by the stakeholders to cover their concerns.
- Definition or development of a method to cover the activities normally structured in analysis, assessment, and evaluation tasks.
- Analysis of the architecture concepts and properties, assessment of the value and utility for the stakeholders, and formulation of findings and recommendations in evaluation reports.

3.2.6 Patterns

Introduction to Patterns The scientific disciplines, whether concerned with phenomena at a molecular, global, or astronomical scale, are based upon discovery and effective modeling of *patterns*. Patterns are recurrences—repeated regularities observed across time, space, or other dimensions. Patterns lie at the heart of physical sciences and the related engineering disciplines, as laws of nature whose mathematical representation and engineering exploitation have transformed the nature and possibilities of human life. In SE, recurring patterns are observable in engineered system requirements, solution architectures, stakeholder value, missions, fitness and trade spaces, parametric couplings, failure modes and risks, markets, system phenomena, principles, and the socio-technical systems of engineering and life cycle management. For example, there are patterns for requirements for refrigerators, patterns for design of coolant compressors, patterns for refrigerant failures, and patterns for maintaining refrigerators. Patterns are visible for products developed for commercial markets and systems engineered under defense contracts, as well as for the socio-technical systems that produce them, such as methodology patterns for eliciting and validating requirements. Whether the patterns are only implicitly and informally recognized and used, or explicit and formal, they can be found across the System of Innovation Pattern shown in Figure 1.6, where they are the basis of group learning. Explicitly modeled patterns help us surface and more efficiently share (learn, teach, practice) what earlier generations of SE practitioners treated as expertise and intuition only obtainable over decades of personal practice. As in the physical sciences, engineering patterns of all kinds are also subject to issues of credibility, validity, applicability, and trust as a basis for decision-making and action. Patterns are not "one size fits all", but instead have both fixed (recurring) and variable (parameterized) aspects, distilled by abstraction across individual instances. Depending on how they are recognized, represented, managed, and applied, patterns may be *informal* or *formal*.

Informal Patterns   The most informally described patterns are those implicit in the expertise or judgement of individual practitioners and teams (as in tribal knowledge), when subject matter experts recognize new occurrences of past experiences. Examples are Jean's expertise in packaging systems, or Jose's expertise in risk assessment. Because of the high value of this experience and interest in making it available to others, historical efforts have been made to explicitly capture and record such patterns, even in informal form, so that they can be transmitted to others. SE *principles* and *heuristics*, often captured as informal prose, illustrate such explicit but informal patterns (see Sections 1.4.3 and 1.4.4). The informal but explicit prose representation of engineering patterns has created popular followings in civil and software engineering communities of practice (Alexander, et al., 1977) (Gamma, et al., 1995). These patterns typically include a prose template description of a problem and an informal description of a design pattern suited to such a problem. Examples of these explicit, informal, but effective patterns include building structural patterns and city layout patterns (in civil architecture patterns), as well as sorting algorithms and graphic user interface designs (in software design patterns). SE practitioners and leaders should not underestimate the value of explicit informal patterns for transmitting knowledge.

Formal Patterns   The sciences' transition from informal prose to formal models powered much of the Science, Technology, Engineering, and Math (STEM) revolution's transformative impact, where model-based representations of patterns are the heart of the related physical sciences. These models have also enabled several generations of powerful automation tools for design, simulation, and production across the engineering disciplines, and more recently this is also impacting SE.

The practice of SE has increased use of explicit formal system descriptive models as central to SE methods, described as Model-Based Systems Engineering (MBSE) (see Section 4.2.1). This also enables the shift to formal model-based representation of patterns and their application in SE, because patterns based in models can be readily transformed (including automated assistance) into configured models specific to an application or project. Likewise, such patterns can be used in automated conformance-checking of other models. Provided the credibility of the patterns for the uses intended is managed, this not only shortens time to a trustable specific model, it also helps shift the language and perspective of multiple systems practitioners and teams into common semantic frameworks specific to a domain or specialty, for improved compatibility and interoperability. For example, do designers of tractors and trailers have a common perspective on the interactions between these engineered products? Can their work be readily checked for consistency? These issues have major impacts on SE effectiveness and productivity.

Formal patterns, particularly when model-based, appear under different names and "flavors" across SE practice and this handbook. Among these are ontologies, Architectural Frameworks, schemas, and Product Line Engineering (PLE) datasets. For more on these, refer to INCOSE S*Patterns Primer (2022) and the other sections of this handbook. Formal patterns also include general and domain-specific system modeling languages.

The power of models in the STEM revolution was not simply that they reflected agreements across groups (as in standards), but also agreements with observed natural phenomena, reduced to simplest form in the patterns of physical laws. These phenomena-based patterns continue to provide the theoretical basis for the individual engineering disciplines, as well as for the foundations of SE (Schindel, 2016, 2020). The central question they address is: What is the *smallest* system model content necessary to represent a SoI, across its life cycle, for purposes of engineering and life cycle management (Schindel, 2011)? This question has practical implications, but is also rooted in the foundations of SE:

- The *practitioner* has an interest in keeping things as simple as possible, but not simpler. "Too large" a model implies the burden of more information than is needed, including redundancies which often include inconsistencies. "Too small" a model implies that information needed during the life cycle is missing.
- *Foundations of an engineering discipline* include representing recurring phenomena fundamental to its corresponding science. The smallest set of elements generating a discipline identifies its foundations (e.g., Newton's Laws generating Mechanics, Maxwells' Equations generating Electrical Science). A definition of a system's mathematical complexity is the size of its smallest generating representation (Li and Vitanyi, 2009).

The *Systematica Metamodel (S*Metamodel)* is a formal pattern describing a neutral (independent of specific modeling languages or tools) answer to the above "smallest model" question, mapped into contemporary model tooling and languages, such as SysML, simulations, or modeling frameworks. An *S*Model* is any model, expressed in any modeling language or tooling, that is mapped to the reference S*Metamodel. The S*Metamodel spans disciplines, tooling, and languages, and is rooted in the phenomena-based models of the physical sciences.

Modern word processing tools are powerful, but varying writer composition skills and practices allow authoring that may produce valuable literature or faulty descriptions and broken semantics. Similarly, observed methods of use of contemporary modeling languages and automated tools allow the generation of system models that are both too small (are missing important elements) and too large (contain undetected redundancies and contradictions) at the same time. Fortunately for formal models, the history of the physical sciences provides patterns about the nature of phenomena and their models, and these can guide the users of contemporary tools and languages to more effective models than bare languages and tools alone. Accordingly, the S*Metamodel provides that guidance in any language or tooling into which it is mapped. Three examples from the S*Metamodel are:

- *All behavior is interaction-based*: Physics has made it clear that there is no "naked" behavior in the absence of interactions, although system modelers sometimes create models that incorrectly assert otherwise. Interactions are the heart of system phenomena, emergence, SE, and S*Models. Failure to understand and represent interactions leads to well-known engineering problems such as overlooking the impact of "external" actor behavior on the performance of an in-service engineered system (Schindel, 2013, 2016).
- *Requirement statements are transfer functions*: Models can help make it clear that requirement statements are not simply prose, but always represent input-output relationships parameterized by state variables. S*Models make this clear and enable improved auditing for problematic or overlooked requirements (Schindel, 2005).
- Stakeholder value trade space, failure effects in risk analysis, and configurability of product line families are all manifestations of the same variables: These are frequently treated as relatively independent specialties and dimensions, greatly over-complicating system representation and understanding, when that apparent dimensionality can be substantially reduced by S*Models (Schindel, 2010).

Other aspects of the S*Metamodel, and the S*Models generated from it, are described in Schindel (2011), INCOSE S*MBSE (2022), and INCOSE S*Patterns Primer (2022).

S*Patterns   S*Patterns are reusable S*Models of families of systems, often domain-specific, configurable to represent multiple individual applications, market segments, or other configurations (Schindel, 2022a). There are also more generic S*Patterns, such as the S*Metamodel itself (INCOSE S*MBSE, 2022), the System Innovation Ecosystem Pattern introduced in Figure 1.6 (Schindel and Dove, 2016) (INCOSE Innovation Ecosystems, 2022), and the Model Characterization Pattern used to generate requirements and metadata for unified characterization of virtual models of all types (INCOSE S*MCP, 2019). Pattern-based MBSE using S*Patterns involves authoring of system patterns and their configuration to application and project-specific S*Model instances, as summarized by Figure 3.17. Part of the "minimality" of the S*Metamodel is its sufficiency for such representations, including configuration rules. Instructional examples of system pattern representations may be found in Schindel and Peterson (2013). System patterns have been used in automotive, heavy equipment, aerospace, medical device, diesel and gas turbine engines, advanced manufacturing, consumer product, cybersecurity, and other domains (Bradley, et al., 2010) (Cook and Schindel, 2015) (Schindel and Smith, 2002) (Schindel, 2012).

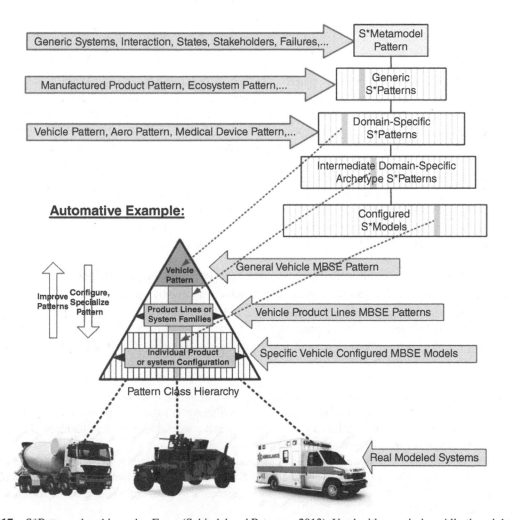

FIGURE 3.17   S*Pattern class hierarchy. From (Schindel and Peterson, 2013). Used with permission. All other rights reserved.

There is more to pattern-based methods than just representing the patterns. Historical descriptions of SE processes can appear to describe all the processes practitioners ought to perform in order to discover, validate, and utilize all the information the system life cycle requires. However, those descriptions have by volume had less to say on the question of "what about what we *already* know?" Such descriptions might be viewed as relying on practitioners to separately work out informal means of exploiting existing knowledge within what the process specifies. To address such questions, the System Innovation Ecosystem Pattern shown in Figure 1.6 describes the curation and mixing of information believed already credible with required new information extraction and validation. Schindel and Dove (2016), INCOSE S*Patterns Primer (2022) and others describe multiple additional levels of detail decomposition of processes, information, ecosystem capabilities, and limitations. Those details show how life cycle processes of ISO/IEC/IEEE 15288 (2023) and this handbook are incorporated to manage group learning and controlled sharing, and especially pattern credibility and uncertainty, across multiple programs of an enterprise, supply chain, or industry group. The System Innovation Ecosystem Pattern is further concerned with the effective linkage between the processes of pattern learning, validation, and curation versus the execution processes of making use of the content of trusted patterns—often by different people, at different times, in different places or organizations.

3.2.7   Design Thinking

Understanding and leveraging the technical, business, and social relationships to successfully design and manage engineered systems is still a challenge in SE practice. SE solutions to this challenge tend to focus on system components, human activities, machine functionality, and human-system integration . Solution design can take advantage of Design Thinking (Dorst, 2015) as a complementary approach to Systems Thinking (see Section 1.5). Design Thinking explores (1) the human needs, (2) the operational and business processes and reasoning by which design concepts are devised and realized, especially those which are creative in nature, together with (3) the systems being realized, (4) its specialization, and (5) their utilities and value provided for the stakeholders.

In a Design Thinking process (Cross, 2000), (Lawson, 1997), context analysis and problem framing techniques are employed to identify all relevant influences on a problem, explore the given problem, and restructure or revise it to suggest a route to a solution. Solution generation techniques, including approaches to idea generation (ideation), are employed to identify a range of possible design solutions which are based on:

- existing known solutions, possibly in the form of variants, patterns or other adaptations;
- applying different forms of design-related reasoning to achieve innovative solutions;
- iterating between decomposing functional requirements and design solutions to achieve optimal design – see, for example, Axiomatic Design (Suh, 2001); and,
- using successive divergent and convergent phases of design synthesis and analysis with respect to the value provided for the stakeholders resulting from business and operational processes.

Design Thinking enables SE practitioners and other team members to understand the stakeholders, challenge assumptions, redefine problems, and realize innovative solutions by drawing upon logic, imagination, intuition, and systemic reasoning. Design Thinking can also be utilized for anticipating and addressing emergent features of systems, and in technical management and organization of engineering processes.

As Design Thinking approaches use solution-based methods, they can be used in various system life cycle stages. Examples are to support business or mission analysis (see Section 2.3.5.1), to identify and validate stakeholder or system requirements (see Sections 2.3.5.2 and 2.3.5.3), or to define the system architecture or its design solution (see Sections 2.3.5.4 and 2.3.5.5).

FIGURE 3.18 Examples of natural systems applications and biomimicry. INCOSE SEH original figure created by McNamara and Anway derived from Studor (2016) and Hoeller, et al. (2016) using NASA images. Usage per the INCOSE Notices page. All other rights reserved.

3.2.8 Biomimicry

Definition   *Natural systems* include living and non-living systems—anything that is not human-made. Natural systems differ from engineered systems (see Section 1.1), which are the primary focus of this handbook.

"*Biomimicry* is a practice that learns from and mimics the strategies found in nature to solve human design challenges—and find hope" (Biomimicry Institute, 2022).

Purpose   *Nature inspired SE* and biomimicry can improve processes, practices, and products through the understanding of how nature is structured, behaves, adapts, interacts, accomplishes functions, and recovers from disturbance. Applying natural systems thinking and engineering can improve system capability, efficiency, and performance, while benefiting operations, support, and the effects on external environments. Examples include optimized information processing and sensing, operation in extreme environments, innovative materials application, distributed architectures, understanding of how emergence arises, lowered environmental impact, and system resilience. Nature has strategies to improve performance in all these areas, including circular approaches to materials and energy. To utilize nature-inspired solutions, SE looks to a universal solution space and asks regularly, "Can nature help me solve this problem?" and, "How can nature help me improve my SoI, product, or process?"

Examples   Examples of successful natural systems applications and biomimicry abound. Select examples are shown in Figure 3.18: Velcro® inspired by burdock (Velcro, 2023); an impeller inspired by the calla lily and nautilus shell (Pax Water Technologies, 2022); grippers inspired by the gecko (NASA JPL, 2013, 2014, 2015); and a sensor inspired by an insect's compound eye (Frost, et al, 2016).

Description   Over time, natural systems have developed a very close fit to their surroundings and other systems. The result is that they exhibit optimized attributes that often exceed the performance of engineered systems. In addition, they often have positive impacts on the environment. The study of natural systems includes forms, structures, materials, behaviors, processes, regenerative strategies, and interactions. Studying natural systems will increase an SE practitioners' repertoire of solutions, architectural variations, and strategies. The SE practitioner on a project is ideally suited to explore opportunities for application of natural systems across all life cycle stages.

To develop natural systems solutions, the SE practitioner uses a systematic process that:
- Begins by being open to alternate solutions;
- Defines requirements in terms of abstract functions or goals, including specific relevant metrics whenever practical;

- In the early stages of solution exploration, uses the abstracted functions to search for and identify multiple natural systems that could satisfy the desired function and examines characteristics of each;
- Selects one or more candidate natural systems;
- Abstracts the strategy that accomplishes the function in nature;
- Explores architectural variations that translates the strategy and generates alternate system element alternatives;
- Transfers the strategy to the SoI;
- Evaluates system element performance at the system level; and
- Evaluates the environmental impact of the system production, operation, support, and retirement.

Partnering with and supporting natural systems scientists can be essential to a successful implementation. An SE team gains from the in-depth knowledge provided by a cross-disciplinary team.

For more information, see INCOSE NS Primer (2023).

4

TAILORING AND APPLICATION CONSIDERATIONS

This section provides considerations for the application of SE with respect to different methodologies, approaches, system types, product sectors, and application domains.

4.1 TAILORING CONSIDERATIONS

There are many standards and handbooks that address life cycle models and SE processes. However, in most cases, these cannot be directly applied to a given organization or project. There is usually a need to tailor them for the specific project, organization, environment, or other situational factors.

The principle behind tailoring is to adapt the processes to ensure that they meet the needs of an organization or a project while being scaled to the level of rigor that allows the system life cycle activities to be performed with an acceptable level of risk. In general, all system life cycle processes can be applied during all stages of the system life cycle, tailoring determines the process level that applies to each stage. Additionally, processes are applied iteratively, recursively, and concurrently as shown in Figure 2.10.

Tailoring scales the rigor of application to an appropriate level based on risk. Figure 4.1 is a notional graph for balancing formal process against the risk of cost and schedule overruns (Salter, 2003). Insufficient SE effort is generally accompanied by high risk of schedule and cost overruns. If too little rigor is applied, the risk of technical issues increases. However, as illustrated in Figure 4.1, too much formal process may also lead to increased cost. If too much rigor is applied or unnecessary process activities or tasks are performed, the risk of cost and schedule slips increases. Tailoring occurs dynamically over the system life cycle depending on risk and the situational environment. Therefore, it should be continually monitored and adjusted as needed.

This section describes the process of tailoring the system life cycle models and SE processes to meet organization and project needs.

INCOSE Systems Engineering Handbook: A Guide for System Life Cycle Processes and Activities, Fifth Edition.
Edited by David D. Walden, Thomas M. Shortell, Garry J. Roedler, Bernardo A. Delicado, Odile Mornas, Yip Yew-Seng, and David Endler.
© 2023 John Wiley & Sons Ltd. Published 2023 by John Wiley & Sons Ltd.

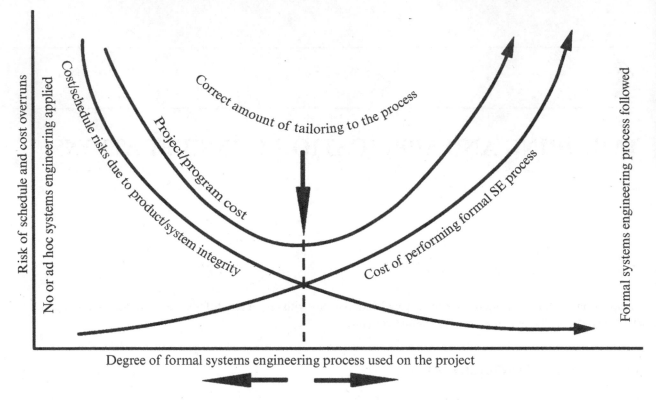

FIGURE 4.1   Tailoring requires balance between risk and process. INCOSE SEH original figure created by Krueger adapted from Salter (2003). Usage per the INCOSE Notices page. All other rights reserved.

Tailoring Process

Overview

Purpose As stated in ISO/IEC/IEEE 15288,
[A.2.1] The purpose of the Tailoring process is to adapt the processes of ISO/IEC/IEEE 15288 (and of this handbook) to satisfy particular circumstances or factors that:

a) Surround an organization that is employing ISO/IEC/IEEE 15288 in an agreement;
b) Influence a project that is required to meet an agreement in which ISO/IEC/IEEE 15288 is referenced;
c) Reflect the needs of an organization in order to supply products or services.

Description At the organization level, the tailoring process adapts external standards in the context of the organizational processes to meet the needs of the organization. At the project level, the tailoring process adapts organizational processes for the unique needs of the project.

Inputs/Outputs Inputs and outputs for the Tailoring process are listed in Figure 4.2. Descriptions of each input and output are provided in Appendix E.

Process Activities The Tailoring process includes the following activities:

- *Identify and record the circumstances that influence tailoring.*
  - Identify the strategic, programmatic, and technical risks for the organization or project.
  - Identify the level of novel concepts or the complexity of the system solution.

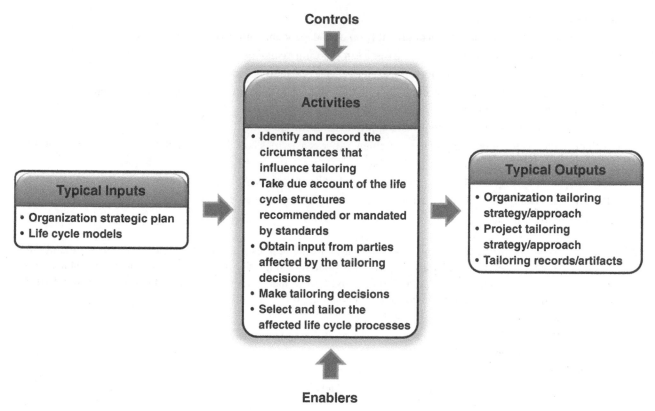

FIGURE 4.2   IPO diagram for Tailoring process. INCOSE SEH original figure created by Shortell, Walden, and Yip. Usage per the INCOSE Notices page. All other rights reserved.

- Identify administrative effects (e.g., geographic distribution, organizational distribution, team size) that may impact tailoring.
- Record the relative importance of circumstances that influence tailoring.
- Identify tailoring criteria for each stage.
- *Take due account of the life cycle structures recommended or mandated by standards.*
  - Identify relevant standards.
  - Evaluate impact on tailoring and implementation across the life cycle.
- *Obtain input from parties affected by the tailoring decisions.*
  - Determine the level of expertise and experience of project members for the processes.
  - Determine the expectations or requirements from stakeholders.
  - Determine the risk tolerance of the stakeholders.
- *Make tailoring decisions.*
  - Assess candidate life cycle models.
  - Record assessment results.
- *Select and tailor the affected life cycle processes.*
  - Capture and maintain rationale for selected life cycle processes.
  - Establish means to continuously evaluate performance of tailored processes.

Note: Tailoring can include the deletion, modification, or addition of outcomes, activities, tasks, typical inputs, or typical outputs.

Common approaches and tips:

- Base decisions on facts and obtain approval from an independent authority.
- Use the Decision Management process to assist in tailoring decisions.
- Constrain the tailoring based on agreements between organizations.
- Control the extent of tailoring based on issues of compliance to stakeholder, customer, and organization policies, objectives, and legal requirements.
- Have different organizations propose and approve the tailoring.
- Influence the extent of tailoring of the agreement process activities based on the methods of procurement or intellectual property.
- Remove extra activities as the level of trust builds between parties.
- Identify the assumptions and criteria for tailoring throughout the life cycle to optimize the use of formal processes.
- Document the rationale for tailoring decisions.

Elaboration *Organizational Tailoring* Organizational tailoring applies specifically to the creating and maintaining organizational-level processes used by all projects. It is done in conjunction with the Life Cycle Model Management process (see Section 2.3.3.1). When contemplating if and how to incorporate a new or updated external standard into an organization, the following should be considered (Walden, 2007):

- Understand the organization;
- Understand the new standard;
- Adapt the standard to the organization (not vice versa);
- Institutionalize standards compliance at the "right" level;
- Allow for tailoring.

Project Tailoring Project tailoring applies specifically to the work executed through projects. It is done in conjunction with the Project Planning process (see Section 2.3.4.1). Factors that influence tailoring at the project level typically include, but are not limited to:

- Stakeholders and acquirers (e.g., number of stakeholders, quality of working relationships);
- Project budget, schedule, and requirements;
- Risk tolerance;
- Complexity and precedence of the system;
- The need for horizontal and vertical integration (see Section 2.3.5.8).

As mentioned in SE principle #12 (see Section 1.4.3), complex systems are engineered by complex organizations. Consequently, today's systems are more often jointly developed by many different organizations. Cooperation must transcend the boundaries of any one organization. Harmony between multiple organizations is often best maintained by agreeing to follow a set of consistent processes, methods, and tools.

Traps in Tailoring Common traps in the tailoring process include, but are not limited to, the following:

- Reuse of a tailored baseline from another system without repeating the tailoring process;
- Using all processes and activities "just to be safe";
- Assuming there is a single set of measures, risks, or other controls that apply to all projects without tailoring;
- Using a pre-established tailored baseline;
- Failure to include relevant stakeholders.

Tailoring for Very Small Enterprises The ISO/IEC/IEEE 29110 series defines Very Small Enterprises (VSEs) as enterprises, organizations, departments, or projects with up to 25 people. In many cases, VSEs find it difficult to apply international standards to their business needs and to justify the application of standards to their business practices. Typical VSEs do not have a comprehensive infrastructure, and the limited personnel usually are performing multiple roles. This may also happen in a large organization when the task is to perform a small project with less than 25 people involved. In this case, it can be extremely challenging to downscale the organization's life cycle model that is designed for much larger projects.

The ISO/IEC/IEEE 29110 series defines guides for VSEs based on a set of VSE characteristics (e.g., business models, situational factors, risk levels). From that, four profiles were derived:

- *Entry* (less than 6 people or start-ups);
- *Basic* (single application by a single work team);
- *Intermediate* (more than one project in parallel with more than one work team); and
- *Advanced* (for VSEs that want to sustain and grow as an independent competitive system developer).

These profiles cover the needs of most VSEs. Each of the profiles defines subsets of international standards (e.g., ISO/IEC/IEEE 15288) relevant to the VSE's respective context. For critical projects, such as mission critical or safety critical, these profiles do not apply, since the criticality of the projects would dictate a much greater level of rigor and comprehensive SE.

4.2   SE METHODOLOGY/APPROACH CONSIDERATIONS

The system definition activities, especially the partitioning of the system, requires that integration is considered throughout the development stage of the life cycle. The integration considerations may also require refinement based on when and how the work is performed. The choice of SE methodology or approach, along with the chosen life cycle model, often affects the sequence of work, which can help determine the focus of resources to address the unique challenges of the project.

This section introduces considerations for the following SE methodologies and approaches:

- Model-Based Systems Engineering (MBSE);
- Agile Systems Engineering;
- Lean Systems Engineering;
- Product Line Engineering (PLE).

Note that other types of SE methodologies and approaches exist.

4.2.1   Model-Based SE

This section provides an overview of the Model-Based Systems Engineering (MBSE) approach and includes a summary of its benefits relative to a more document-based approach. It also references a set of MBSE methodologies and provides a brief description of one representative methodology called the Object-Oriented Systems Engineering Method (OOSEM).

MBSE Overview   The *INCOSE Systems Engineering Vision 2020* (2007) defines MBSE as:

The formalized application of modeling to support system requirements, design, analysis, verification, and validation activities beginning in the [concept stage] and continuing throughout development and later life cycle [stages].

MBSE is often contrasted with a document-based approach to SE. In a document-based SE approach, there is often considerable information generated about the system that is contained in documents and other artifacts such as specifications, interface control documents, system description documents, trade studies, analysis reports, and verification plans, procedures, and reports. The information contained within these documents is often difficult to synchronize and maintain, and difficult to assess in terms of its quality (correctness, completeness, and consistency). Although many systems have been developed using a traditional document-based approach, a model-based approach is becoming essential to address the increasing complexity of systems and support approaches that can more effectively and efficiently adapt to requirements and design changes.

MBSE enhances the ability to capture, analyze, share, and manage the information associated with the specification of a product that can result in the benefits listed below. There is some quantitative data (Rogers and Mitchell, 2021) and considerable qualitative data (OMG MBSE Wiki, 2023, MBSE Events and Related Meetings) from industry papers and presentations that support the following benefits of MBSE:

- *Improved communications* among the development stakeholders (e.g., the acquirer, project management, SE practitioners, hardware and software developers, testers, quality characteristic disciplines).
- *Increased ability to manage system complexity* by enabling the system to be viewed from multiple perspectives.
- *Improved product quality* by providing an unambiguous and precise model of the system that can be evaluated for consistency, correctness, and completeness.
- *Reduced cycle time* by enabling better control of the technical baseline, more rapid impact analysis, improved specification and design reuse, early insight for design decisions, and early discovery of potential defects.
- *Reduced risk by surfacing requirements and design issues early.*
- *Enhanced knowledge capture and reuse* of the information by capturing information in more standardized ways and reducing redundancy of information.
- *Improved ability to teach and learn SE fundamentals* by providing a clear and unambiguous representation of systems and system concepts.

MBSE Methodologies   In an MBSE approach, much of the information that has been traditionally captured in informal diagrams, text, and tables is captured in a *descriptive system model* (see Section 3.2.1). This includes information about the system context, the requirements on the system and its elements, the system architecture including its structure and behavior, the critical parameters needed to specify the analysis of the system, and information about how the system is verified to satisfy its requirements. Modeling languages such as SysML™ (OMG SysML, 2021) are often used to capture this information in a standard way (see Section 3.2.1). The system descriptive model is augmented by other models, such as models to capture the system geometric configuration and various analytical models, to analyze the performance and other quality characteristics of the system. Each kind of model captures different kinds of information about the system. The different models must be managed as the design evolves to ensure a coherent representation of the overall system.

In an MBSE approach, the system descriptive model is a primary artifact of the SE process. MBSE formalizes the application of SE by creating the system descriptive model and integrating it with the other kinds of models. The kind of information and the level of detail of the information that is captured in models and maintained throughout the life cycle depends on the scope of the MBSE effort. An effective MBSE methodology supported by appropriate tools and a team with the requisite SE skills and knowledge are essential to fully realize the benefits of MBSE.

An *MBSE methodology* describes how MBSE is performed to capture the required information in the system descriptive model and related artifacts. Like any methodology, it must be tailored to the particular need of the organization and/or project (see Section 4.1). This includes defining the appropriate life cycle model, tailoring the activities and work products to align with the project scope and modeling objectives, and selecting the appropriate tools to create and manage the models and other relevant data. Estefan (2008) published a survey of candidate MBSE methodologies under the auspices of an INCOSE technical publication. Information on these methodologies is available on the Methodology and Metrics web page of the INCOSE MBSE Wiki (2022). These methodologies and others continue to evolve based on their application to real world projects. OOSEM is summarized below as a representative MBSE method.

OOSEM Summary   OOSEM is an MBSE method intended to help architect systems that satisfy evolving mission and system requirements and can accommodate changes in technology and design. OOSEM is generally consistent with processes in this handbook. It can be adapted to different life cycle models to support the specification, analysis, design, and verification of systems. The method enables the flow-down of requirements from mission, to system, to system element levels, which are realized by applicable hardware, software, data, and other discipline-specific design methods.

OOSEM describes fundamental SE activities whose outputs are model-based artifacts. The modeling artifacts are captured in a system descriptive model using the Systems Modeling Language (SysML™) along with other analytical models. A process model for OOSEM can be downloaded from the INCOSE OOSEM Working Group website (2022).

The OOSEM supports a development process that includes the following subprocesses and activities:

- *Manage the system development*—activities include plan and control the technical effort, including planning, risk management, configuration management, and other project monitoring and control activities;
- *Specify and design the system*—activities include analyze stakeholder needs, specify the system requirements, develop the system architecture, and allocate the system requirements to system elements;
- *Develop the system elements*—activities include design the elements, implement the elements, and verify the elements satisfy the allocated requirements; and
- *Integrate and verify the system*—activities include integrate the system elements and verify that the integrated system elements satisfy the system requirements.

This OOSEM process can be applied at each level of the system hierarchy to specify the requirements of the system and its elements. Applying the process recursively at successive levels of the hierarchy may involve multiple iterations throughout the development process. To be effective, the fundamental tenets of SE must be applied including the use of multi-disciplinary teams and a disciplined management process.

4.2.2   Agile Systems Engineering

The knowledge of requirements for an effective system often continues to change during the system life cycle. Common causes include insufficient initial knowledge, new knowledge revealed during development and utilization, and continual evolution in the targeted operational environment of the system. When evolution of the system's operational environment doesn't stop with initial deployment, a system's functional capabilities must evolve if it is to remain viable. Under these circumstances system engineering is virtually never ending, and retirement is generally an issue of safe and secure functional capability disposal rather than system decommissioning.

Agile Systems Engineering is a principle-based approach for designing, building, sustaining, and evolving systems when knowledge is uncertain, or environments are dynamic. Thus, Agile System Engineering is a what, not a how. As stated in Section 2.2, there are many life cycle models (e.g., Vee, Incremental Commitment Spiral Model (ICSM), DevSecOps (Development, Security, Operations)). Some of them are targeted on a single engineering domain (e.g., XP (Extreme Programming), Scrum, DevOps (Development, Operations), and various scaled approaches such as SAFe (Scaled Agile Framework) in the software engineering domain). Most of them have a strong focus on the development stage.

Agile Systems Engineering is best understood when contrasted to the sequential life cycle approach and in how the two relate to the system life cycle spectrum. Figure 4.3 shows extreme forms of these two life cycle approaches in

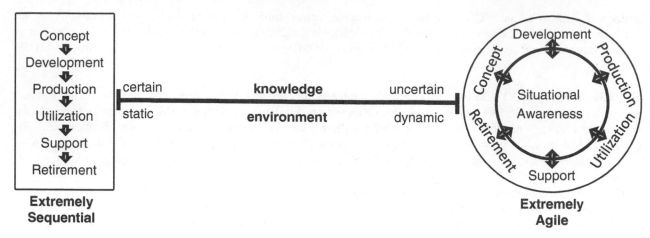

FIGURE 4.3   SE life cycle spectrum. From Dove (2022). Used with permission. All other rights reserved.

terms of their activity stages and data flows. All life cycle approaches fall somewhere between the two ends of the spectrum, depending upon the process-encoded degree of attentiveness and responsiveness to dynamics in knowledge and environment. It is unlikely that either depicted extreme would be effective in actual practice.

An Agile Systems Engineering process is based on strategies for timely and continual knowledge development and affordable application of new knowledge in system development activity. Virtually all forms of Agile Systems Engineering employ incremental or evolutionary development in some way (see Sections 2.2.2 and 2.2.3) as a means to produce demonstrable and/or usable work in process that provokes feedback for real time learning and subsequent application.

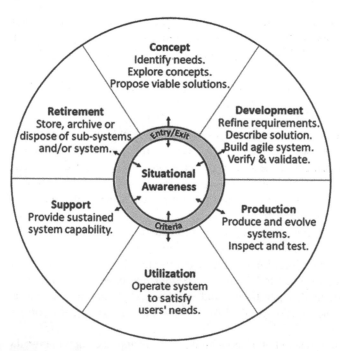

FIGURE 4.4   Agile SE life cycle model. From Dove and Schindel (2019). Used with permission. All other rights reserved.

Software plays an increasingly major role in most systems today. Codified agile software development methodologies offer relevant approaches for rapid knowledge discovery and deployment in the software domain. Patterns from these software approaches can inform agile engineering approaches in other domains and in the encompassing domain of SE; but each domain has unique differences (e.g., external dependencies, fabrication techniques, development cycle time constraints, development support tools).

Agile Systems Engineering Life Cycle Model   The Agile Systems Engineering life cycle model is depicted on the far right of Figure 4.3 and with more detail in Figure 4.4. The six system life cycle stages run around the perimeter, with situational awareness featured in the center. The agile life cycle model can accommodate activities in any and all of the stages concurrently without progressive sequencing. Figure 4.4 depicts a life cycle model (Dove and Schindel, 2019), not to be confused with the Vee model (see Section 2.2.1) which depicts relationships among SE activities applicable in sequential, iterative, and evolutionary approaches.

Situational awareness has no entry or exit criteria, as it should, in principle, be a continuous activity. Entry criteria for all of the other stages begins with a decision to act upon specific triggering awareness, and may require process-prudent, or contract-required, engagement criteria for a stage or stages to be entered. An Agile Systems Engineering process is predicated upon real-time experimentation and learning in all stages, and as such, the entry criteria may be as simple as the decision to enter a stage for experimental knowledge development that may or may not produce artifacts for use in other stages. On the other hand, exit criteria for a stage that produces artifacts for use in other stages should have some fixed requirements for satisfactory stage completion, with recognition that the outcome of stage activity may simply be valuable learned knowledge that aborts the need for producing artifacts of use in other stages.

The retirement stage deals with system elements and older system versions that are retired frequently, as the "current" system evolves. This has implications for safe and secure maintenance, disposal, and reversion processes.

Fleshing out a generic Agile SE Life Cycle Model for a specific project likely starts with default standard processes in each stage, tailored and augmented for specific agile SE differences. Adapting the generic model to a specific organization's process will tailor and augment the generic model as the organization's standard process evolves.

Agile Systems Engineering Life Cycle Operational Considerations   As described in Section 1.3.4, Figure 1.6 depicts the life cycle operational "pattern" as three nested systems:

- **System 1 – The Engineered System** is the target system under development.
- **System 2 – The Life Cycle Project Management System** includes the basic SE development and maintenance processes, and their operational domains that produces System-1.
- **System 3 – The Enterprise Process and Innovation System** is the process improvement system that learns, configures, and matures System-2. System 3 is responsible for situational awareness, evolution, and knowledge management, the provider of operational agility. Intent is continuous, not episodic, information flow among the three systems. Principles and strategies that facilitate operational agility in action include:

Sensing:

- External awareness (proactive alertness);
- Internal awareness (proactive alertness);
- Sense making (risk analysis, trade space analysis).

Responding:

- Decision making (timely, informed);
- Action making (invoke/configure process activity to address the situation);
- Action evaluation (verification and validation).

Evolving:

- Experimentation (variations on process ConOps);
- Evaluation (internal and external judgement);
- Memory (evolving culture, response capabilities, and process ConOps).

The architecture and structural principles that enable system agility are covered in Section 3.1.3.

Agile Systems Engineering Examples   Agile system engineering methodologies are project-context dependent. What is common across methodologies are certain fundamental strategies that get tailored for a specific context. Four published examples illuminate this tailoring in four different contexts:

- As shown in Figure 2.9, Rockwell Collins employed a product line engineering (PLE) (see Section 4.2.4) approach for a large family of radio products composed of software, firmware, and electronic circuit boards with a continuous integration platform that accommodated asynchronous evolution of mixed-domain system element work in process (Dove, et al., 2017).
- The US Navy SpaWar delivered innovative off-road autonomous vehicle technology in continuous six-month development increments with parallel tracks of integration, test, and architecture evolution (Dove, et al., 2016).
- Northrop Grumman evolved user capabilities in six month increments for a software systems-of-systems single-point hub that provided access to 22 independent logistics data bases, with three successive generations under active life cycle control at all times (Dove and Schindel, 2017).
- Lockheed Martin evolved F16, F22, and F35 weapons capabilities with internally developed software and externally subcontracted hardware in roughly six month development increments in a tailored SAFe approach; featuring a continuous integration and demonstration platform with asynchronous evolution of system element simulations, low fidelity proxies, work in process, and completed system elements (Dove, et al., 2018).

4.2.3   Lean Systems Engineering

SE is regarded as an established, sound practice, but is not always delivered efficiently. US Government Accountability Office (GAO, 2008) and NASA (2007a) studies of space systems document major budget and schedule overruns. Similarly, studies by the MIT-based Lean Advancement Initiative (LAI) have identified a significant amount of waste in government projects, averaging 88% of charged time (LAI MIT, 2013; McManus, 2005; Oppenheim, 2004; Slack, 1998). Most projects are burdened with some form of *waste*: politicization, poor coordination, premature and unstable requirements, quality problems, and management frustration. This waste represents a vast productivity reserve in projects and major opportunities to improve project efficiency.

Lean system development and the broader methodology of *lean thinking* have their roots in the Toyota "just-in-time" philosophy, which aims at "producing quality products efficiently through the complete elimination of waste, inconsistencies, and unreasonable requirements on the production line" (Toyota, 2009). *Lean SE* is the application of lean thinking to SE and related aspects of organization and project management. SE is the discipline that enables flawless development and integration of complex technical systems. Lean thinking is a holistic paradigm that focuses on delivering maximum value to the customer and minimizing waste. A popular description of lean is "doing the right job right the first time." Lean thinking has been successfully applied in manufacturing, healthcare, administration, supply chain management, and product development, including engineering.

Lean SE is the area of synergy between lean thinking and SE, with the goal to deliver the best life-cycle value for technically complex systems with minimal waste. The early use of the term lean SE is sometimes met with concern that this might be a "repackaged faster, better, cheaper" initiative, leading to cuts in SE at a time when the profession is struggling to increase the level and quality of SE effort in projects. Lean SE does not take away anything from SE and it does not mean *less* SE. It means *better* SE with higher responsibility, authority, and accountability, leading to better, waste-free workflows with increased mission assurance.

Three concepts are fundamental to the understanding of lean: value, waste, and the process of creating value without waste (captured into lean principles).

Value   The value proposition in engineering projects is often a multiyear, complex, and expensive acquisition process involving numerous stakeholders and resulting in hundreds or even thousands of requirements, which, notoriously, are rarely stable. In lean SE, *value* is defined simply as mission assurance (i.e., the delivery of a flawless complex system, with flawless technical performance, during the product or mission development life cycle) and satisfying the customer and all other stakeholders. This implies completion with minimal waste, minimal cost, and the shortest possible schedule.

Waste in Product Development   *Waste is* "the work element that adds no value to the product or service in the eyes of the customer. Waste only adds cost and time" (Womack and Jones, 1996). The LAI classifies waste into seven categories (McManus, 2005). An eighth category, the waste of human potential, is increasingly added. These categories are defined and illustrated as follows.

- *Overprocessing*—Processing more than necessary to produce the desired output; excessive refinement, beyond what is needed for value.
- *Waiting*—Waiting for people, material or information, or people waiting for information or material; late delivery of material or information, or delivery too early—leading to eventual rework.
- *Unnecessary movement*—Moving people (or people moving) unnecessarily to access or process material or information; unnecessary motion in the conduct of the task; lack of direct access; manual intervention.
- *Overproduction*—Creating too much material or information; performing a task that nobody needs; information over-dissemination and pushing data.
- *Transportation*—Moving material or information unnecessarily; unnecessary hand-offs between people; incompatible communication—lost transportation through communication failures.
- *Inventory*—Maintaining more material or information than needed; too much "stuff" buildup; complicated retrieval of needed "stuff"; outdated, obsolete information.
- *Defects*—Errors or mistakes causing the effort to be redone to correct the problem.
- *Waste of human potential*—Not utilizing or even suppressing human enthusiasm, energy, creativity, and ability to solve problems and general willingness to perform excellent work.

Lean Principles and Lean Enablers for Systems Engineering   Womack and Jones (1996) captured the *process of creating value without waste* into six *lean principles* described in (Oppenheim, 2011), as follows:

The *value principle* promotes a robust process of establishing the value of the system to the customer with crystal clarity early in the project. The process should be customer-centric, involving the customer frequently and aligning employees accordingly.

The *value stream principle* emphasizes detailed project planning and waste-preventing measures, solid preparation of the personnel and processes for subsequent efficient workflow, and healthy relationships between stakeholders (e.g., acquirer, contractor, suppliers, and employees); project frontloading; and use of leading indicators and quality measures. SE practitioners should prepare for and plan all end-to-end linked actions and processes necessary to realize streamlined value, after eliminating waste.

The *flow principle* promotes the uninterrupted flow of robust quality work and first-time-right products and processes, broad steady competence instead of hero behavior in crises, excellent communication and coordination, concurrency, frequent clarification of the requirements, and making project progress visible to all.

The *pull principle* is a powerful guard against the waste of rework and overproduction. It promotes pulling tasks and outputs based on internal and external customer needs (and rejecting others as waste), and better coordination between the pairs of employees handling any transaction before their work begins so that the result can be first-time right.

The *perfection principle* promotes excellence in the SE and organization processes, utilization of the wealth of lessons learned from previous projects into the current project, the development of perfect collaboration policy across people and processes, and driving out waste through standardization and continuous improvement. Imperfections should be made visible in real time, and continuous improvement tools (Six Sigma) should be applied as soon as possible. It calls for a more important role of SE practitioners, with responsibility, accountability, and authority for the overall technical success of the project.

Finally, the *respect-for-people principle* promotes the enterprise culture of trust, openness, honesty, respect, empowerment, cooperation, teamwork, synergy, and good communication and coordination and enables people for excellence.

In 2011, a project undertaken jointly by PMI, INCOSE, and the LAI at MIT developed the Lean Enablers for Managing Engineering Programs (LEfMEP, 2012), adding lean enablers for project management and holistically integrating lean project management with lean SE. A major section of the book is devoted to a rigorous analysis of challenges in managing engineering projects. They are presented under the following 10 challenge themes:

1. Firefighting—reactive project execution;
2. Unstable, unclear, and incomplete requirements;
3. Insufficient alignment and coordination of the extended enterprise;
4. Processes that are locally optimized and not integrated for the entire enterprise;
5. Unclear roles, responsibilities, and accountability;
6. Mismanagement of project culture, team competency, and knowledge;
7. Insufficient project planning;
8. Improper metrics, metric systems, and key performance indicators;
9. Lack of proactive project risk management; and
10. Poor project acquisition and contracting practices.

4.2.4   Product Line Engineering (PLE)

Rarely does anyone build just one edition, just one flavor, just one point solution of anything. In many cases, SE is performed in the context of a product line—a family of similar systems with variations in features and functions. *Product Line Engineering* (PLE) addresses this mismatch by providing models, tools, and methods for holistic engineering of system families.

A note on terminology: Where the PLE field and standards refer to "product," "product line," and "product line engineering," the equivalent terms in SE are "system," "system family," and "system family engineering," respectively. These terms can be used interchangeably (Krueger, 2022).

Challenges with Early Generation System Family Engineering Approaches   When systems in a *system family* are engineered as individual point solutions, techniques such as *clone-and-own reuse* or *branch-and-merge* result in ever-growing duplicate and divergent engineering effort. Trying to manage the commonality and variability among these individually engineered systems in the family has traditionally relied on tribal knowledge and high bandwidth, error-prone interpersonal communication. Furthermore, when each engineering discipline adopts a different ad hoc technique for managing variations among the members of the system family, the result is error prone dissonance when trying to translate and communicate across the different life cycle disciplines.

This is a self-inflicted complexity, over and above the complexity inherent in the systems being engineered. It consumes engineering teams with low-value, trivial, replicative work that deprives them of time and energy that would be better spent on high-value innovative work that advances system and business objectives.

Feature-based Product Line Engineering   *Feature-based PLE* is the modern digital engineering industry good practice for PLE, as defined in the INCOSE PLE Primer (2019) and ISO/IEC 26580 (2021). Feature-based PLE offers significant improvements and benefits in effort, cost, time, scale, and quality by exploiting system similarity while formally managing variation.

Feature-based PLE is used to engineer a system family as a single holistic system rather than a multitude of individual systems. Engineering assets in each engineering discipline are consolidated to eliminate duplication and divergence. A single authoritative variation management model is applied consistently across all assets in all engineering disciplines to eliminate that source of dissonance across the life cycle and to enable organizations to make informed and deliberate cost-benefit decisions about the variations designed into their system family.

Key Elements of a Feature-based PLE Factory   Feature-based PLE uses a *PLE Factory* metaphor, as illustrated in Figure 4.5. See ISO/IEC 26580 (2021) for a full description.

- **Feature Catalogue,** as shown in the upper left, captures a formal model of the distinguishing characteristics about how the members of the system family differ from each other and provides a common language and single authoritative source of truth about variation throughout the engineering organization.
- **Bill-of-Features,** as shown in the upper right, specifies the features selected from the Feature Catalogue for each system in a system family portfolio.
- **Shared Asset Supersets**, as shown in the lower left, are the engineering artifacts that support the creation, design, implementation, deployment, and operation of systems in a system family. They contain *variation points*, which are pieces of content that can be included, omitted, generated, or transformed for a system instance, based on the features selected in a Bill-of-Features for that system.
- **PLE Factory Configurator,** shown in the center, is an automation that applies a Bill-of-Features for a system to each variation point in the Shared Asset Supersets, to determine each variation point's content for the system instance.
- **Product Asset Instances**, shown in the lower right, each contain only the shared asset content suited for that one system in the system family.

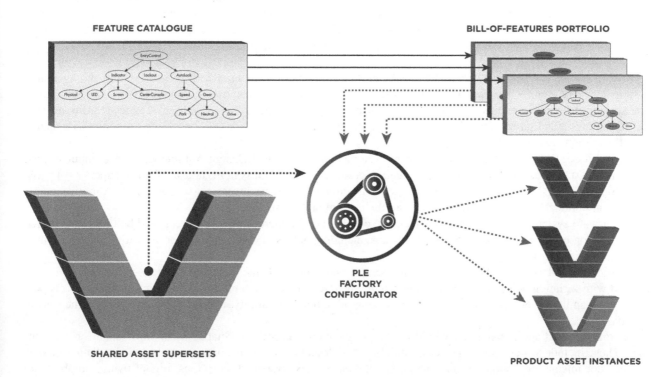

FIGURE 4.5   Feature-based PLE factory. From INCOSE PLE Primer (2019). Usage per the INCOSE Notices page. All other rights reserved.

With Feature-based PLE, engineers now work in the PLE Factory on the Shared Asset Supersets, the Feature Catalogue, and the Bills-of-Features rather than on the individual system instances. Once the PLE Factory is established, engineering assets for the individual systems are automatically instantiated rather than manually engineered. Feature-based PLE transforms the task of engineering a plethora of individual systems into the much more efficient task of producing a single system: The PLE Factory itself. This consolidation also means that change management and configuration management are performed on the single PLE Factory rather than separately on each of the system instances.

Shared Asset Supersets and Variation Points   To work in a PLE Factory, engineers must learn how to create and maintain Shared Asset Supersets, including variation points, for their discipline. For example, requirements engineers learn how to create requirements Shared Asset Supersets with variation points, test engineers learn how to create verification and validation Shared Asset Supersets with variation points, and software engineers learn how to create source code Shared Asset Supersets with variation points.

A Shared Asset Superset contains a single copy of all content used in any system—that is, there is no duplication of content. Content that appears in every system is said to be *common* content, while content that varies from system to system is encapsulated in a variation point. Consistent treatment of variation points in Shared Asset Supersets across all disciplines is a hallmark of Feature-based PLE. Variation points are places in an asset that denote content that is configured according to feature selections in a Bill-of-Features for a particular system instance. Variation point configuration mechanisms typically include selection or omission of the content; selection from among mutually exclusive content alternatives; generation of content based on feature specifications; and feature-based transformation of content from one form into another.

Perhaps one of the easiest examples of Shared Asset Supersets to understand is requirements. A superset of requirements combines individual system requirements to establish all of the system family requirements. Variation points express inclusion and omission, define mutual exclusion, and transform requirement wording in the system specification—all based on feature selections. Requirement transformation can replace numbers, units, or other text with information that is derived from the Bill-of-Features. Requirements that have no variation are common and apply to every system.

MBSE models can be developed as Shared Asset Supersets and instrumented with variation points. For example, system design or architecture models using SysML™ include variation points to specify optional, mutually exclusive, and varying structural elements such as blocks, ports, relationships, objects, classes, activities, states, transitions, use-cases, packages, and others, as well as generation or transformation of values, attributes, and constraints associated with those elements.

Shared Asset Supersets for Electronic Design Automation, Mechanical Design Automation, and Computer-aided Design (CAD) for electronic, mechanical, mechatronic, and cyber-physical systems take the form of supersets of parts, properties, relationships, assemblies, system elements, circuit boards, wiring harnesses, and more. Variation points instrument optional, mutually exclusive, and varying content in these models.

In software systems, Shared Asset Supersets are constructed for source code, resources, and build scripts. Source code variation points can be defined in several ways, including blocks of code, optional or mutually exclusive source files, and macro substitutions.

Verification and validation Shared Asset Supersets for automated and manual test plans and test cases can be instrumented with variation points to identify and configure the tests for each system, based on feature selections. It is possible to streamline or even eliminate redundant testing of common capability across multiple systems in the system family.

A broad array of additional assets with digital representations can serve as Shared Asset Supersets in system families. These include system budgets or cost models, schedules and work plans, user manuals and installation guides, process documentation, marketing brochures, simulation models, system descriptions, digital twins, supply chain orders, manufacturing specs, contract proposals, and much more. Feature-based PLE can be applied to all SoI types defined in Section 4.3.

Organizational Change and Return-on-Investment with Feature-based Product Line Engineering   For many organizations, Feature-based PLE represents a shift in engineering approach that requires organizational change, along with commitment from engineering and business leadership to make that change. The ROI to justify the organizational change is in most cases compelling, based on the elimination of low-value, mundane, replicative work, with doubling, tripling and larger improvements in engineering metrics such as: lowering engineering complexity; reducing overall engineering time, cost, and effort; increasing portfolio scalability; and improving system quality (Gregg, et al., 2015 and McNicholas, 2021). In consideration of this ROI, the question to leadership is, "What if your engineers could do their normal day's work before lunch; what would you have them do in the afternoon?" There are many answers to this question, all of them good.

4.3   SYSTEM TYPES CONSIDERATIONS

The concept of SoI was introduced in Section 1.3.1. The type of SoI has significant implications on SE. This section introduces SE considerations for the following types of SoIs:

- Greenfield/Clean Sheet Systems
- Brownfield/Legacy Systems
- Commercial-off-the-Shelf (COTS)-Based Systems
- Software-Intensive Systems
- Cyber-Physical Systems (CPS)
- System of Systems (SoS)
- Internet of Things (IoT)/Big Data-Driven Systems
- Service Systems
- Enterprise Systems

Note that other types of SoIs exist.

4.3.1   Greenfield/Clean Sheet Systems

"Greenfield" and "brownfield" are terms used in real estate. Greenfield land is previously undeveloped space, such as a (green) farmer's field. Brownfield land has been previously developed, typically has existing structures and services in place, and may contain undesirable or hazardous materials (also known as waste) that must be remediated. Greenfield SE, also known as "clean sheet" or "blank slate" SE, involves systems that are new designs and have no, or limited, legacy systems constraints, other than system interfaces. Given the incremental and spiral development life cycle approaches of today, a greenfield system may evolve toward brownfield even before it is delivered, from the developer's perspective.

Traditionally, SE has been taught by considering systems from a greenfield perspective. One starts with a "clean or blank sheet of paper" and determines the set of stakeholders and their needs and requirements, translates them into system requirements, architects and designs a system solution, implements the system elements, and then integrates, verifies, and validates the system elements and the system solution. While this is an effective way to teach SE and to prepare practitioners with skills applicable to the entire system life cycle, few system development efforts are truly greenfield. Greenfield, therefore, is an almost theoretical situation that is rarely seen in practice. The remaining considerations provide different perspectives and implications beyond greenfield SE.

Sometimes, it can be quite traumatic for organizations that make brownfield updates to its legacy products over a long period of time to transition from brownfield development back to greenfield (Axehill, 2021). They may need to "relearn" how to do greenfield.

4.3.2 Brownfield/Legacy Systems

As described in Section 4.3.1, brownfield (and greenfield) are terms used in real estate. Brownfield land has been previously developed, typically has existing structures and services in place, and may contain undesirable or hazardous materials (also known as waste) that must be remediated. Brownfield SE, also known as "legacy" SE, involves significant modifications, extensions, or replacement of an existing "as-is" system in an existing environment to an updated "to-be" system. Brownfield systems often contain waste (e.g., technical debt) that may need to be remediated (Seacord, et al., 2003) (Hopkins and Jenkins, 2008). "In-service" systems are another example of brownfield system (Kemp, 2010) (Van De Ven, et al., 2012). Brownfield systems typically have explicit continuity requirements, where the operation of the as-is system needs to continue, resulting in a deliberate transition to the updated system.

The nature of greenfield and brownfield systems drives different life cycle approaches that reflect different areas of emphasis. Table 4.1 lays out some of the key differences across a set of aspects important to SE (Walden, 2019) (Baley and Belcham, 2010). This impacts not only the system solution, but also the team that is put in place to develop the system. As with all development efforts, SE processes need to be tailored to fit the needs of a given project (see Section 4.1). SE in a brownfield environment augments the SE life cycle processes described in this handbook with site surveys and reconstruction activities to understand the as-is systems, identify gaps, and engineer the to-be system (Walden, 2019).

TABLE 4.1   Considerations of greenfield and brownfield development efforts

Aspect	Greenfield	Brownfield
Life Cycle Stage(s) (of Initial SoI)	Concept, Development	Utilization, Support
Focus	New or novel features	Maintenance or adding new features while retaining select legacy functionality
Maturity (of Initial SoI)	Low to Moderate	High for maintenance; Mix for existing system and environment, plus new development for upgrade or replacement
Architecture and Design Review	Reviewed and modified at multiple levels	Reviewed only when significant updates are made/performed
Verification	The entire SoI typically needs to verified	Only the updated and impacted parts of the system need to be verified (there may be regression testing for the unchanged parts)
Validation	The entire SoI typically needs to validated with the customer/user	The entire SoI (including changes) typically needs to validated with the customer/user to check for new emergent behavior
Manufacturing/ Production	May be in place if using the existing line, or is developed (or tailored) as development progresses	Mostly in place, reverse engineering of existing designs may be required if the original design can no longer be produced (e.g., due to as-is use of banned materials)
Maintenance and Logistics	Developed (or tailored) as development progresses	Mostly in place, but may need changes or upgrades depending on the replacement system elements
Practices and Processes	Developed (or tailored) as work progresses	Mostly in place, though not necessarily relevant to the new team
Team Composition	Newly formed group	Mix of old and new, bringing both historical biases and fresh ideas

From Walden (2019) derived from Baley and Belcham (2010). Used with permission. All other rights reserved.

4.3.3   Commercial-off-the-Shelf (COTS)-Based Systems

One of the key trade-off studies SE practitioners perform is the "make vs. buy" decision on system elements. "Make" represents custom-built solutions; "buy" represents outsourced development and commercial-off-the-shelf (COTS) solutions. Directed use also can result in COTS. Most systems have some COTS content. The following characteristics can be useful when deciding if a particular system or system element can be characterized as COTS (Oberndorf, et al., 2000) (Tyson, et al., 2003):

- Sold, leased, or licensed to the general public;
- Offered by a vendor trying to profit from it;
- Supported and evolved by the vendor, who retains the intellectual property rights;
- Vendor (not acquirer) controls the frequency of the product's maintenance and updates;
- Available in multiple, identical copies;
- Used without hardware or source code modification.

The promise of COTS is to save development time, reduce technical risk, reduce time-to-market, reduce cost-to-market, and take advantage of latest technology. However, often these promises are not realized. Considerations for COTS-based systems include (Long, 2000):

- COTS products not built to your specific requirements (including missing functionality, extra functionality, and unwanted behaviors).
- Unique, or different than expected, interfaces, including a vendor's use of proprietary data formats and/or communications protocols, may occur.
- Vendor claims and decisions may impact schedule.
- The details needed to understand how COTS products may impact the safe and secure operation of the SoI may not be readily available, including the trustworthiness of the vendor, the use of open-source software, and the use of third-party software of unknown origin.
- COTS product may be insufficiently documented.
- "Not Invented Here" (NIH) syndrome may deter engineers from using COTS.
- Delivery times may not be met.
- Special integration challenges may occur.
- COTS products are often not verified to your specific requirements and may lack verification data for the operating environment for the SoI.
- "Sole Source" suppliers result in more risk.
- Need to consider the entire life cycle cost (LCC) of maintenance and technology rolls/refresh due to COTS obsolescence and diminishing manufacturing sources and material shortages (DMSMS) (Note: IEC 62402 (2019) provides guidance for establishing a framework for obsolescence management process which is applicable through all stages of system life cycle).

There are differences in approaches to SE for COTS-based systems development. Some of the key COTS-based SE considerations are shown in Table 4.2. Effective use of COTS generally requires COTS evaluation starting during needs analysis. In some circumstances, an "internal sales pitch" for each viable candidate COTS-based system needs to be developed, highlighting which requirements are met, which are partially or not met, and what additional capabilities and cost advantages (now and potentially in the future) each possible system provides. For an organization which

TABLE 4.2   Considerations for COTS-based development efforts

Aspect	Traditional Systems Engineering	COTS-Based Considerations
Focus	The SoI	The SoI as well as how potential COTS products in the marketplace could be assembled to meet most/all the needs
Stakeholder Needs and Requirements	Fairly explicit stakeholder requirements	Flexible and prioritized capabilities stated in broad terms
System Requirements and Functionality	Requirements and functionality are defined and allocated based on technical considerations	COTS capabilities and functionality form the basis for the system requirements allocation and evolution and COTS may introduce additional system-level constraints
System Element Requirements	Extra or missing system element requirements are typically bad	Need to strike a balance between what the system needs and what the market can provide, missing or extra COTS requirements may be a reality due to the marketplace and they may also necessitate extra COTS wrappers and glue
System Architecture and Design	Focus is on optimizing the SoI	Focus is on optimizing the set of COTS and custom components that make up the SoI
Integration, Verification, and Validation	Done with known (internal) system element owners	Can typically get an early version of the system up and operating dramatically sooner than with a "make" system, criteria more difficult to establish since COTS performance must satisfy the market requirements while balancing the needs of the system, execution and defect resolution more difficult due to external COTS element owners
Technical Management	Well understood set of processes	More challenging decision environment, additional risks are present for COTS, and potentially increased configuration and information management (CM and IM) activities
Agreements	Acquisition agreements are primarily outsourced development efforts	Acquisitions also include COTS items and must consider other aspects such as licensing, additional COTS vendor support, and obsolescence management
Quality Characteristics	Known with internal team support and data, but the full picture may not be obtainable until after system deployment	For proven COTS, can be determined up front, may have to rely on COTS vendors for some data, consideration of life cycle cost (LCC) is critical for COTS

has only done "make" system development in the past, moving to a COTS-based development requires a different mind-set and different development skills, including the new role of COTS integrator. These skills are often different than those needed for non-COTS-based system development.

4.3.4   Software-Intensive Systems

ISO/IEC/IEEE 42010 (2022) defines a software-intensive system as:

> Any system where software contributes essential influences to the design, construction, deployment, and evolution of the system as a whole.

Software, like physical entities, is encapsulated in system elements. Software elements often contribute to system functionality, behavior, quality characteristics, interfaces, and observable performance indicators for software-intensive systems. Encapsulated software is sometimes custom-built and is sometimes obtained by using software elements from libraries. Reused software elements may be modified by tailoring them for intended use. Also, software elements may be licensed from software vendors. In most cases, licensed software packages cannot be modified. Software engineers sometimes encapsulate licensed software packages in software shells or wrappers that provide the interfaces needed to integrate the needed capabilities of the software into a system while masking unwanted capabilities.

Software is widely incorporated in systems and provides "essential influences" because software is a malleable entity composed of textual and iconic symbols that in many cases, but not always, can be constructed, modified, or procured sooner and at less expense than fabricating, modifying, or procuring physical elements that have equivalent capabilities. In some cases, but not always, software elements can provide capabilities that would be difficult to realize in hardware.

But in some cases, a physical element of equivalent capability may be preferred to a software element because software may not provide certifiable safety, security, or performance at the necessary levels of assurance. Making tradeoffs between physical elements and software elements is an essential aspect of developing and modifying software-intensive systems. Trade studies are best conducted when SE practitioners consult physical engineers and software engineers who are working collaboratively.

Interfaces provided by software in a software-intensive system can provide passive pass-through connections or, if desired, software included in an interface can actively coordination interactions among the connected elements. Software interfaces may be internal and/or external to a system. Internal software interfaces can provide connections that coordinate interactions among physical elements, among software elements, and between physical elements and software elements. External software interfaces can provide passive and active connections to entities in the physical environment, including humans and other sentient entities that are software-enabled, plus connections to the external interfaces of other physical-only, software-only, and software-intensive systems. A "software-only" system is a software application implemented on a stable computing platform. The computing platform includes hardware and software that support the software application. The platform is often implemented using commodity items.

Sometimes the essential influences of software are observable in a system's external interfaces, as in human-user interface displays, and sometimes not, as in the interfaces for direct interaction with an external system. In the latter case, the software is said to be embedded in the SoI because performance of the interface is not directly observable by a human, although the effects of interface performance may be observable. Software is also used to provide interfaces among the systems that constitute an SoS. Software can contribute to functionality, behavior, quality attributes, external interfaces, and performance indicators for a composite SoS.

Developing and modifying software-intensive systems presents challenges for SE practitioners when partitioning system requirements and elements of the architecture, allocating performance parameters to physical elements and to software elements, establishing and controlling physical-software interfaces, and facilitating integration of physical elements with software elements (Fairley, 2019). These challenges sometimes arise because the culture, terminology, processes, and practices of software engineering are unfamiliar to SE practitioners and conversely, the various aspects of SE may be unfamiliar to software engineers. Techniques for improving communication between SE practitioners and software engineers are presented in Section 5.3.1 and in Fairley (2019).

4.3.5 Cyber-Physical Systems (CPS)

CPS are the integration of physical and cyber (software) processes in which the software monitors and controls the physical processes and is, in turn, affected by them. CPS are enabled by sensors and feedback loops and their provenance has increased because of significant advances in sensor technology and affordability. Figure 4.6 illustrates the behavior of a CPS: sensors may be deployed in the system hardware and/or its environment. Sensor data is used by software algorithms to control the hardware in responses to changes in the environment and/or in the hardware itself. The algorithms control the dynamic behavior of the CPS to achieve one or more goals, which could include homeostasis (maintaining equilibrium). An example might be an automobile with sensors to detect obstacles (in the environment) and take evasive action if an obstacle is detected ahead (the actuators would be steering or braking). Other

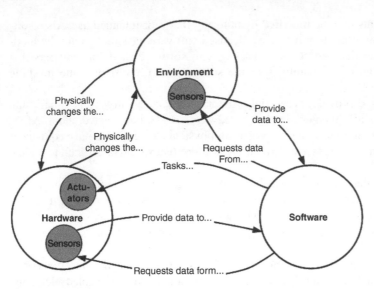

FIGURE 4.6 Schematic diagram of the operation of a Cyber-Physical System. INCOSE SEH original figure created by Henshaw. Usage per the INCOSE Notices page. All other rights reserved.

sensors may be deployed to monitor the health of the vehicle, for instance, wear of the brake pads may trigger an action to modify the driving or notify the owner of the need for maintenance. Digital twins are also an important concept in this regard that refers to a digital surrogate that is a dynamic physic-based description of physical assets (physical twin), processes, people, places, systems, and devices that can be used for various purposes. The digital representation provides both the elements and the dynamics of how an Internet of Things (IoT) device operates and lives throughout its life cycle (see Section 4.3.7).

The CPS concept is closely aligned to Industry 4.0, an initiative to revolutionize industry through so-called smart systems (Kagermann, 2013). CPS always include both software and hardware and are almost always networked, in which case they are Cyber-Physical Systems of Systems (CPSoS). If the individual CPS are networked using Internet Protocols, then they form part of an IoT, but they may interact through other mechanisms (e.g., mechanical, electromagnetic, thermal). The relationships between the concepts of SoS, CPS, and IoT are illustrated in Figure 4.7. The "Things" in IoT are constituent systems and they are always networked, and therefore, always an SoS.

CPS are a feature of almost every industry and many aspects of society. They provide automated, and even autonomous, control of technologies ranging from robotic manufacture to SMART cities and from automated insulin delivery to control of critical infrastructure. They may form a contribution to business models that include physical systems and can even underpin a business concept by providing resilience or cost savings.

There are many implications of CPS for engineers. Two of the most significant are complexity and ethics. The increased level of complexity is due to both extensive networks and the problem of modeling the combination of physical dynamics with computational processes. Lee (2015) has pointed out that such models are nearly always nondeterministic due to the lack of temporal semantics in the cyber and physical modeling programs. This has significant implications for modeling and verification within the SE processes. The lack of determinism, together with questions concerning the transfer of decision making from humans to machines have ethical challenges that engineers must face in terms of definition and realization of CPS (European Parliament Research Service, 2016).

The scope and dimensions of CPS is well-illustrated by the CPS Concept Map (Asare, et al., 2012), which essentially defines CPS as feedback systems that are applicable across a wide range of applications such as infrastructure, healthcare, manufacturing, military and many more. One might consider them relevant to any application in which the system is required to be dynamic and the control thereof is managed by software. The concept map also highlights the explicit need for security and safety considerations in design as well as the challenges of verification and validation in large complex systems. It points toward the need for improved modeling and, indeed, progress has been made in the use of MBSE for CPS development through a framework to implement suitable tool chains (Lu, 2019).

Given that software forms an integral part of many systems and devices, it can reasonably be stated that SE practitioners are very often concerned with CPS and usually CPSoS.

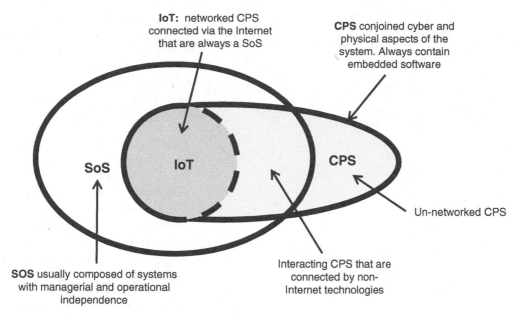

FIGURE 4.7   The relationship between Cyber-Physical Systems (CPS), Systems of Systems (SoSs), and an Internet of Things (IoT). From Henshaw (2016). Used with permission. All other rights reserved.

4.3.6   Systems of Systems (SoS)

ISO/IEEE/IEC 21839 (2019) defines a System of Systems (SoS) as:

> A set of systems or system elements that interact to provide a unique capability that none of the constituent systems can accomplish on its own.

Constituent systems can be part of one or more SoS. Each constituent system is a useful system by itself, having its own development, management goals, and resources, but interacts within the SoS to provide the unique capability of the SoS (ISO/IEEE/IEC 21839, 2019).

The following characteristics can be useful when deciding if a particular SoI can better be understood as an SoS (Maier, 1998):

- Operational independence of constituent systems;
- Managerial independence of constituent systems;
- Geographical distribution;
- Emergent behavior;
- Evolutionary development processes.

Of these, operational independence and managerial independence are the two principal distinguishing characteristics for applying the term SoS.

Figure 4.8 illustrates the concept of an SoS. The air transport system is an SoS comprising multiple aircraft, airports, air traffic control systems, and ticketing systems, which along with other systems such as security and financial

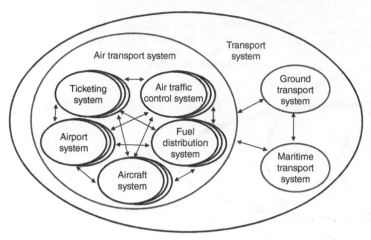

FIGURE 4.8 Example of the systems and systems of systems within a transport system of systems. From ISO/IEC/IEEE 21841 (2019). Used with permission. All other rights reserved.

systems facilitate passenger transportation. There are equivalent ground and maritime transportation SoS that are all in turn part of the overall transport system (an SoS in the terms of this description).

There are three international standards that provide useful guidance on SoS:

- ISO/IEC/IEEE 21839 (2019)—"System of systems (SoS) considerations in life cycle stages of a system" focuses on the SE of an individual constituent system and identifies considerations to be addressed as the engineering of the systems progresses from concept through retirement.

- ISO/IEC/IEEE 21840 (2019)—"Guidelines for the utilization of ISO/IEC/IEEE 15288 in the context of system of systems (SoS)" provides guidance on the application of the processes to the special case of SoS, including considerations for how constituent systems relate within the SoS.

- ISO/IEC/IEEE 21841 (2019)—"Taxonomy of systems of systems" lays out a taxonomy of SoS types based on authority relationships as (shown in Table 4.3)

ISO/IEC/IEEE 21840 (2019) guidance on application of ISO/IEC/IEEE 15288 (2023) life cycles processes (see Section 2.3) is based on the differences between systems and SoS and the impact on the SE processes as shown in Table 4.4.

Dahmann (2014) identified the following challenges that influence the engineering of an SoS:

- *SoS authorities*—In an SoS, each constituent system has its own local "owner" with its stakeholders, users, business processes, and development approach. As a result, the type of organizational structure assumed for most traditional SE under a single authority responsible for the entire system is absent from most SoS. In an SoS, SE relies on crosscutting analysis and on composition and integration of constituent systems, which in turn depend

TABLE 4.3 SoS types

Directed	The SoS is created and managed to fulfill specific purposes and the constituent systems are subordinated to the SoS. The constituent systems maintain an ability to operate independently; however, their normal operational mode is subordinated to the central managed purpose.
Acknowledged	The SoS has recognized objectives, a designated manager, and resources for the SoS; however, the constituent systems retain their independent ownership, objectives, funding, and development and sustainment approaches. Changes in the systems are based on cooperative agreements between the SoS and the constituent systems.
Collaborative	The constituent systems interact more or less voluntarily to fulfill agreed upon central purposes. The central players collectively decide how to provide or deny service, thereby providing some means of enforcing and maintaining standards.
Virtual	The SoS lacks a central management authority and a centrally agreed upon purpose for the SoS. Large-scale behavior emerges-and may be desirable-but this type of SoS must rely on relatively invisible mechanisms to maintain it.

From ISO/IEC/IEEE 21841 (2019) derived from SEBOK. Used with permission. All other rights reserved.

TABLE 4.4   Impact of SoS considerations on the SE processes

SE Process	Implementation as Applied to SoS
Agreement Processes	Because there is often no top level SoS authority, effective agreements among the systems in the SoS are key to successful SoSE.
Organizational Project Enabling Processes	SoSE develops and maintains those processes which are critical for the SoS within the constraints of the system level processes.
Technical Management Processes	SoSE implements Technical Management Processes applied to the particular considerations of SoS engineering - planning, analyzing, organizing, and integrating the capabilities of a mix of existing and new systems into a system of systems capability while systems continue to be responsible for technical management of their systems.
Technical Processes	SoSE Technical Processes define the cross-cutting SoS capability, through SoS level business or mission analysis and stakeholder needs and requirements definition. SoS architecture and design frame the planning, organization, and integration of the constituent systems, constrained by system architectures. Development, integration, verification, transition, and validation are implemented by the systems. with SoSE monitoring and review. SoSE integration, verification, transition and validation applies when constituent systems are integrated into the SoS and performance is verified and validated.

on an agreed common purpose and motivation for these systems to work together toward collective objectives that may or may not coincide with those of the individual constituent systems.

- *Leadership*—Recognizing that the lack of common authorities and funding poses challenges for SoS, a related issue is the challenge of leadership in the multiple organizational environments of an SoS. This question of leadership is experienced where a lack of structured control normally present in SE requires alternatives to provide coherence and direction, such as influence and incentives.

- *Constituent systems*—SoS are typically composed, at least in part, of in-service systems, which were often developed for other purposes and are now being leveraged to meet a new or different application with new objectives. This is the basis for a major issue facing the application of SE to SoS, that is, how to technically address issues that arise from the fact that the systems identified for the SoS may be limited in the degree to which they can support the SoS. These limitations may affect initial efforts at incorporating a system into an SoS, and systems' commitments to other users may mean that they may not be compatible with the SoS over time. Further, because the systems were developed and operate in different situations, there is a risk that there could be a mismatch in understanding the services or data provided by one system to the SoS if the particular system's context differs from that of the SoS.

- *Capabilities and requirements*—Traditionally (and ideally), the system engineering process begins with a clear, complete set of initial user requirements and provides a disciplined approach to develop and evolve a system to meet these and emerging requirements. Typically, SoS are composed of multiple independent systems with their own requirements, working toward broader capability objectives. In the best case, the SoS capability needs are met by the constituent systems as they meet their own local requirements. However, in many cases, the SoS needs may not be consistent with the requirements for the constituent systems. In these cases, SE of an SoS needs to identify alternative approaches to meeting those needs either through changes to the constituent systems or through the addition of other systems to the SoS. In effect, this is asking the systems to take on new requirements with the SoS acting as the "user."

- *Autonomy, interdependence, and emergence*—The independence of constituent systems in an SoS is the source of a number of technical issues when applying SE to an SoS. The fact that a constituent system may continue to change independently of the SoS, along with interdependencies between that constituent system and other constituent systems, adds to the complexity of the SoS and further challenges SE at the SoS level. These dynamics can lead to unanticipated effects at the SoS level leading to unexpected or unpredictable behavior in an SoS even if the behavior of the constituent systems is well understood.

- *Testing*—The fact that SoS are typically composed of constituent systems that are independent of the SoS poses challenges in conducting end-to-end SoS testing, as is typically done with systems. First, unless there is a clear understanding of the SoS-level expectations and measures of those expectations, it can be very difficult to assess the level of performance as the basis for determining areas that need attention or to ensure users of the capabilities and limitations of the SoS. Even when there is a clear understanding of SoS objectives and metrics, testing in a traditional sense can be difficult. Depending on the SoS context, there may not be funding or authority for SoS testing. Often, the development cycles of the constituent systems are tied to the needs of their owners and original ongoing user base. With multiple constituent systems subject to asynchronous development cycles, finding ways to conduct traditional end-to-end testing across the SoS can be difficult if not impossible. In addition, many SoS are large and diverse, making traditional full end-to-end testing with every change in a constituent system prohibitively costly. Often, the only way to get a good measure of SoS performance is from data collected from actual operations or through estimates based on modeling, simulation, and analysis. Nonetheless, the SoS SE team needs to enable continuity of operation and performance of the SoS despite these challenges.
- *SoS principles*—SoS is a an area where there has been limited attention given to ways to extend systems thinking to the issues particular to SoS. The community is beginning to identify and articulate the crosscutting principles that apply to SoS in general and to develop working examples of the application of these principles. There is a major learning curve for the average SE practitioner moving to an SoS environment and a problem with SoS knowledge transfer within or across organizations.

Beyond these general SE challenges, in today's environment, SoS pose particular issues from a security perspective. This is because constituent system interface relationships are rearranged and augmented asynchronously and often involve COTS elements from a wide variety of sources. Security vulnerabilities may arise as emergent phenomena from the overall SoS configuration even when individual constituent systems are sufficiently secure in isolation.

The SoS challenges cited in this section require SE approaches that combine both the systematic and procedural aspects described in this handbook with holistic, nonlinear, iterative methods. There is a growing set of approaches to applying SE to SoS (Cook and Unewisse, 2019). These include SoS life cycle engineering approaches such as the SoS Wave Model (Dahmann, et al., 2011) and the Designing for Adaptability and evolutioN in System of Systems Engineering (DANSE). These approaches address both functionality of constituents to create coherent aggregate SoS capability (Axelsson, 2020) as well as management of interfaces among constituents (Hoehne, 2020).

4.3.7   Internet of Things (IoT)/Big Data-Driven Systems

SE is based on engineering requirements, engineering calculations, testing, modeling, and simulations—and all are based on data or data generation. SE practitioners often make decisions based on intuition, previous experience, or qualitative assessments. The 4th Industrial Revolution, with its proliferation of sensors of various types and big data analytics, creates an opportunity for SE, as a discipline, and for SE practitioners, as professionals and decisions makers, to be more data-driven. The following recommendations apply to modern SE tasks and decisions:

- Bring as much diverse data and as many diverse viewpoints to maximize the generation of information quality.
- Use data to develop a deeper understanding of the business context and the problem at hand.
- Develop an appreciation for the impact of variation, both in data and in the overall business.

- Deal with uncertainty, which means that SE also recognizes mistakes.
- Recognize the importance of high-quality data and invest in trusted sources and in making improvements.
- Conduct good experiments and research to supplement existing data and address new questions.
- Recognize the criteria used to make decisions and adapt under varying circumstances.
- Realize that making a decision is only the first step; SE practitioners must keep an open mind and revise decisions if new data suggests a better course of action.
- Work to bring new data and new data technologies into the organization.
- Learn from mistakes and help others to do so, by applying lessons-learned processes.
- As SE practitioners, strive to be a role model when it comes to data, working with leaders, peers, and subordinates to help them become data driven.

There are three general goals in analyzing data:

1. *Prediction*: To predict the response to future values of the input variables.
2. *Estimation*: To infer how response variables are associated with input variables.
3. *Explanation*: To understand the relative contribution of input variables to response values.

Predictive modeling is the process of applying models and algorithms to data for the purpose of predicting new observations. In contrast, explanatory models aim to explain the causality and relationship between the independent variables and the dependent variables. Classical statistics focuses on modeling the stochastic system generating the data. Statistical learning, or computer age statistics, builds on big data and the modeling of the data itself. If the former aimed at properties of the model, the latter is looking at the properties of computational algorithms. SE practitioners need to be educated in data sciences to enable them to practice the above tools and methods as an integral part of SE.

Data analytics and IoT are wide-scope revolutions of digital surroundings. They create complex CPS that add new functionalities and capabilities to the existing physical environment. Designing an IoT system that has analytic capabilities involves "multi stack" layers, addressing SoS and network of networks.

SE practitioners should view a system as interconnected system elements performing the system functions. To meet this challenge, the SE practitioner who leads data-driven designs needs interdisciplinary knowledge of the main aspects of IoT: computing, sensors/actuators, software, network, analytics and data science.

4.3.8   Service Systems

OASIS (2012) defines a service as:

> A mechanism to enable access to one or more capabilities, where the access is provided using a prescribed interface and is exercised consistently with constraints and policies as specified by the service description.

It involves application of specialized competences (knowledge and skills) through deeds, processes, and performances for the benefit of another entity or the entity itself in real world. The entity involved with the service can be technical, socio-technical, or strictly social.

For service systems, understanding the integration needs among loosely coupled systems and system elements, along with the information flows required for both governance and operations, administration, maintenance, and provisioning of the service, presents major challenges in the definition, design, and implementation of services (Domingue, et al., 2009; Maier, 1998). Cloutier, et al. (2009) presented the importance of Network-Centric Systems (NCS) for dynamically binding different system entities in engineered systems rapidly to realize adaptive SoSs that, in the case

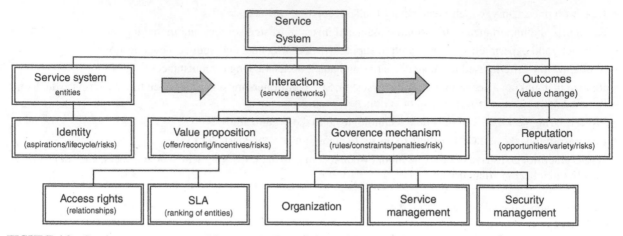

FIGURE 4.9   Service system conceptual framework. From Spohrer (2011). Used with permission. All other rights reserved.

of service systems, are capable of knowledge emergence and real-time behavior emergence for service discovery and delivery.

Figure 4.9 illustrates the conceptual framework of a service system. Typically, a service system is composed of service system entities that interact through processes defined by governance and management rules to create different types of outcomes in the context of stakeholders with the purpose of providing improved customer interaction and value cocreation.

Services not only involve the interaction between the service provider and the consumer to produce value, but have other intangible attributes like quality of service (e.g., ambulance service availability, response time to an emergency request). The demand for service may have loads dependent on time of day, day of week, season, or unexpected needs (e.g., natural disasters), and services are rendered at the time they are requested. Thus, the design and operations of service systems "is all about finding the appropriate balance between the resources devoted to the systems and the demands placed on the system, so that the quality of service to the customer is as good as possible" (Daskin, 2010).

In many cases, taking a service SE approach is imperative for the service-oriented, customer-centric, holistic view to select and combine service system entities to define and discover relationships among service system entities to plan, design, adapt, or self-adapt to cocreate value. Typically, five types of resources need to be considered: people; tangible products and environment infrastructure; organizations and institutions; protocols; and shared information and symbolic knowledge in the service delivery process. Major challenges include the dynamic nature of service systems evolving and adapting to constantly changing operations and/or business environments and the need to overcome silos of knowledge. Interoperability of service system entities through interface agreements must be at the forefront of the service SE design process for the harmonization of operations, administration, maintenance, and provisioning procedures of the individual service system entities (Pineda, 2010). In addition, service systems require open collaboration among all stakeholders, but recent research on mental models of multidisciplinary teams shows integration and collaboration into cohesive teams has proven to be a major challenge (Carpenter, et al., 2010).

In summary, in a service system environment, SE practitioners should bring a customer focus to promote service excellence and to facilitate service innovation through the use of emerging technologies to propose creation of new service systems and value cocreation. SE practitioners must play the role of an integrator, considering the interface requirements for the interoperability of service system entities—not only for technical integration but also for the processes and organization required for optimal customer experience during service operations.

4.3.9 Enterprise Systems

This section illustrates the applications of SE principles and concepts when the SoI is an enterprise. The aim is to continuously improve and help transform the enterprise to better deliver value and to survive in a globally competitive environment. Enterprise SE is an emerging discipline that focuses on frameworks, tools, and problem-solving approaches for dealing with the inherent complexities of the enterprise including exploitation of new opportunities that can facilitate achievement of enterprise goals. A good overall description of enterprise SE is provided in Rebovich and White (2011). For more detailed information on this topic, please see the Enterprise SE articles in Part 4 of SEBoK (2023).

Enterprise   An enterprise consists of a purposeful combination (e.g., a network) of interdependent resources (e.g., people, processes, organizations, supporting technologies, and funding) that interact with each other to coordinate functions, share information, allocate funding, create workflows, and make decisions, and that interact with their environment(s) to achieve business and operational goals through a complex web of interactions distributed across geography and time (Rebovich and White, 2011).

An enterprise must do two things: (1) develop things within the enterprise to serve as either external offerings or as internal mechanisms to enable achievement of enterprise operations, and (2) transform the enterprise itself so that it can more effectively and efficiently perform its operations and survive in its competitive and constrained environment.

It is worth noting that an enterprise is not equivalent to an "organization." As shown in Figure 4.10, an enterprise has organizations that participate in it, but these organizations are not necessarily "part" of the enterprise. The organizations that participate in the enterprise will manage a variety of resources for the benefit of the enterprise, such as people, knowledge, and other assets such as processes, principles, policies, practices, culture, doctrine, theories, beliefs, facilities, land, and intellectual property. These organizational resources will consume or produce money, time, energy, and material when acting on behalf of the enterprise.

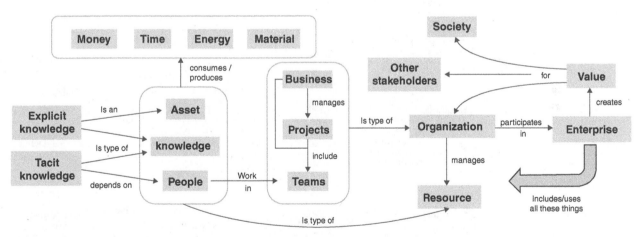

Notes:
1. All entities shown are decomposable, except people. For example, a business can have sub-businesses, a project can have subprojects, a resource can have sub-resources, an enterprise can have sub-enterprises.
2. All entities have other names. For example, a program can be a project comprising several subprojects (often called merely projects). Business can be an agency, team can be group, value can be utility, etc.
3. There is no attempt to be prescriptive in the names chosen for this diagram. The main goal of this is to show how this chapter uses these terms and how they are related to each other in a conceptual manner.

FIGURE 4.10   Organizations manage resources to create enterprise value. From SEBoK (2023). Used with permission. All other rights reserved.

Creating Value   As shown in Figure 4.10, an enterprise creates value for society, for other stakeholders, and for the organizations that participate in that enterprise. It also shows other key elements that contribute to the value creation process. There are many types of organizations to implement value-creating enterprises: businesses (companies), networks of companies, programs and projects, virtual organizations, etc. A typical business may participate in multiple enterprises through its portfolio of projects. A large SE project can be an enterprise in its own right (implemented as a virtual organization), with participation by many different businesses, and may be organized as a number of interrelated subprojects. In many cases, enterprises find themselves in a rapidly changing environment where stakeholder needs change over time. Therefore, an enterprise must constantly adapt its capabilities to meet the enterprise strategic goals and objectives.

Capabilities in the Enterprise   As shown in Figure 4.11, the enterprise acquires or develops systems or individual elements of a system. The enterprise can also create, supply, use, and operate systems or system elements. Since there could possibly be several organizations involved in this enterprise venture, each organization could be responsible for particular systems or perhaps for certain kinds of elements. Each organization brings their own organizational capability with them, and the unique combination of these organizations leads to the overall operational capability of the whole enterprise.

The word "capability" is used in SE in the sense of "the ability to do something useful under a particular set of conditions." This section discusses three different kinds of capabilities: organizational capability, system capability, and operational capability. It uses the word "competence" to refer to the ability of people relative to the SE task. Individual competence (sometimes called "competency") contributes to, but is not the sole determinant of, organizational capability. This competence is translated to organizational capabilities through the work practices that are adopted by the organizations. New systems (with new or enhanced system capabilities) are developed to enhance enterprise operational capability in response to stakeholder's concerns about a problem situation.

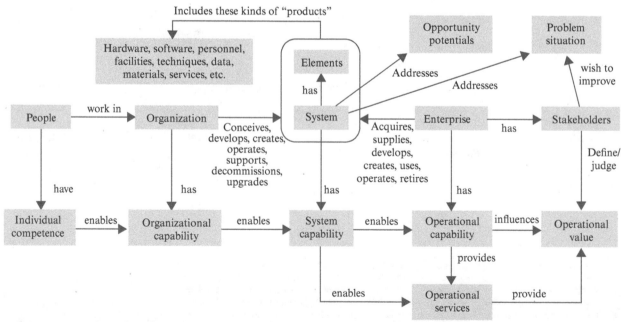

FIGURE 4.11   Individual competence leads to organizational, system, and operational capability. From SEBoK (2023). Used with permission. All other rights reserved.

As also shown in Figure 4.11, operational capabilities provide operational services that are enabled by system capabilities. These system capabilities are inherent in the system that is conceived, developed, created, and/or operated by an enterprise. Enterprise SE concentrates its efforts on maximizing operational value for various stakeholders, some of whom may be interested in the improvement of some problem situation.

Enterprise SE, however, addresses more than just solving problems; it also deals with the exploitation of opportunities for better ways to achieve the enterprise goals. These opportunities might involve lowering operating costs, increasing market share, decreasing deployment risk, reducing time to market, and any number of other enterprise goals. The importance of addressing opportunity potentials should not be underestimated in the execution of enterprise SE practices.

The operational capabilities of an enterprise will have a contribution to operational value (as perceived by the stakeholders). Notice that the organization or enterprise can deal with either the system as a whole or with only one (or a few) of its elements. These elements are not necessarily hard items, like hardware and software, but can also include "soft" items, like people, processes, principles, policies, practices, organizations, doctrines, theories, beliefs, and so on.

Enterprise Drivers and Outcomes   An enterprise needs to consider its own needs that relate to enabling assets (e.g., personnel, facilities, communication networks, computing facilities, policies and practices, tools and methods, funding and partnerships, equipment and supplies) when addressing the stakeholders' needs. The purpose of the enterprise's enabling assets is to effect state changes to relevant elements of the enterprise necessary to achieve targeted levels of performance. The enterprise "state" shown in Figure 4.12 is a complex web of past, current, and future states (Rouse, 2009). The enterprise work processes use these enabling assets to accomplish their work objectives to achieve the desired future states.

Since a high degree of complexity is to be assumed, it is advisable to apply formalized modeling methods to achieve the enterprise strategic goals and objectives. It has proven useful to use enterprise architecture analysis to model these states and the relative impact each enabling asset has on the desired state changes. This analysis can be used to determine how best to fill capability gaps and minimize the excess capabilities (or "capacities"). The needs and capacities are used to determine where in the architecture elements need to be added, dropped, or changed. Each modification represents a potential benefit to various stakeholders, along with associated costs and risks for introducing that modification.

Enterprise Opportunities and Opportunity Assessments   The potential modifications that are identified represent opportunities for improvement. Usually, these opportunities require the investment of time, money, facilities, personnel, and so on. There might also be opportunities for "divestment," which could involve selling of assets, reducing capacity, canceling projects, and so on. Each opportunity can be assessed on its own merits, but usually these opportunities have dependencies and interfaces with other opportunities, with the current activities and operations of the enterprise, and with the enterprise's partners. Therefore, the opportunities may need to be assessed as a "portfolio," or, at least, as sets of related opportunities. Typically, a business case assessment is required for each opportunity or set of opportunities. If the set of opportunities is large or has complicated relationships, it may be necessary to employ portfolio management techniques. The portfolio elements could be bids, projects, products, services, technologies, intellectual property, etc., or

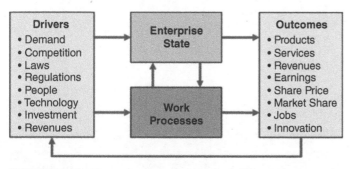

FIGURE 4.12   Enterprise state changes through work process activities. From Rouse (2009). Used with permission. All other rights reserved.

any combination of these items. Examples of an enterprise portfolio captured in an architecture modeling tool can be found in Martin (2005), Martin et al. (2004), and Martin (2003).

The results of the opportunity assessment can be compiled and laid out in an enterprise plan (sometimes conveyed as an enterprise ConOps) that considers all relevant factors, including system capabilities, organizational capabilities, funding constraints, legal commitments and obligations, partner arrangements, intellectual property ownership, personnel development and retention, and so on. The plan usually goes out to some long horizon, typically more than a decade, depending on the nature of the enterprise's business environment, technology volatility, market intensity, and so on. The enterprise plan needs to be in alignment with the enterprise's strategic goals and objectives and leadership priorities.

Practical Considerations   When it comes to performing SE at the enterprise level, Rebovich and White (2011) provide several good practices:

- Set enterprise fitness as the key measure of system success. Leverage game theory and ecology, along with the practices of satisfying and governing the commons.
- Deal with uncertainty and conflict in the enterprise through adaptation: variety, selection, exploration, and experimentation.
- Leverage the practice of layered architectures with loose couplers and the theory of order and chaos in networks.

Enterprise governance involves shaping the political, operational, economic, and technical landscape. One should not try to control the enterprise like one would in a traditional SE effort at the project level.

4.4   APPLICATION OF SYSTEMS ENGINEERING FOR SPECIFIC PRODUCT SECTOR OR DOMAIN APPLICATION

This chapter presents how SE is applied in different product sectors or application domains. For each of these, unique and domain-specific terms, concepts, activities, methods, and practices are introduced.

The following domains are presented in alphabetical order:

- Automotive Systems;
- Biomedical and Healthcare Systems;
- Commercial Aerospace Systems;
- Defense Systems;
- Infrastructure Systems;
- Oil & Gas Systems;
- Power & Energy Systems;
- Space Systems;
- Telecommunication Systems;
- Transportation Systems.

Note that the application of SE is not limited to the product sectors and application domains listed above. SE is a generic discipline that can be applied in most situations and domains (with varying levels of value). However, the details of how the practice is applied will vary in different product sectors or application domains.

4.4.1 Automotive Systems

Overview of SE Applications within the Automotive Domain   The automotive industry has a long history of engineering complicated and more and more complex consumer products, characterized by diverse product ranges, high production volumes, and a very competitive market. To make vehicles attractive, manufacturers have to balance efficiency for mass production with driving performance. Strong, well-orchestrated processes across the supply chain are key to meeting this challenge. Quality is highly dependent on good processes being followed rigorously, and economic performance relies on optimization. Electrification, connectivity, autonomy, and consumer choice are driving complexity, creating opportunities for SE to enter the mainstream in coming years. The INCOSE Automotive SE Vision 2025 document (2020) provides an excellent summary of the current and future trends in this domain. See Table 4.5 for a comparison of automotive with two other domains considering some of the characteristics that may affect SE approaches.

Emergence of SE in the Automotive Domain   Since General Motors vehicles first shared a common chassis in 1908, the automotive industry has employed extensive reuse in combination with variant management techniques to manage costs and keep delivery cycles short. Until the late 1960s, activities familiar to the SE domain centered on parts development, promoting reuse through standards and matrix organizations. Architects mainly addressed geometry until the on-board electronics and software revolution started in the 1970s. Standalone mechatronic and automated control systems appeared first, delivering applications such as engine control, anti-lock braking, and automatic heating and air conditioning. During the 1990s, vehicles acquired extensive networks of interconnected electronic control units. In the same period, many manufacturers invested heavily in engineering teams that focused on elicitation and refinement of stakeholder needs to address emergent properties such as safety, environmental impact, dynamic performance, and occupant comfort. This subset of activities (e.g., stakeholder requirements, electrical and electronic architecture) provided the ingredients for the emergence of SE in automotive.

Contemporary SE in the Automotive Domain   Increasing complexity, driven by the parallel trends of electrification, on- and off-board networks, automation, and autonomy has led to growing interest in automotive SE since 2000. Electronics and software began to outstrip mechanical design as a means for manufacturers to provide distinctive products. As shown in Table 4.6, the automotive industry is very standards-driven and the last decade has seen many new developments. These standards address, in particular, architectural approaches, software elements, safety, and security processes. Practices have evolved in isolated pockets, leading to significant variation in how the discipline is interpreted and executed. Convergence of capabilities at different maturity levels, led by engineering executives with different worldviews and backgrounds, has led to the adoption of many different SE paradigms with varying enthusiasm. SE practitioners might draw the system boundary at the vehicle level, or be restricted to applying their techniques to individual features, system elements, or domains, sometimes focusing on electronics and software. Approaches differ too, with some being requirements-led, others architecture-centric. SE has entered the automotive domain in incremental steps, and continues to do so, due to the need to accommodate the special features and legacy of each company. Disruptive change is hard for established players because their business models have low margins, and rely on trusted, repeatable processes that leverage extensive reuse of system elements. Similarly, production needs efficient processes to continuously deliver high volumes of system elements and assemblies on a just-in-time basis, making interruptions to manufacturing difficult to tolerate.

The explosion of electrical, electronic, and software systems is driving awareness that engineering based on assembling parts from suppliers without a systems approach is no longer enough. Full autonomy is on the horizon, and advanced driver assistance features like lane keeping and emergency braking are well established. Many new vehicles have high-integrity system elements linked to the outside world. Managing the complexity and cybersecurity risks this creates is a great challenge (see Section 6.5). Cultural and methodological changes are ongoing in industry incumbents. They increasingly face competition from newcomers with backgrounds in software-intensive industries (see Section 4.3.4), who in turn must adapt their culture and scale their business models to the realities of high-volume automotive manufacturing.

TABLE 4.5   Comparison of automotive, aerospace/defense, and consumer electronics domains

	Automotive	Aerospace/ Defense	Consumer Electronics
Customer requirements	Assumed by manufacturer	Defined by customer	Assumed by manufacturer
Legislative environment	Certification/inspection of product design, auditing of development and production processes (e.g., ISO/TS 16949 (2009)/IATF 16949, ISO 26262 (2018), E-marking, conformity of production, many regulatory Inc) standards (UNR, FMRSS, etc.) applicable to the overall system and its elements. Process regulations/standards are a new development.	Certification (e.g., DO-178C from RTCA Inc) in process, tooling and product. Auditing of development and production processes	CE/UL/FCC marking, according to a small number of standards (typically <10 per product)
User skill level	Somewhat trained	Highly trained	Untrained
Complexity (PLE/ component reuse)	High	Low	Medium
Complexity (sociodynamic)	High	High	Low
Product development cycle time	Medium (3–7 years)	Long (10+ years)	Short (1–2 years)
Delivery cadence	Annual	Decade	Annual
Product design life	Medium (10–20 years)	Long (30+ years)	Short (1–2 years)
Approach to maintenance	Repair	Repair	Replacement
Connectivity need	Medium (trend to high)	Low-medium (defined in stakeholder requirements)	Medium-high
Number of external integration interfaces	Medium (trend to high)	Medium (trend to high)	Medium
Safety/ cybersecurity criticality	High	High	Medium-high
Typical industry operating margin	4–6%	6–8%	8–10%
Annual production volume	10k–900k	100–100k	100k–100M

New Eco-Systems Involving the Automotive Industry: The Example of "Mobility as a Service"   In many urban areas, local governments implement policies to reduce the number of private cars and to foster the deployment of mobility services as a complement to public transport. The number of privately owned, individual cars has decreased dramatically in some big cities. For instance, in Paris, capitol of France, the percentage of cars as a mode of transportation has decreased from 46% in 2002, to only 13% in 2022. Vehicle manufacturers should not expect that this trend

TABLE 4.6   Representative organizations and standards in the automotive industry

Organization/standard	Description
SAE International, formerly the Society of Automotive Engineers	One of the main organizations that coordinate the development of technical standards for the automotive industry. Currently, SAE International is a globally active professional association and standards organization for engineering professionals in various industries, whose principal emphasis is placed on transport industries such as automotive, aerospace, and commercial vehicles
Japan Society of Automotive Engineers (JSAE)	An organization that sets automotive standards in Japan, analogous to the SAE
Association for Standardization of Automation and Measuring	An incorporated association under German law whose members are primarily international car manufacturers, suppliers, and engineering service providers from the automotive industry. The ASAM standards define protocols, data models, file formats, and application programming interfaces (APIs) for the use in the development and testing of automotive electronic control units
AUTomotive Open System ARchitecture (AUTOSAR)	An open and standardized automotive software architecture, jointly developed by automobile manufacturers, suppliers, and tool developers. Some of its key goals include the standardization of basic system functions, scalability to different vehicle and platform variants, transferability throughout the network, integration from multiple suppliers, maintainability
The GENIVI Alliance	A nonprofit consortium whose goal is to establish a globally competitive, Linux-based operating system, middleware, and platform for the automotive in-vehicle infotainment (IVI) industry. GENIVI specifications cover the entire product life cycle and software updates and upgrades over the vehicle's lifetime
ISO/TS 16949 (2009)/ IATF 16949	An international standard for particular requirements for the application of ISO 9001 quality management systems for automotive production and relevant service part organizations
IEC 62196 (2022)	An international standard for set of electrical connectors and charging modes for electric vehicles maintained by the International Electrotechnical Commission (IEC)
ISO 26262 (2018)	Road vehicles - Functional safety
ISO/SAE 21434 (2021)	Road vehicles - Cybersecurity engineering

will change. Vehicles they used to sell are more and more being replaced by public transport, biking, and walking, as well as services still involving vehicles like car-sharing or ridesharing. All these service offers can be integrated into a larger service enabling them to combine and thus making on-demand mobility faster and easier. This is called Mobility as a Service (MaaS), with a lot of initiatives around the world triggered by the Sustainable Development Goals introduced in Section 3.1.10. However, MaaS is not a silver bullet nor standard yet. MaaS may be considered as the mission for an SoS involving mobility service operators, both public and private. They have to cooperate in order to offer a more attractive user experience and at the same time to make the conditions for local mobility more sustainable.

The Future of SE in the Automotive Domain   Dealing with massively expanding complexity in an extremely challenging environment where standards and regulation trail fast-paced innovation is a challenge for this process-driven industry, but an opportunity for SE. As connectivity and autonomy become the norm, vehicles are built on highly configurable software platforms for providing mobility as a service. Development cycles that took five years are being compressed, where service updates that took a year will be expected in weeks. Delivering change like this means fundamental shifts in thinking are required across the board: from new business models, through service-centric architectures, to security-informed safety paradigms. SE is the means by which this can be achieved.

4.4.2 Biomedical and Healthcare Systems

Overview of SE Applications within the Biomedical and Healthcare Domain   SE has become more important to the healthcare industry (SEBoK, 2023), especially as systems and processes get more complex and quality characteristics such as safety, security, reliability, and human systems integration become more challenging. SE offers numerous benefits to biomedical and healthcare systems including the following:

- Supports design and development of healthcare systems using well-defined processes and standards,
- Offers well-defined approaches to design and implement architectures for proper interfacing, networking and communications using open industry standards,
- Enables operators and enterprises to scale up without compromising quality of operations,
- Enables better insights and control of many production systems including quality assurance, inventory, and cost control, and
- Augments user experience of various stakeholders like doctors, and surgeons by system level integration of emerging digital platforms like augmented reality, virtual reality, and robotics.

In the medical industry, especially for medical devices, it is important to understand that "risk management" is generally centered around product (user safety) risk and (called system safety in this handbook—see Section 3.1.11) rather than project (technical or business) risk (called risk management in this handbook—see Section 2.3.4.4).

Unique Considerations for Healthcare Delivery   SE applied to healthcare delivery differs significantly from conventional SE as applied in traditional fields such as defense, aerospace, and automotive. Most healthcare delivery projects involve improvement of an imperfect workflow or care process or the design of a limited scope new workflow or care process in a local clinic, hospital, laboratory, or in population health. If successful, solutions are shared with peer institutions in the same medical organization. As a result, most SE projects in healthcare delivery involve only a few stakeholders and a handful of requirements. Approaches leveraging lean SE have shown to be successful in many cases (Oppenheim, 2021) (see Section 4.2.3). Healthcare delivery operations have a critical need for the SE process to address pervasive healthcare problems such as care fragmentation (e.g., the systemic misalignment of incentives) or lack of coordination that spawn inefficient allocation of resources or harm to patients. Just as in medical device development, SE in healthcare delivery also strongly emphasizes patient safety. Methods such as the Systems Engineering Intervention for Patient Safety (SEIPS) (Carayon, 2006) focus on tailoring SE processes to the specific context of patient-centered medicine.

Unique Considerations for Medical Devices   In contrast to healthcare delivery systems, some medical device and healthcare IT companies use a more traditional form of SE. However, some are heavily tailoring SE approaches to incrementally demonstrate the value of SE. Many devices must work in harsh environments, including inside the human body. Interoperability, interconnectivity, and transportability are increasingly critical for medical devices and SaMDs (Software as Medical Devices). During audits and submissions, regulators require device developers to follow standard quality system processes (e.g., ISO 13485). Standards such as ISO 14971 (application of risk management to medical devices), IEC 60601 (medical device safety), IEC 62304 (Medical Device software—Software life cycle processes), and IEC 62366 (application of usability engineering to medical devices) are driving medical device organizations to take a deeper look into system safety and the engineering practices behind it. Thus, SE practitioners are increasingly being brought on board to leverage their life cycle management skills and support validating that the final product does indeed meet the needs of its stakeholders. In addition to an emphasis on systems safety, the medical device sector is seeing an increasing need for several SE methodologies including, but not limited to, SoS management, stakeholder management, agile systems development, trade analysis, MBSE, and PLE.

Unique Activities, Methods, and Practices   Healthcare Systems are often broad in context including a population of diverse patients, many healthcare professionals, many medical devices, many insurance companies, many delivery systems, regulators, and the government. One emphasis of SE practitioners in the biomedical and healthcare domain is patient safety risk, often more so than technical or business risks (see Section 6.1). Traceability is often a key factor in regulatory submissions and audits. Organizations that have strong SE practices are therefore in a better position to avoid pitfalls and to effectively defend their decisions if a regulatory audit does occur. In general, applicable standards do not need to be excessively tailored, although organizations with new or maturing practices may want to focus on lean implementations to obtain early and effective system adoption. Carefully balancing the trades between healthcare costs, better health outcomes for populations, and profits for shareholders is an ongoing challenge for Healthcare SE practitioners. On a larger scale, healthcare SE practitioners that can influence policy and incentives will become even more valuable to their organizations.

4.4.3   Commercial Aerospace Systems

Overview of SE Applications within the Commercial Aerospace Domain   SE is part of the strategies for the development of solutions and products in the commercial aerospace system domain. Commercial aerospace systems are complex, and their complexity continues to increase. The increased use of software makes it possible to implement more functions than before, which contributes to a further increase in complexity. At the same time, the expectation is raised that the increased use of software will make solutions and products available more quickly than the historic mechanical systems of the past. As shown in Figure 4.8, commercial aerospace systems are often part of larger SoSs. Future commercial aerospace systems will include autonomy, artificial intelligence, neural networks, novel propulsion, advanced human system integration (HSI), and cybersecurity.

Commercial aerospace systems use sequential as well as incremental and evolutionary life cycle models, including agile methods with smaller cycles. Thus, the processes in this handbook can be used to address and help organizations manage these new factors derived from complexity settings. The adoption of new technologies and perspectives emphasize some concepts such as the systemic approaches and use of SoS approaches to support organizations by putting them on the forefront of the market with competitive products, adapted to the new reality of increasing interoperability.

Unique Terms & Concepts   The commercial aerospace organizations of many countries have specific policies, standards, and guidebooks to guide the application of SE in their organizational environment. For example, ARP 4754A (2010) describes the standard practices for verifying commercial aircraft requirements.

There are many other systems related to this domain. For example, in the aviation domain, there are systems that go far beyond the aircraft itself according to the interaction characteristic of system elements described on system concept definition. Examples include an air traffic control system.

New applications of commercial aerospace systems are being continually introduced. For example, some organizations specifically created to address the new aerospace segment of flying cars have started a great race in a totally different way. By using new methodologies and approaches, these systems are being developed by considering their integration in completely new context. The same is happening with unmanned and autonomous vehicles. These new applications require an understanding of the ecosystem of the new operational contexts, as well as the lifestyles of its users.

Unique Activities, Methods, and Practices   SE may help the realization of effective commercial aerospace systems through the following activities, methods, and practices:

- **Stakeholders.** Stakeholders vary greatly and can range from federal government services, to aircraft manufacturers, to passengers.
- **Design and construction practices.** Model-based design is generally used from construction model specifications, which enables and maintains traceability between requirements and models.
- **Interfaces.** Because commercial aerospace systems' system elements are developed in various parts of the world and brought to a single (or multiple) location for assembly, adherence to interface management principles is critical.
- **Risk management.** Risk management is essential, especially for the introduction of new technologies.
- **Safety.** Finally, it is important for SE management to assure that safety is not compromised by organizational factors, as described by Paté-Cornell (1990).

Examples of how SE helps is resolving unique domain challenges include:
For aircraft original equipment manufacturers (OEMs), SE:

- Helps in design and manufacturing of aircraft subsystems, assembly and integration testing using well-defined process, standards, and quality standards.
- Offers well-defined approaches to create designs or architectures, processes, and roadmaps for proper interfaces, instrumentation, and communications that enable better visibility of the static and dynamic operational data and status of the subsystems.
- Enables operators and enterprises to scale up without compromising quality of production using a well-defined SE framework, tools, and emerging technologies.
- Enables better insights and control of congestion and traffic control of many schedules like flights, passenger, luggage, and food.
- Augments user experience of various stakeholders by system level integration of emerging digital technologies like augmented reality and virtual reality for enriched cockpit and instruments.

For airlines, SE:

- Helps in the support stage, to maintain the fleet.
- Helps balance performance and environmental impacts.
- Offers a set of procedures and activities to manage the services that consider human resources, information, and operation data.

Other Unique Considerations   SE is increasingly being applied in commercial practice. Petersen and Sutcliffe (1992), for example, discuss the principles of SE as applied to aircraft development. Life cycle functions of the commercial aerospace industry gives SE its own unique characteristics.

4.4.4   Defense Systems

Overview of SE Applications within the Defense Domain   While SE has been practiced in some form from antiquity, what has now become known as the modern definition of SE has its roots in defense systems of the twentieth century. It became recognized as a distinct activity in the late 1950s and early 1960s due to technological advances taking place that led to increasing levels of system complexity and systems integration challenges, and the need for SE further increased with the large-scale introduction of digital computers and software.

SE within defense evolved to address systemic approaches to issues such as the widespread adaptation of COTS technologies and the use of SoS approaches. It offers well-defined designs/architecture, processes and roadmaps for proper interfacing, networking, and communications. This enables better integrity and interoperability of real-time intelligence data across various devices, from various vendors, and platforms using open industry standards. Today, with increasing emphasis on networks and capabilities the defense organizations of many countries are recognizing the criticality of end-to-end SoS performance and increasing focus on integration to deliver these capabilities.

Unique Considerations   Defense systems have numerous characteristics and consequently, a huge complexity, making SE essential for their development:

- They are complex technical systems with many stakeholders and compressed development timelines.
- The systems must be highly available and work in extreme conditions all over the world—from deserts to rain forests and to arctic outposts.
- There are long system life cycles, so logistics is of prime importance.
- There is typically a strong human interaction, so usability/human systems integration is critical for successful operations.
- There is at times a need for defense operators and enterprises to accelerate development and production (e.g., quick response in event of national emergency or increased threats) without compromising quality of operations using a well-defined SE framework, tools, and emerging technologies

Unique Activities, Methods, and Practices   SE has a strong heritage in defense, and much of the SE processes in this handbook can be used as is in a straightforward manner, with normal project tailoring to address unique aspects of the project. It is important to note that as ISO/IEC/IEEE 15288 (2023) has evolved into a more domain- and country-neutral SE standard, so care must be taken to ensure that the defense focus is reasserted upon application. An example of specific implementation of ISO/IEC/IEEE 15288 when utilized for US Department of Defense projects is provided in IEEE 15288.1 (2014). This standard provides the basis for selection, negotiation, agreement, and performance of necessary SE activities and delivery of products. Additionally, the standard allows flexibility for both innovative implementation and tailoring of the specific SE processes to be used by system suppliers, either contractors or government system developers, integrators, maintainers, or sustainers. The defense organizations of many countries also have specific policies, standards, and guidebooks to guide the application of SE in their environment.

4.4.5   Infrastructure Systems

Overview of SE Applications within the Infrastructure Domain   This section addresses physical capital projects infrastructure including public works, transport, complex buildings, and industrial facilities. Within the infrastructure domain, SE practices are more developed in the high-technology system elements that involve software development, control systems, system security, or system safety. Infrastructure projects tend to define the high-level design solution without requirements decomposition, allocation, or interface identification. Architectures, traceability, and relationships within the project are often implied rather than specified. Infrastructure owners can benefit by applying SE to provide systematic, formal, verifiable connections between the business needs and the final product.

Unique Terms and Concepts   Infrastructure projects are distinguished from manufacturing and production, as they usually focus on unique, large physical systems where construction takes place on site rather than in a factory. These projects are adapted and integrated to existing environments, and are often characterized by loosely defined boundaries, evolving system architectures, multiphase implementation efforts which can exceed a few decades, and multiple-decade asset life cycles. As a result, stakeholders' expectations and design solutions evolve over an extended

TABLE 4.7   Infrastructure and SE definition correlation

Systems Engineering Term	Infrastructure Term	Recommendation
Acquirer	Owner or Agency	
Acquisition	Contracting phase; Procurement	Share good practices and lessons learned to improve procurement documents to enable better owner control of the project.
Business requirements	Project need; Business case	Derives contractor requirements from the business requirements, hold requirement reviews and include in the contractors' scope.
Configuration control	Versioning	Configuration identification, change management, status accounting, configuration audit according to ISO 10007 (2017).
Decision gate	Milestone	Clearly define entry and exit criteria for decision gates
Life cycle	Project life cycle	Include how the infrastructure will deliver its intended function and long-term asset management. Add in contractor's scope expectations that will benefit the entire project life cycle.
Performance requirement	Often found in Technical Specifications	Allocate top-level system performance requirements to system elements, defining performance requirements. Best performed by the acquiring entity unless the procurement method is a PPP.
Requirements	Design Criteria; Scope of Work; or Specifications	Integrate full life cycle considerations into design criteria, including operations, maintenance, and disposal/replacement planning.
Supplier	Design Consultant; Contractor	Use requirements management to strengthen procurement language and enforce contract requirements during the project. Clearly define acceptance criteria and performance measures.
System architecture	Context diagram; Schematics; Process and Instrumentation Diagrams	Consider creating early in project life cycle to support requirement allocation and interface management. Use ICDs or N^2 diagrams to complement the system architecture.
Verification	Design Review; Quality Control (QC)/Quality Assurance (QA)	Provide sufficient schedule and budget for both QC and Q A activities, including specific audit periods and "pens down" dates for each milestone. In the design phase, confirm the design meets requirements (QC), and procedures were followed (QA)
Validation	Construction Inspection; Quality Control (QC) / Quality Assurance (QA)	Include sufficient budget and authority for QA to ensure compliance. Ensure acceptance testing refers back to stakeholder needs and includes a focus on whether it meets its intended use.

timeframe. Unlike other SE domains, most infrastructure projects cannot be standardized and do not involve a prototype.

Many of the processes described in this handbook can be used to manage infrastructure projects but in some cases with different terminology, as illustrated in Table 4.7. There are some areas where existing infrastructure practices could be adjusted slightly to better align with SE practices.

Unique Activities, Methods, and Practices   SE may help the realization of effective infrastructure systems through the following activities, methods, and practices:

- **Stakeholders.** Stakeholders can range from governmental legislators who control funding for the project, to local/regional agencies that add beautification needs, to landowners with adjacent property impacted by a proposed project. In all government-funded projects, segments of the public may also be a stakeholder group. The wide array of potential stakeholders makes requirements gathering, cost, and schedules volatile. Public and political pressure can cause premature initiation of projects, with incomplete project scope and ill-defined metrics.

- **Design and construction practices.** Within infrastructure, the engineering disciplines have well-established, traditional practices and are guided by independent industry codes and standards that are not shared between disciplines. Design requirements are generally dissociated from construction specifications, therefore limiting traceability between design and construction.
- **Interfaces.** Infrastructure projects have external, often uncontrollable interfaces that can impact the project. Interfaces can include existing built systems, natural systems, environmental, and other internal and external dynamics.
- **Risks.** The contractual framework and allocation of liability and commercial risk are major factors impacting procurement and contracting processes. SE practices may therefore help to manage risk associated with cost estimating, changing scope, system integration, and verification. They may also improve construction productivity, making infrastructure development more cost-effective.

Other Unique Considerations   SE concepts are relatively newly applied in the infrastructure domain. As the application of SE grows within the infrastructure domain, an effort should be made to train engineering discipline specialists in SE concepts. Four key SE processes are useful to introduce SE on infrastructure projects: requirements management, interface management, verification, and validation. These processes can improve infrastructure project delivery, and total life cycle view that integrates design, construction, and asset management.

4.4.6   Oil and Gas Systems

Overview of SE Applications within the Oil and Gas Domain   The emergence of SE within the Oil and Gas (O&G) domain is relatively new compared to other sectors of similar complexity. Most applications of SE have occurred within the past decade to varying levels of implementation. Due to fluctuation in oil prices, new systems with increasing complexity and efforts to reduce greenhouse gas emissions have motivated a risk-averse industry to adapt to, and in some cases drive, change. This has encouraged an entire culture known to resist change to challenge assumptions and traditional ways of working, especially working in a document-centric environment.

The greatest SE-related need has been in the system requirements definition and requirements management space. With a domain-wide focus on digitalization, the change in how requirements are defined and transmitted throughout the supply chain has benefited from an SE approach. The industry leaders have either switched, or are switching, to data-centric requirement sets. There has been collaboration between suppliers and acquirers to improve the quality and traceability of requirements and create metrics for measurement of progress. INCOSE and American Petroleum Institute (API) cooperated on some trials in 2017 and 2018 that explained the aims and elements of good requirement writing to a panel of experts involved in updating a standard. They then supported the engineers in the writing and recrafting of the content, resulting in higher quality requirements and clearly separating between instruction, information, and verification (IOGP, 2021).

Beyond requirements, additional SE practices are also being introduced in the O&G domain. For example, SE practitioners take advantage of requirement definition to develop system architectures and systematically define interfaces. By using requirement management tools, configuration management and change management of requirements can be implemented in projects. In other cases, systems thinking tools, such as context diagrams and functional trees, are used as a foundation for SE practices, such as functional modeling. Technical requirements are also leading to conversations and implementation of verification and validation strategies and realization. With the companies incorporating digital design data across disciplines and throughout the life cycle, the digitalization of requirements has led to the auto-creation of specifications and test plans.

Unique Considerations for SE within Oil and Gas   One of the challenges when considering SE in O&G is that it is difficult to evaluate the entire domain as one. The long and complicated supply chain includes diverse and segmented companies across the globe. And it is not solely crude oil or natural gas. With the current energy transition affecting all aspects of engineering, most O&G companies have been shifting focus from fossil-based systems to include renewable sources that are efficient solutions with net-zero emissions. Many oil and gas companies have targets of net-zero greenhouse gas emissions for operations by 2050.

Another challenge for the domain is that since SE has not been implemented as a holistic approach, there are pockets of SE maturity that do not always intersect. This is a result of most O&G companies following a well-established and practiced sequential (waterfall) stage gate process. Therefore, SE implementation is generally only implemented where a clear case for change is needed and demonstrated, or when all other approaches have been exhausted. To help with this, the INCOSE O&G working group developed a scalable presentation for various high-level conversations and presenting success case studies from participating O&G companies. One area where SE continues to gain traction and show value is in new product development projects. By introducing SE principles and methods early in the project, the project team can see the benefits of applying SE and embrace the changes to the traditional ways of working.

4.4.7   Power & Energy Systems

Overview of SE Applications within the Power and Energy Domain   During the first two decades of the twenty-first century, the global energy system has been subject to a complex set of requirements stemming from the Paris Agreement, the United Nations Sustainable Development Goals (see Section 3.1.10), reduction in greenhouse gas emissions, and other efforts to avoid degradation of our social foundations and ecological ceiling (Raworth, 2017). To provide an effective solution for such a complex set of problems demands the realization, or modification, of many new systems, elements, and enabling systems supported by a holistic systems approach. The United Kingdom's (UK) Council for Science and Technology stated with respect to the UK's Net Zero ambitions that "by drawing on SE principles, a detailed and credible plan can provide the framework required to drive change, give reassurance to businesses, investors and consumers, and engage the whole of society in delivering this change" (CST, 2020).

At the heart of the sustainability transition is the convergence of business, technology, and socio-politics to guide innovation around what is viable, feasible, and desirable. This new intersection of disciplines can be enabled through SE. Yet many incumbent organizations and legacy approaches dominate the power and energy landscape, meaning a change in thinking or practice is resisted, or even directly opposed. As a result, the application of SE in power and energy remains largely immature in the first quarter of the twenty-first century compared to more established sectors such as defense, space, and transportation. A cultural shift toward shared knowledge management systems, portfolio management, and organization infrastructure is required to fully embrace and adopt SE.

Unique Terms and Concepts   The language used in SE is made effective through SE heuristics or when translated to real-world examples in the power and energy context. For example, an SoS may be considered an abstract SE term, yet it perfectly describes the nature of distributed energy resources.

Unique Activities, Methods, and Practices   SE may help the realization of effective power and energy systems through the following activities, methods, and practices:

- **Architecture and design.** Provides robust architecture, design, and development processes for higher integrity and interoperability across the supply chain infrastructure from upstream (e.g., solar plant, nuclear, wind farms), mid-stream (e.g., large-scale storage, energy vector processing, transmission and distribution networks) and downstream (e.g., retail outlets, local area networks, domestic microgeneration, private storage).
- **Risks.** Enables better identification and handling of risks associated with energy security and resilience.

- **Portfolio management.** Provides the platform for joined up roadmaps and communication channels which enables stakeholder acceptance, transition, and utilization of emerging paradigms such as smart grids, district heat networks, renewable technologies, demand side response, and electric vehicles.
- **Sustainability.** Enables better management of reductions in greenhouse gas emissions through robust technical and management processes with a whole life cycle perspective, open industry standards, quality management, and assurance.

Other Unique Considerations   Power and energy systems are typically at the SoS level, so activities follow the key characteristics and challenges associated with an SoS. As an example, achieving the goal of energy security requires as much understanding of geo-politics as it does the evolution of cybersecurity as digitalization grows.

The application of SE needs to transcend geographic boundaries and domain silos. For example, consideration for how SE supports the implementation of the Clydebank declaration, which calls for the establishment of green shipping corridors for zero-emission maritime transport between shipping ports, presents an energy, transport, and logistics challenge on an international scale.

On a global scale, power and energy systems must remain persistently operational for billions of system users whilst maintaining a constant state of equilibrium with demand balanced by supply (augmented by flexibility solutions such as demand side response and storage). In addition, power and energy systems typically have a life cycle of 30–100 years or even into the thousands of years for end-of-life decommissioning and waste storage from nuclear fission facilities. These considerations form complexity multipliers for the sustainable energy transition.

Adapting the mental models and behaviors of system users will be crucial to effecting change. This demands elements of social sciences, systems science, and systems thinking to complement the rigor of SE processes. To support this, there is a need to provide feedback mechanisms to influence the micro-behavior of the human actors in the system in such a way as to maintain the macro-stability of the system (Sillitto, 2010) achieved through HSI (see Section 3.1.4). But SE cannot focus on change in human behaviors without consideration for market dynamics and political levers. SE can help provide the coherence and joined-up thinking necessary to make energy policy an enabling system for delivering the overarching goals of energy decarbonization, digitalization, decentralization, and democratization. We must also avoid the trap of pushing technology solutions that deliver undesirable user experience. Considerations range from consumer price point, to acoustic noise of technologies, to retrofit disruptions, to the availability of energy in remote or isolated communities. Our challenge, as future ancestors, will be to find ways to support growing energy demands whilst delivering a sustainable energy supply chain that is available and affordable to everyone on a global scale.

4.4.8 Space Systems

Overview of SE Applications within the Space Domain   Space systems are systems that are designed to operate and perform tasks into and within the space environment. This may consist of: spacecraft (and their associated payloads and instruments); mission packages(s); ground stations; data links between spacecraft and ground, launch systems; and directly related supporting infrastructure. Due to the relatively high costs of deploying assets into earth orbit or beyond, space systems typically require high reliability with little maintenance other than software changes (note that designing for maintainability in space one of the many trade-offs that need to be considered during conceptual design). This makes it necessary for all system elements to work the first time or be compensated by operational workarounds; this can impact the risk posture of the system being developed.

The space domain has evolved into three main areas of interest, with some overlap:

- Civil,
- Commercial, and
- National Security Space.

Each of these areas have their own motivations that can influence the way they develop systems.

Unique Activities, Methods, and Practices   Key emphases of SE in the space domain are integration, verification (including testing), and validation of highly reliable, well-characterized systems. Risk management is also key in determining when to incorporate new technologies and how to react to changing requirements through multiyear developments and programmatic challenges. SE provides coordination for multi-disciplinary engineering expertise that enables optimized designs.

Civil systems are typically acquired by government agencies, which typically focus on performance risk, determining when to incorporate new technologies, and how to react to changing requirements through multiyear developments and programmatic challenges. This lends itself to the use of the sequential approaches, such as the SE Vee model, or, in some cases, the waterfall model.

Commercial systems strive for profitability (cost and schedule) and are more amenable to using incremental and evolutionary approaches. This allows them to deploy systems faster, and rapidly gain experience that can improve later iterations of their product.

National Security Space, much like Civil, may emphasize performance over cost and schedule, and have traditionally used the Vee and waterfall models for development. They typically are more tolerant of accepting risk from the injection of new technologies.

Unique Standards   Overall, proper application of SE in the space domain helps in design and development of space elements for easier manufacturing and lowering maintenance cost using well-defined processes and standards. SE offers well-defined architecture and design processes and roadmaps for proper interfacing, networking, and communications that enable better integrity and interoperability of space systems and elements provided by various contractors, and across the supply chain using open industry standards.

Civil, commercial, and national security entities have their own drivers that determine how standards are created and adopted. Most space-faring nations and international consortiums have developed and adopted their own standards that are specific for space systems. Some examples include:

- In the United States, the Department of Defense has created Military (or "Mil") Standards that have been readily adopted by both Civil and Commercial primarily due to the need for high reliability and survivability in a hostile environment. NASA has also created a set of technical standards, as has the AIAA and other organizations.
- The European Cooperation for Space Standardization is an initiative established to develop a coherent, single set of user-friendly standards for use in all European space activities.
- The Euro-Asian Council for Standardization, Metrology and Certification, a regional standards organization operating under the auspices of the Commonwealth of Independent States, has developed GOST (Russian: ГОСТ),a set of spacecraft certification standards that is commonly used by Russia.
- JAXA (Japan Aerospace Exploration Agency) has developed a library of standards known as JERG (JAXA Engineering Requirement, Guideline).
- International organizations such as ISO and IEEE have also been involved in the development of standards that enhance interoperability.

Other Unique Considerations   An additional challenge in the space domain occurs when humans are integrated into the system. Typically, it includes the incorporation of design features and capabilities that accommodate human interaction with the system to enhance overall safety and mission success. The system needs to ensure that the human needs are addressed in terms of effectively utilizing human capabilities and performance, hazards are controlled to a level considered safe for human operations, and provide, to the maximum extent practical, the capability to safely recover the crew from hazardous situations. At the time of this writing, only missions led by three nations (Russia, the United States , and China) have sent humans into space. Each country has developed their own set of standards:

- Rovcosmos for Russia, and
- Human Space Flight Requirements for Civil, typically governed by NASA, for the United States,
- CNSA for China.

Of the three, only the United States has begun to explore human commercial space flight, where such requirements are governed by the US Federal Aviation Administration (FAA).

4.4.9 Telecommunication Systems

Overview of SE Applications within the Telecommunication Domain   Telecommunication systems are defined by having a route to transfer information across and to distinct endpoints that are used to share (send and/or receive) information. They differ from postal systems in that the information shared is in the form of electronic media (applications or services) rather than transporting packages or handwritten letters.

Telecommunication systems are enablers for other services. Almost all modern systems either make use of telecommunication technologies provided by other systems, or contain telecommunication technologies within them (e.g., digital signage and ticketing systems within public transport systems; battlespace communication systems used by the military; environmental monitoring systems such as those used to monitor/predict the weather or detect and provide advance warning of earthquakes and other environmental events). Lives and livelihoods depend on telecommunication systems.

Communication network complexity and the social cost of telecommunication failures will only increase (White and Tantsura, 2016). It is therefore opportune for telecommunication leaders and practitioners to advocate, apply, and extend the best telecommunication SE approaches to cope with this complexity and risk.

Unique Terms and Concepts   Telecommunication systems are built on a wide range of technologies: satellite communication, cellular networks, land mobile radio, microwave, radio, television, Wi-Fi, Bluetooth, and global positioning systems. They are increasingly software-intensive systems (Donovan and Prabhu, 2017). Telecommunications includes communication networks owned by carriers, internet service providers, government agencies, and other enterprises, as well as broadcast networks (e.g., radio, cable, television) and over-the-top service provider applications (e.g., messaging, video conferencing, social media applications) (Adkins, et al., 2020) (Birman, 2012). Some telecommunication systems, like the internet and the public switched telephone network, have no single owner; their design depends upon collaboration in international telecommunication standards bodies.

The telecommunication transport network may be dedicated for a specific purpose (service or application) or shared for multiple services or applications. Communication networks may be employed for emergency services, defense, transportation, health, financial, industrial supervisory control, and data acquisition purposes. Many of these are considered to be national critical infrastructure (Lewis, 2019).

Telecommunication systems typically have some of the following characteristics:

- Diverse geographical distribution;
- Multi-party ownership and management (network domains or applications);
- Multiple constituent systems with independent life cycles that continuously evolve over many decades;
- A small stable set of functions, allocated across system elements, to achieve the common purpose of enabling communication;
- Many nodes and types of nodes; and
- Strong interdependence between nodes (failures within one node may cause other nodes to become isolated unless the specific failure mode is anticipated and the network is designed to withstand the failure).

Unique Activities, Methods, and Practices   Network specifications, architectures, and models have a small number of functions and node types at their core, but they must also support many function and node variants. Engineering planning and architecture must be flexible enough to accommodate change in parts of the network owned and operated by others. Design and verification activities depend on a thorough analysis of failure modes and interactions across the network. Scale and variation lead to significant configuration management challenges (Xu and Zhou, 2015). These characteristics affect engineering activities throughout the life cycle. Like other types of networks, communication networks can be represented by a network model that defines a grouping of nodes and links to help understand how resources flow from one node to another.

While some telecommunication systems in slowly changing environments can survive on implicit engineering practices, such approaches have been found to be ineffective and inefficient. They typically fall short when one or more of the following exist:

- New and/or complex stakeholder needs;
- New operating models;
- Significant safety or security risks;
- High complexity; or
- Constrained, high-cost operating environments.

Other Unique Considerations   Many vendor services and technologies are mature and are slow to evolve. Telecommunication networks typically comprise COTS vendor equipment/applications using industry interface standards and semi-standardized architectures (see Section 4.3.3). This equipment is typically integrated together without the use of SE. The promise from COTS vendors is that their equipment is suitable for rapid configuration and integration, and implicitly can survive with simpler requirements/business analysis and engineering processes, including outsourced vendor standardized engineering. This can lead to situations where there are unexpected outcomes. Without an appropriate set of engineering disciplines, such as those based on SE, it is difficult to assess, let alone manage, the risks.

4.4.10 Transportation Systems

Overview of SE Applications within the Transportation Domain   Ground transportation systems, such as highways, busses, people-movers, mass transit, and rail involve complex capital programs for fleet acquisition and/or building of related infrastructures and are invariably within a SoS on an operational level. The system life cycle for ground transportation assets is 25 to 100+ years and often involves public funding and a related fiducial public trust. During this life cycle, operational processes are continuously optimized and improved using block changes.

The ground transportation industry segment is an emerging SE practice area, largely consisting of transit authorities, railroad operators rolling stock manufacturing industries, and civil engineering construction firms. Globally, some geographic regions are progressing the deployment and acceptance of SE more rapidly. They have created effective SoS approaches that are in concert with emergent societal trends such as smart cities, an embedded safety culture, and a through life approach to service delivery. For example, in the UK, the Institution of Civil Engineers has made solid progress in moving toward an SE approach within the civil engineering domain. Other regions are at the early stages of integrating SE into their ground transportation systems processes. These imbalances can be overcome in time through global collaboration.

Unique Considerations   Additional mandates for the use of SE methods, skills, and competencies are generated from rapidly increasing system complexities in modern transportation systems, as evidenced by Intelligent Transportation System (ITS) concepts including regional, national, and local smart cities initiatives. Train control automation, bus scheduling, and ride-share coordination, coupled with autonomous vehicles, micro-grid hybrid traction power, passenger fare collection, and related trip planner smartphone apps provide emerging complexities within modern ground transportation capability. However, public transportation must improve as public needs change, new technologies are introduced, or as environmental concerns and quality-of-life concerns encourage greater use of public transit. Ground transportation services are typically "in-service" brownfield systems and must migrate to an updated version, while still maintaining service, so that assets may be cost effectively managed in the public's interest (see Section 4.3.2). Some transit assets in large cities run 24/7 service as a matter of public safety and demand, requiring careful planning. A "whole" organization approach is needed for success.

Unique Terms and Concepts   The transportation domain uses the same terms as the infrastructure domain, see Table 4.7.

Unique Activities, Methods, and Practices   Some examples of early initiatives on the application of SE are in the United Kingdom, the European Union, and, specifically, the Netherlands. In the UK, Network Rail established the Governance for Railway Investment Projects (GRIP) to help manage projects and to align them with SE processes. Leading EU train operating companies and the main rolling stock manufacturers seek to develop a common, open architecture such as EULYNX and the Reference Command and Control System Architecture to support and optimize trackside and rolling stock acquisitions and upgrades. In the Netherlands, SE is advocated by the network management institution, ProRail, as part of their project management approach for delivering rail infrastructure projects. ProRail created a SE Handbook for their project life cycle approach which is comparable to GRIP.

5

SYSTEMS ENGINEERING IN PRACTICE

5.1 SYSTEMS ENGINEERING COMPETENCIES

The terms "competence" and "competency" are two distinct terms used to define the personal attributes of individual Systems Engineering (SE) practitioners. Competence is the ability to perform an activity or task. Competency is the set of skills required in the performance of a job. The competence reflects the total capacity of the individual, whereas a competency is a set of skills that the individual will be required to perform for a job. The sum of an individual's competencies will make up their competence. Those competencies are measured and assessed to provide an estimate or a picture of the overall competence of a SE practitioner.

The INCOSE Systems Engineering Competency Framework (SECF) (2018) provides a set of competencies that identify knowledge, skills, abilities, and behaviors important to effective SE that can be applied in any domain context. The INCOSE SECF spans a wide range of competencies organized into five themes: core, professional, technical, management, and integrating.

- The *core competencies* underpin engineering, as well as SE.
- The *professional competencies* are primarily based in behavior as a SE practitioner (see Section 5.1.2).
- The *management competencies* relate to performing tasks associated with controlling and managing SE activities.
- The *technical competencies* are associated with technical processes to accomplish SE.
- The *integrating competencies* recognize SE as an integrating discipline, considering activities from other specialists from project management, logistics, quality, and finance to create a coherent whole.

The INCOSE Systems Engineering Competency Assessment Guide (SECAG) (2022) provides guidance on how to evaluate individuals for proficiency in the competencies and how to differentiate between proficiency at each of five levels defined within the INCOSE SECF. For each competency, the INCOSE SECAG provides a description, why it

INCOSE Systems Engineering Handbook: A Guide for System Life Cycle Processes and Activities, Fifth Edition.
Edited by David D. Walden, Thomas M. Shortell, Garry J. Roedler, Bernardo A. Delicado, Odile Mornas, Yip Yew-Seng, and David Endler.
© 2023 John Wiley & Sons Ltd. Published 2023 by John Wiley & Sons Ltd.

matters, and possible contributory evidence. Indicators of competence are provided along with examples of relevant knowledge, experience, and possible objective evidence of personal involvement in activities or professional behaviors applied.

Both the INCOSE SECF and INCOSE SECAG are comprehensive resources that are globally accepted and tailorable to the needs of the organization or individual. The INCOSE SECF and INCOSE SECAG are intended for use in hiring, assessing, training, and advancing SE practitioners; and may be used for other purposes as deemed appropriate.

Given the complexity and multidisciplinary nature of today's systems and system of systems, it is not possible for a single person to know everything about a system of interest (SoI). As shown in Figure 5.1 SE practitioners are often referred to as "T-shaped" in describing their professional expertise (Delicado, et al., 2018). SE practitioners must have both a depth of knowledge of a fundamental engineering discipline, while at the same time develop and maintain a breadth of knowledge about systems and the multiple disciplines involved.

5.1.1   Difference between Hard and Soft Skills

"The complexity of modern system designs, the severity of their constraints, and the need to succeed in a high tempo, high-stakes environment where competitive advantage matters, demands the highest levels of technical excellence and integrity throughout the life cycle" (INCOSE SECF, 2018, Page 47). Interactions with stakeholders, including customers, project managers, all types of engineers, operations, marketing, and various departments contribute to the overall success of a system and to the life cycle cost (LCC) to conceive, produce, utilize, support, and retire that system. SE practitioners and SE teams need both hard skills and soft skills to meet these challenges. Table 5.1 contrasts the differences between hard skills and soft skills. As examples in the hard skills category, consider typical engineering technical aspects like structural, hydrodynamic, reliability, or electrical analyses, because they require science, mathematics, and quantitative modeling in order to solve engineering-related problems. Examples of hard skills for SE practitioners include requirements analysis, architectural evaluation, and risk management. As examples in the soft skills category, consider influencing a peer, motivating a team, and resolving a conflict. They require emotional intelligence and appropriate behaviors to solve people-related problems.

Soft skills, which are also known as interpersonal skills, intrapersonal skills, people skills, professional skills, and other terms, are those skills related to aspects such as teamwork, collaboration, and facilitation. Understanding the nuances of an organization, the dynamics of a team, or the experiences of an individual requires soft skills. It is not uncommon for an engineer to question the value of soft skills, and question how these skills apply to a specific engineering problem. However, soft skills often make the difference between a smooth application of SE and one riddled with challenges. Both the INCOSE SECF and INCOSE SECAG outline areas of professionalism and ethics that provide insights and information related to important soft skills for ensuring long-term sustainability for the life cycle of a system.

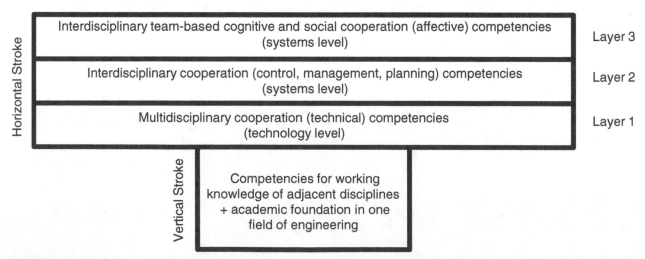

FIGURE 5.1   The "T-shaped" SE practitioner. From Delicado, et al. (2018). Used with permission. All other rights reserved.

TABLE 5.1   Differences between the hard skills and soft skills

Hard Skills	Soft Skills
• Concrete definitions	• Subjective definitions
• Measurable	• Difficult to measure
• Testable	• Difficult to test
• Individual application	• Social application
• Low self-awareness required	• High self-awareness required
• Personal affect has low impact	• Personal affect has high impact

INCOSE SEH original table created by McCoy and Whitcomb on behalf of the INCOSE Professional and Soft Skills Working Group. Usage per the INCOSE Notices page. All other rights reserved.

The field of engineering brings a unique perspective toward developing competencies in these hard and soft skills. Early in the career of an engineer, there is a strong emphasis toward competency in the hard skills, such as those related to engineering, math, physics, chemistry, industrial processes, and technical management. The INCOSE SECF outlines these skills in the areas of Technical Competencies, Management Competencies, and Core Competencies. As the career of an engineer develops, the emphasis moves toward a balance between hard and soft skills. For an SE practitioner, a higher demand is often placed on these soft skills due to interactions with stakeholders, team members, and senior managers. The INCOSE SECF outlines these skills in the Professional Competencies.

5.1.2   System Engineering Professional Competencies

Professionalism can be summarized as a personal commitment to professional standards of behavior, ethics, obligations to society, the profession, and the environment. SE practitioners are trusted to apply reasoning, judgment, and problem solving to reach unbiased, informed, and potentially significant decisions because of their specialized knowledge, skills, abilities, and behaviors. SE professionalism includes consideration of personal behaviors beyond using methods and tools. SE practitioners recognize the benefits of behaviors and outcomes related to professional competencies from ethics, professionalism, and technical leadership to communications, negotiation, team dynamics, facilitation, emotional intelligence, coaching, and mentoring. These Professional Competencies are documented in the INCOSE SECF (2018). The evaluation of an individual's Professional Competencies can be accomplished using the INCOSE SECAG (2023).

5.1.3   Technical Leadership

Leadership can be generally defined as: "The act (or art!) of enabling people to produce results or achieve outcomes they would not have on their own." Technical Leadership is leadership in situations that involve technology. As illustrated in Figure 5.2, Technical Leadership exists at the intersection of Technical Expertise and Leadership Skills.

Strong technical leadership is critical for the successful development, operation, sustainment, and evolution of engineered systems. To successfully lead technical teams and technical enterprises, technical leaders must possess all the leadership skills required of any effective leader. Technical leaders must also possess, and be recognized as possessing, significant technical expertise. A good technical leader must possess expertise in one or more technical areas and have some level of understanding across a wide range of disciplines to earn credibility for leading technical teams. They must also be aware of the limits of their own knowledge so they know when to seek the expertise of others.

Why Is Technical Leadership Important for SE Practitioners?   SE practitioners are responsible for the success of the system as a whole. They must understand the needs of a broad range of stakeholders and ensure those needs are met. They must work across traditional engineering disciplines to ensure the individual contributions of each integrate harmoniously to produce the desired outcome. Because they seldom have the positional authority to ensure these outcomes, they must lead through influence, leveraging their technical knowledge and their personal qualities to create an environment in which the individuals and teams accomplish the desired goals.

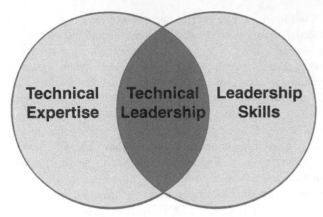

FIGURE 5.2 Technical leadership is the intersection of technical expertise and lead-ership skills. INCOSE SEH original figure created by Gelosh and Pennotti on behalf of the INCOSE Technical Leadership Institute. Usage per the INCOSE Notices page. All other rights reserved.

What Does It Take to Be a Good Technical Leader? The INCOSE Technical Leadership Institute (TLI) developed a model that identifies and describes six interrelated behaviors that technical leaders must master to successfully lead through influence. Each behavior is described in Table 5.2, and each includes a question technical leaders should continually ask themselves as they seek to lead.

5.1.4 Ethics

There will always be pressure to cut corners to deliver projects faster or at lower costs, especially for a profession such as SE. As stated in the INCOSE Code of Ethics (2023),

> The practice of SE can result in significant social and environmental benefits, but only if unintended and undesired effects are considered and mitigated.

TABLE 5.2   Technical leadership model

Technical Leadership Behavior	Description	Question for the Technical Leader
Holding the Vision	A vision is an aspirational statement that defines who we are and where we want to go. It provides an impelling purpose. It must be the start of a continual and ongoing conversation that SE practitioners are well positioned to support, reinforce, and encourage.	*What outcome are we striving to achieve and how am I advancing that vison?*
Thinking Strategically	Using Strategic Thinking, technical leaders formulate a hypothesis before acting, treat the action as an experiment to test their hypothesis, and, based on the results they observe, continue along the path they are on or formulate a new hypothesis and begin the testing anew.	*What patterns are emerging from my actions, and what are the implications for what I should do next?*
Fostering Collaboration	Complex problems cannot be solved by individuals working alone; their solution requires the efforts of many. The goal is to foster collaboration that allows new ideas to emerge through creative conflict and experimentation.	*What relationships am I building today for myself and for others?*
Communicating Effectively	Effective communication is not just about speaking, it is also about actively listening which requires attention to both the content being spoken and the emotion with which it is communicated.	*Who am I trying to influence and what is their greatest challenge?*
Enabling Others to Succeed	The technical leader's role is to influence, guide, encourage, and support those who can produce desired outcomes. The leader's success will derive from their success, and their acknowledgment that the leader contributed to it.	*What obstacles are preventing others from acting, and how can I help remove those obstacles?*
Demonstrating Emotional Intelligence	In order to lead others, technical leaders must continually seek feedback to decrease their blind spots and be willing to reveal things that help others know them better. While the former can make them uncomfortable and the latter make them more vulnerable, the payoff will be more than worth the effort.	*That am I afraid of admitting to others and how might disclosing it improve our relationship?*

INCOSE SEH original table created by Gelosh and Pennotti on behalf of the INCOSE Technical Leadership Institute. Usage per the INCOSE Notices page. All other rights reserved.

Part of the role of the SE practitioner as a leader and professional is knowing when unacceptable risks or trade-offs are being made, knowing how to influence key stakeholders, and having the courage to stand up for stakeholders, the community, and the profession when necessary. The INCOSE Code of Ethics contains sections on "Fundamental Principles," "Fundamental Duties to Society and Public Infrastructure," and "Rules of Practice" to help the SE practitioner in practical applications of ethics to their work and daily lives.

5.2   DIVERSITY, EQUITY, AND INCLUSION

The following definitions are taken from the Accreditation Board for Engineering & Technology (ABET) (2017) and provide a reference point for conversations and materials about diversity, equity, and inclusion.

- *Diversity* is the range of human differences, encompassing the characteristics that make one individual or group different from another.
- *Equity* is the fair treatment, access, opportunity and advancement for all people, achieved by intentional focus on their disparate needs, conditions, and abilities.
- *Inclusion* is the intentional, proactive, and continuing efforts and practices in which all members respect, support, and value others.
- INCOSE uses the compound term Diversity, Equity, and Inclusion (abbreviated to DEI) when referring to the broad subject matter.

Diversity encompasses a wide range of characteristics. Figure 5.3 shows a representative set of characteristics grouped into five areas: intrinsic, employment, environment, interaction, and family (Harding and Pickard, 2019).

FIGURE 5.3   Categorized dimensions of diversity. From Harding and Pickard (2019) derived from SEBoK (2023). Used with permission. All other rights reserved.

DEI is vital to successful SE because of the wide range of contexts in which SE is applied and the consideration of multiple stakeholder viewpoints at the heart of the systems approach. SE practitioners play a pivotal role in integrating DEI concepts into the team's composition and approach and in the system design and development process through:

1. Ensuring that the SE team and its leadership is inclusive, welcomes a diverse range of talent, promotes cognitive diversity and diversity of ideas, and, where necessary, takes deliberate action to provide psychological safety and communication equity.
2. Ensuring that the systems we realize are as accommodating as possible of the differences within the entire stakeholder community.

Failure to address either of these aspects results in sub-optimal outcomes, whether in terms of missed solutions, lower productivity, or delivering a system that does not equitably meet the needs of the full range of the stakeholder community (i.e., essentially failing to meet the ultimate goal of delivering a total optimal system solution for all). Furthermore, in providing total optimal system solutions, one purpose of the system is to provide similar (equal) outcomes to each user. To accomplish this, the SE practitioner must understand the difference between equality and equity. Equality is providing each user exactly the same resources but may not result in similar (equal) outcomes. Equity, however, is proactively addressing disparities that exist between individual users (e.g., different situations, needs, requirements, life experiences, challenges) and changing the system so that each user can experience a similar outcome.

5.3   SYSTEMS ENGINEERING RELATIONSHIPS TO OTHER DISCIPLINES

SE practitioners routinely work within broad multidisciplinary teams. The following sections highlight SE practitioner interactions with some key related disciplines.

5.3.1   SE and Software Engineering (SWE)

This section describes an overview of the relationships between SE and software engineering (SWE), especially when developing and modifying *software-intensive systems*. ISO/IEC/IEEE 42010 (2022) defines a software-intensive system as:

> Any system where software contributes essential influences to the design, construction, deployment, and evolution of the system as a whole.

Section 4.*3.4* discusses software-intensive systems. These systems are also known as software-enabled and cyber-physical systems (see Section 4.3.5).

What is this discipline?
ISO/IEC TR 19759 (2015) defines SWE as:

> The application of a systematic, disciplined, quantifiable approach to the development, operation, and maintenance of software; that is, the application of engineering to software.

SE practitioners need to know that software has become ubiquitous in modern systems and is important in updates to many older systems. Because software elements are malleable, software can thus, in most cases, be tailored, adapted, and modified more readily than can physical elements, especially in the support stage of the system life cycle. Software elements, like physical elements, can be improved and replaced if the interfaces are preserved. Software often provides or facilitates provision of the following system attributes: functionality, behavior, quality characteristics (QCs), and system interfaces (both internal and external). Software embedded in software-intensive systems can also monitor system performance and provide observable performance indicators.

   The differences between software elements and physical elements present significant challenges for SE practitioners when facilitating development and modification of software-intensive systems. Software elements are logical entities composed of textual and iconic symbols that are processed and interpreted by other software that is executed on computer hardware. A single mistyped symbol, including a single mistake in a logical expression, in large system that may include thousands or millions of symbols and logical expressions can result in failure of the entire system. The mistake may not be detected during system development or modification because even small software programs contain large numbers of logical branches and iteration loops that prevent exhaustive testing in reasonable amounts of time. Assuring the quality of a software-intensive system is thus challenging. Detailed concentration on the precise development of software elements and software interfaces may cause SWE practitioners to lose sight of the impacts of their software on larger system issues.

What is its relationship to SE?

The relationship of SWE to SE is becoming increasingly important because software is a large and growing part of modern systems. The issues that arise for *SE* practitioners when developing or modifying software-intensive systems involve partitioning of requirements and architecture so that SWE practitioners (and hardware engineering (HWE) practitioners of various kinds) can design their elements and interfaces and pursue their development practices based on the differing natures of the mediums in which they work (Fairley, 2019). Well-defined and controlled interfaces, both provided and required, are essential so that separately developed elements can be efficiently integrated. Integration of software and hardware elements is typically accomplished in an incremental manner, as the elements become available. Defining and controlling interfaces is an ongoing challenge.

The primary benefit of coordinating SE and SWE during system development and modification is incorporation of software-provided capabilities in a manner that results in efficient and effective systems. SE practitioners can consult with SWE practitioners during system analysis and design to develop options and tradeoffs for configuring software and hardware elements and their interfaces. In addition, SWE practitioners can provide recommendations to SE practitioners for processes that can be used when developing and integrating software and hardware elements and when performing system verification, validation, and deployment (Fairley, 2019). Consultations and recommendations can ameliorate the software problems that sometimes result in late deliveries, insufficient system performance, and difficult system modifications.

How does it impact/is it impacted by SE?

SE practitioners who develop software-intensive systems are, as always, concerned with facilitating development of systems that are delivered in a timely manner, that satisfy performance parameters, and that can be modified efficiently and effectively. SE practitioners can better achieve these goals by taking advantage of the expertise of SWE practitioners, but they may fail to do so because they may not be familiar with the culture, terminology, and practices of SWE. SWE practitioners may not contribute their expertise because they are not consulted or if consulted may have insufficient knowledge of system level issues to provide recommendations. Involving knowledgeable SWE practitioners at the system level can improve communication. Cross-training, shadowing, mentoring, and collaborative workshops that include SE practitioners and SWE practitioners can result in synergetic relationships that will provide better communication and allow more effective and efficient development and modification of software-intensive systems.

More information on SE for software-intensive systems can be found in The Guide to the System Engineering Body of Knowledge (SEBoK) in the Part 6 knowledge area titled "SE and Software Engineering."

5.3.2 SE and Hardware Engineering (HWE)

This section describes an overview of the relationships between SE and hardware engineering (HWE).

What is this discipline?

HWE includes the development and implementation of physical elements for systems, enabling systems, and support equipment for systems. HWE includes mechanical engineering (ME) for mechanical elements and electrical engineering (EE) for electrical and electronic elements.

What is its relationship to SE?

The SE team must assist the hardware team in establishing hardware requirements, physical interface requirements, and establishing and tracking key physical measures (e.g., size, weight, and power (SWaP) budgets) at every level of the system architecture. During the System Architecture Definition and Design Definition processes (see Section 2.3.5.4 and 2.3.5.5), requirements will be allocated and derived across many hardware elements. Initially, these allocations may be required to meet a set of requirements at the system level without a full understanding of the actual values at the element level. This is where a requirements budget can be utilized by the SE team to derive these critical hardware attributes across all the applicable hardware elements and design teams.

These hardware design teams will consist of diverse HWE fields, with a focus on the hardware design. The hardware design teams will accept an initial budget allocation from the SE team based on a summary analysis of the system requirements and design. The SE team gives allocations based on the requirements and hardware design team inputs. The SE team will then track progress to those budgets, determine any impacts at the system level, and relay change requests as needed. Hardware considerations include SWaP, inertia, balance, frequency, phase, and others.

Interface management is an important consideration for SE and HWE. Interface decisions will have an impact on both the interconnectivity of the system and hardware selection. Utilizing open source or standard interfaces, if feasible, can reduce costs and development time. If unique or custom interfaces are necessary, they must be documented in the architecture before design, development, and testing can occur. This requires the SE team to balance interface decisions between capability and hardware standards.

The application of SE to hardware differs from software-only applications in two primary ways: hardware solutions may exist that can meet all or part of a decomposed requirement and the hardware performance requirements generally must be built into the initial hardware deliverable rather than iterated into the design (with some exceptions). A make or buy decision is often performed at each level of the architecture based on the results of a market analysis and tradeoff study of existing solutions that will satisfy the requirements. If the result of the decision is to make, the architecture can be further decomposed or Implementation process initiated. However, if the result of the decision is to buy, then the architecture decomposition for that system element will end (see Section 1.3.5).

How does it impact/is it impacted by SE?
Hardware material selection is supported by existing application or domain-specific standards. On a smaller, less complicated systems, HWE may be able to select all materials without the need of a formal SE process. With the increasing complexity of large systems and SoSs, the role of SE in hardware material selection becomes paramount, especially in applications where human lives could be affected. For instance, when multiple hardware sources could have system elements with mechanical or electrical interactions, the role of SE is to document architecture and design decisions and provide timely input into the material selection process before costly errors are found in verification or production. Sometimes SE acts to balance the software and hardware requirements, because software requirements can influence the hardware requirements, and vice versa. For instance, having a high availability requirement can result in the need for redundant software systems deployed on redundant hardware systems, meaning that the weight and the power budgets are increasing. SE serves as the bridge between software and hardware.

5.3.3 SE and Project Management (PM)

This section describes an overview of the relationships between SE and project management (PM).

What is this discipline?
As defined by the Project Management Institute (PMI), PM is defined as:

> The application of knowledge, skills, tools, and techniques to project activities to meet the project requirements. (PMI, 2022)

PM activities include initiating, planning, executing, monitoring, and closing *projects*. Within this handbook these are primarily distributed across the Technical Management Processes (see Section 2.3.4) for the SE portion of the responsibilities, but also include some activities in the Agreement Processes (see Section 2.3.2) and Organizational Project Enabling Processes (see Section 2.3.3). In the PMI Project Management Body of Knowledge (PMBoK) (2017), they are part of Project Integration Management, Project Scope Management, Project Schedule Management, Project Cost Management, Project Quality Management, Project Resource Management, Project Communications Management,

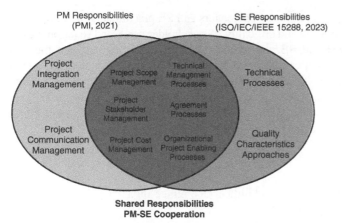

FIGURE 5.4   The intersection between PM and SE. INCOSE SEH original figure created by Roussel on behalf of the INCOSE PM-SE Integration Working Group. Usage per the INCOSE Notices page. All other rights reserved.

Project Risk Management, Project Procurement Management, and Project Stakeholder Management. According to the PMBoK, a *program* is a set of related projects that are viewed as an entity that requires coordinated management.

What is its relationship to SE?

While SE and PM are distinct disciplines, extensive research has shown that effective integration between PM and SE improves project performance, achieving better results in schedule and budget performance, as well as stakeholder requirements satisfaction, when compared with projects with lesser integration. In 2011, INCOSE formed a strategic alliance with PMI and Massachusetts Institute of Technology (MIT) to research and advance the integration of the two disciplines, driven by the vision that better integration would lead to the delivery of better solutions for organizations and their stakeholders. One output of this alliance was *Integrating Program Management and SE* (Rebentisch, 2017), which defines this integration as "a reflection of the organization ability to combine project management and SE practices, tools and techniques, experience and knowledge in a collaborative and systematic approach in the face of challenges in order to be more effective in achieving common goals/objectives in complex project environments." A summary of the respective and shared responsibilities between PM and SE is shown in Figure 5.4. Cooperation between project managers and SE practitioners must exist within all these shared activities.

How does it impact/is it impacted by SE?

A high degree of integration between PM and SE is characterized by collaborative decision making, a shared responsibility toward a common goal, having the project manager involved in technical aspects of the project, and having the SE practitioners involved in programmatic aspects of the project. An understanding of the differences, culture, background, and behavior of the two disciplines is also required. A team's ability to combine PM and SE practices, tools and techniques, experience, and knowledge in a collaborative and systematic approach enables addressing challenges in order to achieve common goals and objectives. Specifically, an integrated team achieves rapid and effective decision making, effective collaborative work, and effective information sharing (Rebentisch, 2017).

PM and SE overlap in the early stages of concept and development but tend to diverge in the later stages of development and production. For example, in the early life cycle stages the SE practitioner focuses on the technical details of the SoI, verification, and validation. The project manager focuses on the overall project performance and delivery of benefits, including high-level finance and budgetary requirements. PM and SE should cooperate on concurrent development of the breakdown structures (see Section 2.3.4.1), and in the management of them through the life cycle. If the different structures are managed separately by the respective teams without coordination, problems may arise in the project. As the project proceeds through the later life cycle stages, greater integration between PM and SE reduces unproductive tension, a cause of project delays, cost increases, and, sometimes, project failure (Rebentisch, 2017).

As stated in Rebentisch (2017), PM delivers the sustainable benefits of the overall project, while SE delivers the technical aspect of the project. These two roles overlap to integrate technical and programmatic aspects of the project and create potential for unproductive tension, if not effectively managed with cooperation. For example, PM and SE share the same objective to satisfy stakeholder needs and requirements. However, project managers tend to focus on project stakeholders, while SE practitioners tend to focus on system stakeholders. The separation between these two types of stakeholders can generate tension and misunderstandings between project managers and SE practitioners. It is important that they jointly identify all stakeholders and agree on the priority and criticality of stakeholder needs and requirements.

Where there is not effective integration between PM and SE, unproductive tension emerges between the two disciplines. This tension produces conflict and works at cross purposes with project success. Tension can be related to practices, techniques, as well as responsibilities. Misaligned measures can cause tension. Common measures are critical to ensuring that each party has the same concerns and information. The maturity level of each party is critical. An immature or inexperienced PM organization can render ineffective a mature, high-performing SE organization (and vice versa). When one or both disciplines perform inadequately, the entire effort is impaired.

5.3.4 SE and Industrial Engineering (IE)

This section describes an overview of the relationships between SE and industrial engineering (IE).

What is this discipline?
Bidanda (2022) defines Industrial Engineering as:

> Optimizing the utilization of human resources, facilities, equipment, tools, technologies, information, and handling of materials to produce quality products and services safely and cost-effectively considering the needs of customers and employers.

The Institute of Industrial and Systems Engineers (IISE) Industrial and SE Body of Knowledge (ISEBoK) (2022) is composed of 14 knowledge areas. The knowledge areas include: Operations Research & Analysis, Economic Analysis, Facilities Engineering & Energy Management, Quality and Reliability Engineering, Ergonomics & Human Factors, Operations Engineering & Management, Supply Chain Management, Safety, Information Engineering, Design and Manufacturing Engineering, Product Design & Development, Systems Design & Engineering.

What is its relationship to SE?
IE is closely related to SE. The IISE states that "Industrial and SE is concerned with the design, improvement, and installation of integrated systems of people, materials, information, equipment and energy. It draws upon specialized knowledge and skill in the mathematical, physical, and social sciences together with the principles and methods of engineering analysis and design, to specify, predict, and evaluate the results to be obtained from such systems." As evidence of the close relationship of IE and SE, Figure 5.5 compares the ISEBoK (2022) with SE topics in this handbook. The numbered topics on the figure are the knowledge areas explicitly identified in the ISEBoK. The figure identifies the knowledge areas that are usually performed by SE practitioners, the ones usually performed by industrial engineers (IE), and the knowledge area descriptions that are used by both disciplines.

As an illustration of the relationships between IE and SE, four knowledge areas are discussed below.

Operations Research and Analysis—Operations Research (OR) includes a variety of techniques to quantify and improve the efficiency of systems and organizational processes using scientific mathematical models. The mathematical techniques include linear programming, transportation models, linear assignment models, network flows, dynamic programming, integer programming, nonlinear programming, metaheuristics, decision analysis, game theory, stochastic modeling, queuing systems, simulation, systems dynamics, and analytics. Industrial engineers use OR to understand, design, and improve the operation of industrial systems and processes. SE practitioners can perform OR analyses or use OR studies performed by others to make system decisions (see Section 5.3.5).

Information engineering—Information engineering is a "methodology for developing an integrated information system based on the sharing of common data, with emphasis on decision support needs as well as transaction-processing (TP) requirements" (Gartner, 2022). Information engineering topics include: data types; information system concepts; information requirements; output design; data processing; database concepts; storage and processing; system analysis; system design; system evaluation; information management; and data analytics. Since information systems are a critical component of modern engineering systems, IE and SE practitioners work closely with information engineers.

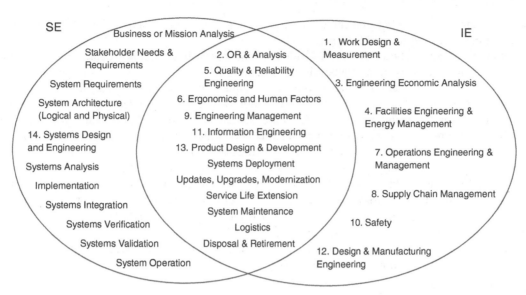

FIGURE 5.5   IE and SE relationships. From Parnell, et al. (2022). Used with permission. All other rights reserved.

Product Design and Development—As stated in ISEBoK (2022), "From an Industrial Engineering knowledge view, it is the processes and analysis employed supporting efficient decision-making during Product Design and Development." Product design and development topics include: design process; preliminary and detailed design; verification and testing; planning for manufacture; metrics for design and development; life cycle costing; and risk and opportunity management. SE practitioners may provide inputs to the product design and development process throughout the system life cycle.

System Design and Engineering—The ISEBoK (2022) lists the SEBoK (2022) as the reference for this section. The topics include the system life cycle concepts from mission engineering to operations.

How does it impact/is it impacted by SE?
There are some significant overlaps. IE and SE practitioners must work closely with each other to accomplish the goals of the project. Different organizations use different job titles for IE and SE, and the same job title may have different responsibilities in different organizations.

5.3.5   SE and Operations Research (OR)

This section describes an overview of the relationships between SE and operations research (OR).

What is this discipline?
The Institute for Operations Research and the Management Sciences (INFORMS) defines OR as:

> The scientific process of transforming data into insights using advanced analytical methods to making better decisions. (INFORMS, 2022)

The name "operations research" originated from the recognition of the successful use of scientific and mathematical modeling techniques for military *operations* during World War II (Gass and Harris, 2001). For example, mathematical modeling assisted in developing strategies to employ bombers as well as arrange convoys against submarine attacks.

OR is sometimes referred to as Management Science, particularly in business applications, as well as Operational Research in some countries (Quade and Boucher, 1968).

A branch of applied mathematics, OR includes wide range of approaches. Common to all is the establishment of a model to represent the system to support analysis for decision making. In some cases, the model may be solved directly, but other common approaches include optimization, simulation, and decision analysis.

Optimization is useful in determining the best combination of decision variables. Optimization models can be linear, integer, and nonlinear depending on the form of the objective function and constraints. Goal programming is useful when thresholds are established as an achievement target on multiple objectives. The approach avoids falling short of goals and incurring a penalty but permits a broader trade-off analysis.

Simulation is useful for assessing how well the system as a whole will perform its role in a given environment. Systems will typically be analyzed using stochastic discrete event simulations. These are often called Monte Carlo simulations, as random numbers are drawn to represent outcomes from distributions for the input uncertainties. These models are run multiple times and the output probabilities of occurrence are established by aggregating the simulation runs and using statistical tools for interpretation (Fishman, 1997).

Decision analysis uses a model of the decision makers' preferences and risk attitudes to analyze tradeoffs between alternatives and is often used for strategic decision making.

What is its relationship to SE?

OR techniques frequently support SE by assisting in understanding stakeholder needs and estimating how a proposed system will behave. Decision analysis is used to elicit stakeholder needs and preferences to construct a value model. The value model is used to understand value tradeoffs between different system or system element alternatives. The performance of the anticipated system alternatives may be represented through a simulation. This has been so successful that model-based SE (MBSE) has become an important area (see Section 4.2.1). Optimization may be applied to improve designs. For example, by minimizing a system element weight while providing required structural strength. Optimization may be used to minimize life cycle costs using inventory theory. Queueing theory may be used to understand system processing times. Forecasting may be used to project forward from historical data. Network flow analysis may be used to improve network capacity.

Blanchard and Fabrycky (2011) discuss the SE morphology for product (system) realization. They note the steps after design synthesis are (1) Estimation and Prediction, and (2) Design Evaluation. Estimation and Prediction often rely heavily on OR. For example, assessing the response times associated with different possible locations for a new firehouse or estimating spare parts requirements as part of a life cycle cost perspective.

Optimization may be applied in the development stage as well as in the production stage. Optimization may be applied at higher levels of the systems hierarchy. For example, a less costly design may support a larger number of employed systems for the same results.

Operations research techniques may be applied to historical and test data. Data analytics may be used to establish system performance as a mathematical function of inputs. Sensitivity analysis may be used to establish acceptable system inputs. Response surface methodology may be used to establish more abstract but tractable models. Forecasting methods may be applied to time series data.

Decision analysis provides the ability to understand decision quality and assess the value or utility delivered under certainty and uncertainty. Decision analysis provides not only the ability to select the best presented alternative but also supports insight in examination of hybrid alternatives, allowing searching for improved designs. It is also transparent and traceable, and so is fully defensible and allows representation of differing preferences of various stakeholders.

How does it impact/is it impacted by SE?

OR and SE practitioners must work closely with each other to accomplish the goals of the project. SE practitioners can perform OR analyses or use OR studies performed by others to make system decisions.

5.4   DIGITAL ENGINEERING

Definitions

The US Defense Acquisition University Glossary (DAU, 2022) defines Digital Engineering as:

> An integrated digital approach that uses authoritative sources of systems' data and models as a continuum across disciplines to support life cycle activities from concept through [retirement].

Digital Engineering is crosscutting: it includes all engineering disciplines using well-formed models to execute their processes and communicate a system's design. Digital Engineering emphasizes continuity of models and their use across the life cycle. Digital Engineering seeks to maximize the use of models and the computer and places emphasis on consistent and rigorous engineering, strong data management practices, and continuous improvement through technologic innovation. It requires a supporting infrastructure and environment and a capable workforce and culture that is committed to working in accordance with process, following methods, and using tool the organization supplies them.

Model-based SE (MBSE) (see Section 4.2.1) is one of core elements of Digital Engineering. In MBSE and Digital Engineering, a *Digital System Model* is a digital representation of a system. It integrates the authoritative MBSE and other Digital Engineering technical data and associated artifacts, defining all aspects of the system throughout the system life cycle. The Digital System Model is composed of a federated set of models that serve as an authoritative source of truth (ASOT) for the system's design.

Digital Engineering leverages MBSE and the Digital System Model to enable digital threads and digital twins. A *digital twin* is a digital surrogate of the system, incorporating models to emulate the actual system or some of its elements. The digital twin evolves through the life cycle with the mission and definition of the system. A *digital thread* is a set of interconnected, cross-discipline model data that seamlessly expedite the controlled interplay of digital artifacts to inform decision makers throughout a system's life cycle. Digital threads can be used to produce *digital artifacts* that are a combination of authoritative data, information, knowledge, and wisdom addressing stakeholders' unique perspective.

Digital Engineering in Projects

Digital Engineering performed on projects needs to be based around the project and SoI requirements, as well as opportunities and risks that are identified from stakeholders including customers, users, organizational leadership, the infrastructure, decision makers, and project managers. It is important to pay special attention to Digital Engineering contributions and inputs when doing the technical planning across the life cycle.

As each life cycle stage is considered, it is critical to research and document opportunities for future project use, as well as for reuse of models, simulations, and data across the life cycle. For example, models and simulations developed during earlier life cycle stages may be reused for verification and training in later life cycle stages.

Digital Engineering for the Enterprise

Digital Engineering is an approach for projects, but is also important as an enterprise digital transformation initiative (see Section 5.5). To achieve maximum benefit both project and enterprise level concerns should be considered together. Enterprise level implementation efforts are necessary to position the engineering infrastructure and environment so that a project will be able to perform their engineering activities in a digital manner. Consistent use of the infrastructure and environment on a project, and across projects, will yield increasingly consistent engineering work products. Consistency across work products will lead to great gains in reuse and will enable greater speed through computer automation.

Digital Engineering Ecosystem

The Digital Engineering ecosystem can be vast, crossing engineering domain and organizational boundaries. The Digital Engineering ecosystem should be treated as an SoS and must be developed using SE good practices. The Digital Engineering ecosystem consists of the models, tools, processes, and people/roles that come together to develop the systems the organization cares about. The ecosystem can be broader than this, depending on the scope. For example, some organizations utilize their Digital Engineering ecosystems to support the operation of their systems. Others include cross-project libraries and methods in their Digital Engineering ecosystems, as well as interconnections with external Digital Engineering ecosystems across their supply chain.

Technological Innovation

Technological innovation is an important part of Digital Engineering. It is the catalyst that drives change into the engineering practice through strategic and planned implementation. It is an essential part of a continuous process improvement program that seeks out and injects technology into the systems that are used to develop the SoI. Technological innovation seeks to optimize the use of computer and information technologies to enhance the speed, agility, quality, and precision of all engineering activities that occur through the development life cycle.

5.5   SYSTEMS ENGINEERING TRANSFORMATION

In discussing SE transformation, there are three important transformations to be considered:

- Transforming from no SE to full use of SE,
- The internal SE from traditional to agile methodologies as appropriate, and
- The internal SE from document-based to model-based disciplines.

Considering the first, for an organization to begin implementing SE requires a cultural shift to understand the basics of the system life cycle and implement the life cycle processes described in this handbook in the organization's set of processes (see Section 2.3.3.1). When performing this type of transformation, organizations will need to prepare for the major cultural shift required for the team members to learn, understand, and perform SE tasks.

The second type of transformation for internal SE from traditional to agile methodologies also requires a cultural shift to understand the basics of agile SE (see Section 4.2.2) in the organization's set of processes. Agile SE requires tools and capabilities that may also cause the organization to need additional infrastructure and human talent. Every organization will need to determine the best way to integrate agile SE practices and methodologies, whether as a wholesale change or a gradual change for their SE processes. In practice, the main reasons why are required these two types of transformations are that the system life cycle is either being compressed during the concept and development stages and/or the life cycle is being shifted to the left with early concept task being anticipated with early involvement of SE.

The third transformation is SE's transformation to a model-based discipline supports transdisciplinary digital engineering through model-based activity, advancement, organizational change, and a broad model community engagement (Peterson 2019). This transformation addresses:

- Knowledge representation and immersive technologies,
- Product (System) modeling,
- Model-based SE (MBSE) approaches and methods,

- Virtual prototyping and virtual product integration at scale,
- Foundational theory, principles, and heuristics, and
- MBSE in support of Digital Engineering.

Knowledge representation and immersive virtual reality technologies (characterized by deep absorption or immersion) enable highly efficient and shared human understanding of systems in a virtual environment that spans the full life cycle. Systems modeling forms the product-centric backbone of the digital enterprise which incorporates a model-centric approach to integrate technical, programmatic, business, regulatory, and governance concerns (see Section 3.2.1). Model-based approaches and sophisticated model-based methods extend beyond product modeling to include systems of systems (SoSs) and enterprise-level modeling and analysis (see Section 4.2.1). Large scale virtual prototyping and virtual product integration based on integrated models enable significant time-to-market reductions (see Sections 3.2.2 and 2.3.5.8). Foundations of theory, principles, and heuristics allow for a better understanding of increasingly complex systems and decisions in the face of uncertainty (see Section 1.4). MBSE in support of Digital Engineering is standard practice and is integrated with other modeling and simulation as well as digital enterprise functions (see Section 5.4).

5.6   FUTURE OF SE

The primary focus of this handbook is the state-of-the-good-practice in SE. This section highlights some emerging areas, looking toward the future of SE. INCOSE's Systems Engineering Vision (2022) is an excellent resource. In addition, the SE Body of Knowledge (2023) is continually updated to reflect both the state-of-the-practice and the state-of-the-art in SE.

INCOSE's Future of Systems Engineering (FuSE) is a systems community initiative to realize the INCOSE Vision 2035 (2022) and to evolve the instruction, practice, and perception of SE to:

- Position the discipline to leverage new technologies in collaboration with allied fields,
- Enhance SE's ability to solve the emerging challenges, and
- Promote SE as essential for achieving success and delivering value in the engineering of socio-cyber-physical systems and SoSs at scale and subject to non-deterministic influences and effects.

An important aspect of SE is to keep current on emerging trends, technologies, and challenges when considering both the SoI and the SE processes themselves. While performing SE, the practitioner needs to consider advances in computing, communications, software, human systems integration, and algorithms such as augmented intelligence and Machine Learning (ML) for both use in systems and SE. Leveraging these technologies will increase the capabilities of systems and the SE practitioner. By *scale*, we mean the challenges in applying SE from smaller (e.g., miniaturization of electronics) through larger scales (e.g., cloud-based systems with millions of users). With the exponential growth of scale and hyperconnectivity of systems and SoSs, scale is becoming even more important to the discipline of SE.

At an early 2019 FuSE workshop hosted by INCOSE, the terms Artificial Intelligence (AI) for SE and SE for AI were first used to describe the dual transformation envisioned for both the SE and AI disciplines. The "AI4SE" and "SE4AI" labels have quickly become symbols for an upcoming rapid evolutionary phase in the SE community. AI4SE applies AI and ML techniques to improve human-driven SE practices. This goal of "augmented intelligence" includes outcomes such as achieving scale in model construction and efficiency. Enhancing and assisting SE processes, methods, and tools, with tangible impacts on the quality of the engineered system as well as on the cycle time for the various life cycle activities, would be some of the primary focus areas of AI4SE (SEBoK, 2023).

The FuSE roadmap drives this evolution of SE to:

- Be increasingly adaptable, evolvable, and fit for purpose,
- Account for human abilities and needs as an integral system element and human interactions with systems and SoSs,
- Be more responsive in resolving increasingly challenging societal needs, and
- Realize and enhance the INCOSE Vision 2035 (2022) and other visionary statements.

Greater understanding of the inter-coupled technical, economic, social, and environmental systems will provide the basis for significantly increased involvement of SE practitioners in the policy arena. In this expanded role, SE practitioners will also make important contributions to the design of viable systems and transition pathways supporting global sustainability transformation. The scope of SE will widen to recognize and include policy, legal, economic, and environmental specializations.

6

CASE STUDIES

Real-world examples that draw from diverse industries and types of systems are provided throughout this handbook, and in this part, five case studies have been selected to illustrate the diversity of systems to which Systems Engineering (SE) principles and practices can be applied: medical therapy equipment, a bridge, a breach of a cybersecurity system, a redesign of a high-tech medical system for low-tech maintenance, and autonomous vehicles. They represent examples of failed and successful systems.

6.1   CASE 1: RADIATION THERAPY—THE THERAC-25

Background   Therac-25, a dual-mode medical linear accelerator (LINAC), was developed by the medical division of the Atomic Energy Commission Limited (AECL) of Canada, starting in 1976. A completely computerized system became commercially available in 1982. This new machine could be built at lower production cost, resulting in lower prices for the customers. However, a series of tragic accidents led to the recommended recall and discontinuation of the system.

The Therac-25 was a medical LINAC, or particle accelerator, capable of increasing the energy of electrically charged atomic particles. LINACs accelerate charged particles by introducing an electric field to produce particle beams (i.e., radiation), which are then focused by magnets. Medical LINACs are used to treat cancer patients by exposing malignant cells to radiation. Since malignant tissues are more sensitive than normal tissues to radiation exposure, a treatment plan can be developed that permits the absorption of an amount of radiation that is fatal to tumors but causes relatively minor damage to surrounding tissue.

Six accidents involving enormous radiation overdoses to patients took place between 1985 and 1987. Tragically, three of these accidents resulted in the death of the patients. This case is ranked in the top ten worst software-related incidents on many lists. Details of the accidents and analysis of the case are available from many sources, including Jacky (1989) and Leveson and Turner (1993).

INCOSE Systems Engineering Handbook: A Guide for System Life Cycle Processes and Activities, Fifth Edition.
Edited by David D. Walden, Thomas M. Shortell, Garry J. Roedler, Bernardo A. Delicado, Odile Mornas, Yip Yew-Seng, and David Endler.
© 2023 John Wiley & Sons Ltd. Published 2023 by John Wiley & Sons Ltd.

Approach   Therac-25 was a revolutionary design compared to its predecessors, Therac-6 and Therac-20, both with exceptional safety records. It was based on a double-pass concept that allowed a more powerful accelerator to be built into a compact and versatile machine. AECL designed Therac-25 to fully utilize the potential of software control. While Therac-6 and Therac-20 were built as stand-alone machines and could be operated without a computer, Therac-25 depended on a tight integration of software and hardware. In the new, tightly coupled system, AECL used software to monitor the state of the machine and to ensure its proper operations and safety. Previous versions had included independent circuits to monitor the status of the beam as well as hardware interlocks that prevented the machine from delivering radiation doses that were too high or from performing any unsafe operation that could potentially harm the patient. In Therac-25, AECL decided not to duplicate these hardware interlocks since the software already performed status checks and handled all the malfunctions. This meant that the Therac-25 software had far more responsibility for safety than the software in the previous models. If, in the course of treatment, the software detected a minor malfunction, it would pause the treatment. In this case, the procedure could be restarted by pressing a single "proceed" key. Only if a serious malfunction was detected was it required to completely reset the treatment parameters to restart the machine.

The software for Therac-25 was developed from the Therac-20's software, which was developed from the Therac-6's software, a brownfield, or legacy, development (see Section 4.3.2). One programmer, over several years, evolved the Therac-6 software into the Therac-25 software. A stand-alone, real-time operating system was added along with application software written in assembly language and tested as a part of the Therac-25 system operation. In addition, significant adjustments had been made to simplify the operator interface and minimize data entry, since initial operators complained that it took too long to enter a treatment plan.

At the time of its introduction to market in 1982, Therac-25 was classified as a Class II medical device. Since the Therac-25 software was based on software used in the earlier Therac-20 and Therac-6 models, Therac-25 was approved by the federal Food and Drug Administration under Premarket Equivalency.

Conclusions   The errors were introduced in the concept and early development stages, when the decisions were made to create the software for Therac-25 using the modification of existing software from the two prior machines. The consequences of these actions were difficult to assess at the time, because the starting point (software from Therac-6) was a poorly documented product and no one except the original software developer could follow the logic (Leveson and Turner, 1993).

The issues from the Therac case are, unfortunately, still relevant, as evidenced by similar deaths for similar reasons in 2007 upon the introduction of new LINAC-based radiation therapy machines (Bogdanich, 2010).

6.2   CASE 2: JOINING TWO COUNTRIES—THE ØRESUND BRIDGE

Background   The Øresund Region is composed of eastern Denmark and southern Sweden and since 2000 has been linked by the Øresund Bridge. The area includes two major cities, Copenhagen and Malmö, has a population of three million, and counts as Europe's eighth largest economic center. One fifth of the total Danish and Swedish Gross National Product (GNP) is produced in the region. The official name of the bridge is translated "the Øresund Connection" to underscore the full integration of the region. For the first time ever, Sweden is joined permanently to the mainland of Europe by a 10-minute drive or train ride. The cost for the entire Øresund Connection construction project was calculated at 30.1 billion DKK (3 billion USD), and the investment is expected to be paid back by 2035.

The Øresund Bridge is the world's largest composite structure, has the longest cable-stayed bridge span in the world carrying motorway and railway traffic, and boasts the highest freestanding pylons. The 7.9 km (5 miles)-long bridge crosses the international navigation route between the Baltic Sea and the North Sea. A cable-stayed high bridge rises 57 m (160 ft) above the surface of the sea, with a main span of 490 m (0.3 miles). Both the main span and the approach bridges are constructed as a two-level composite steel-concrete structure. The upper deck carries a four-lane motorway, and the lower deck carries a two-track railway for both passenger trains and freight trains. The rest of the distance is spanned by the artificial island Peberholm ("Pepper" islet, named to complement the Saltholm islet to the north) and a tunnel on the Danish side that is the longest immersed concrete tunnel in the world. Since completion, Peberholm has become a natural habitat for colonies of rare birds, one of the largest of its kind in Denmark and Sweden.

Nations other than Denmark and Sweden also contributed to this project. Canada provided a floating crane, aptly named Svanen (the swan), to carry prefabricated bridge sections out to the site and place them into position. Forty-nine steel girders for the approach bridges were fabricated in Cádiz, Spain. A specially designed catamaran was built to handle transportation of the foundations for the pylons, which weighed 19,000 tons each.

Approach   As noted in the many histories of the bridge, the development stage of the project began with well-defined time, budget, and quality constraints. The design evolved over more than seven years, from start to delivery of final documentation and maintenance manuals. More than 4,000 drawings were produced. The consortium dealt with changes, as necessary, using a combination of technical competence and stakeholder cooperation. Notably, there were no disputes and no significant claims against the owners at the conclusion, and this has been attributed to the spirit of partnership.

What is not often reported is that the success of the development stage is clearly based on the productive, focused, creative effort in the concept stage that began when the royal families of Denmark and Sweden finally agreed in 1990 to move ahead with a bridge project connecting their two countries. That SE effort shaped the approach to the project with well-defined time, budget, and quality constraints at the transition to the development stage. During the concept stage, the SE team also recognized that the concerns of environmental groups would—and should—impact the approach to the construction of the bridge. The owners took a creative approach by inviting the head of a key environmental group to be part of the board of directors.

From the beginning of the development stage, the owners defined comprehensive requirements and provided definition drawings as part of the contract documents to ensure a project result that not only fulfilled the quality requirements on materials and workmanship but also had the envisioned appearance. The contractor was responsible for the detailed design and for delivering a quality-assured product in accordance with the owners' requirements. The following are representative of the requirements levied at the start of the project:

Schedule: Design life, 100 years; construction time, 1996–2000
Railway: Rail load, International Union of Railways (UIC) 71; train speed, 200 km/h
Motorway: Road axle load, 260 kN; vehicle speed, 120 km/h
Ambient environment: Wind speed (10 min), 61 m/s; wave height, 2.5 m; ice thickness, 0.6 m; temperature, +/− 27°C
Ship impact: To pylons, 560 MN; to girder, 35 MN

In addition to established requirements, this project crossed national boundaries and was thereby subject to the legislations of each country. Technical requirements were based on the Eurocodes, with project-specific amendments made to suit the national standards of both countries. Special safety regulations were set up for the working conditions, meeting the individual safety standards of Denmark and Sweden.

The railway link introduced yet another challenge. In Denmark, the rail traffic is right-handed, as on roadways, whereas the trains in Sweden pass on the left-hand side. The connection needed to ensure a logical transition between the two systems, including safety aspects. In addition, the railway power supply differs between the two countries; thus, it was necessary to develop a system that could accommodate power supply for both railway systems and switch between them on the fly.

The design of a major cable-stayed bridge with approach spans for both road and railway traffic involves several disciplines, including, but not limited to, geotechnical engineering, aerodynamics, foundation engineering, wind tunnel tests, design of piers and pylons, design of composite girders, design of cables and anchorages, design of structural monitoring system, ship impact analysis, earthquake analysis, analysis of shrinkage and creep of concrete, ice load analysis, fatigue analysis, pavement design, mechanical systems, electrical systems, comfort analysis for railway passengers, traffic forecast, operation and maintenance aspects, analysis of construction stages, risk analysis for construction and operation, quality management, and environmental studies and monitoring.

Comprehensive risk analyses were carried out in connection with the initial planning studies, including specification of requirements to secure all safety aspects. Important examples of the results of these studies for the Øresund Bridge were as follows:

Navigation span was increased from 330 to 490 m.

The navigation channel was realigned and deepened to reduce ship groundings.

Pier protection islands were introduced to mitigate bridge/ship accidents.

Risks were considered in a systematic way, using contemporary risk analysis methods such as functional safety analysis using fault tree and "what-if" techniques. Three main issues were considered under the design–build contract:

General identification and assessment of construction risks

Ship collision in connection with realignment of navigation channel

Risks in connection with 5-year bridge operation by contractor

A fully quantified risk assessment of the human safety and traffic delay risks was carried out for a comprehensive list of hazards, including fire, explosion, train collisions and derailments, road accidents, ship collisions and groundings, aircraft collisions, environmental loads beyond design basis, and toxic spillages. An example of a consequence of this analysis was the provision of passive fire protection on the tunnel walls and ceilings.

Both Denmark and Sweden are proud of being among the cleanest industrial countries in the world. Their citizens, and therefore the politicians, would not allow for any adverse environmental impact from the construction or operation of a bridge. The Great Belt and Øresund Strait both constitute corridors between the salty Kattegat and the sweeter water of the Baltic Sea. Any reduction in water exchange would reduce the salt content and, therefore, the oxygen content of the Baltic Sea and would alter its ecological balance. The Danish and Swedish authorities decided that the bridge should be designed in such a way that the flow through of water, salt, and oxygen into the Baltic was not affected. This requirement was designated the zero solution. To limit impacts on the local flora and fauna in Øresund during the construction, the Danish and Swedish authorities imposed a restriction that the spillage of seabed material from dredging operations should not exceed 5% of the dredged amounts. The zero solution was obtained by modeling with two different and independent hydrographical models.

In total, 18 million cubic meters of seabed materials were dredged. All dredged materials were reused for reclamation of the artificial peninsula at Kastrup and the artificial island, Peberholm. A comprehensive and intensive monitoring of the environment was performed to ensure and document the fulfillment of all environmental requirements. In their final status report from 2001, the Danish and Swedish authorities concluded that the zero solution as well as all environmental requirements related to the construction of the link had been fulfilled. Continual monitoring of eel grass and common mussels showed that, after a general but minor decline, populations had recovered by the time the bridge was opened. Overall, the environment paid a low price at both Øresund and the Great Belt because it was given consideration throughout the planning and construction stages of the bridges.

Conclusions   This award-winning bridge is the subject of numerous articles and a doctoral thesis, where details of the construction history and collaboration among all the stakeholders are provided (Jensen, 2014; Nissen, 2006; Skanska, 2013). This project provides a clear example of the benefit of a solid concept stage where the management team was able to resist the customer-driven temptation to jump prematurely into the development stage.

6.3 CASE 3: CYBERSECURITY CONSIDERATIONS IN SYSTEMS ENGINEERING—THE STUXNET ATTACK ON A CYBER-PHYSICAL SYSTEM

Background   As our world becomes increasingly digital, the issue of cybersecurity is a factor that the SE practitioner needs to consider. Both hardware and software systems are increasingly at risk for disruption or damage caused by threats taking advantage of digital technologies. Stuxnet, a cyber-attack on Iran's nuclear capabilities discovered in

2010, illustrates the need for the SE practitioner to be comprehensive in application of secure design principles and methods for assessing and avoiding vulnerabilities, and rigorous in mitigation of attack potential (Failliere, 2011; Langner, 2012).

This case study discusses a high degree of attack sophistication previously unseen—malware complexity at military-grade performance, nearly no side effects, and pinpoint accuracy. However, though the creation and deployment of Stuxnet were expensive undertakings, the strategy, tactical methods, and code mechanisms became openly available for others to reuse and build upon at much less expense. Cyber-physical system attacks are becoming increasingly prevalent, and SE must consider the implications of cybersecurity to reduce the vulnerabilities.

Iran's Natanz nuclear fuel enrichment plant (FEP) is a military-hardened facility, with a security fence surrounding a complex of buildings, which are in turn each protected by a series of concrete walls. The complex contains several "cascade halls" for the production of enriched uranium in gas centrifuges. This facility was further hardened with a roof of several meters of reinforced concrete and covered with a thick layer of earth.

Each of the cascade halls is a cyber-physical system, with an industrial control system (ICS) of programmable logic controllers (PLCs), computers, an internal network with no connections to the outside world, and capacity for thousands of centrifuges. Though the internal network is isolated from the outside world by an "air gap," possible vulnerabilities still include malicious insider collusion, non-malicious insider insertion of memory devices brought in from the outside, visiting service technicians, and supply chain intervention. It has been suggested that all of these breech vectors may have played a role in the massive centrifuge damage that began occurring in 2009 and continued at least through 2010.

Malware, now known as Stuxnet, was introduced into the ICS of at least one of the cascade halls and managed to take surreptitious control of the centrifuges, causing them to spin periodically and repeatedly at rates damaging to sustained physical operation. The net effect of the attack is still unclear, but at a minimum, it ranged from disruption of the production process up to potential permanent damage to the affected centrifuges.

Approach   Many characteristics of Stuxnet were unprecedented and stand as the inflection point that ushered in a new era of system attack methodology and cyber-physical system targeting. Illuminating forensic analysis of the Stuxnet code was conducted by several well-known cybersecurity firms, with detailed postmortems covered in two documents from the Institute for Science and International Security (Albright, et al., 2010, 2011). This analysis is beneficial in expanding the risk landscape that the SE practitioner should consider during design. Below are some concepts that are concerned in the context of Stuxnet:

Knowing what to do (intelligence)—To be successful, a threat has to be able to take advantage of the targeted system(s). It is uncertain how the perpetrators knew what specific devices were employed in what configuration at Natanz; but after the Stuxnet code was analyzed, Natanz was clearly identified as the target. Stuxnet infected many sites other than Natanz, but it would only activate if that site was configured to certain specific system specifications. The perpetrators needed specific system configuration information to know how to cause damage and also to know how to single out the target among many similar but not identical facilities elsewhere. The SE practitioner needs to consider that adversaries will attempt to gain intelligence on a system and must consider methods to prevent this.

Crafting the code—A zero-day attack is one that exploits a previously unknown vulnerability in a computer application, one that developers have had no time to address and patch. Stuxnet attacked Windows systems outside the FEP using a variety of zero-day exploits and stolen certificates to get proper insertion into the operating system and then initiated a multistage propagation mechanism that started with Universal Serial Bus (USB) removable media infected outside the FEP and ended with code insertion into the ICS inside the FEP. SE practitioners need to be prepared for many different attack vectors (including internal threats) and must consider them during system design.

Jumping the air gap—It was widely believed that Stuxnet crossed the air gap on a USB removable media device, which had been originally infected on a computer outside of the FEP and carried inside. But it was also suggested that the supply chain for PLCs and PLC maintenance personnel may have been at least two additional attack vectors. Whatever the methods, the air gap was crossed multiple times. USB removable media could have also affected a bidirectional transfer of information, sending out detailed intelligence about device types connected to the FEP network subsequently relayed to remote servers outside of the control of the facility. The SE practitioner always needs to remember that threats to the system are both inside and outside the system boundary.

Dynamic updating—Analysis shows that the attack code, once inserted, could be updated and changed over time, perhaps to take advantage of new knowledge or to implement new objectives. Stuxnet appears to have been continuously updated, with new operational parameters reintroduced as new air gap crossings occur. The SE practitioner needs to prepare for situations after a successful attack has occurred.

Conclusions   As the complexity and technology of systems change, the SE practitioner's perspective needs to adjust accordingly. The increasing use of digital-based technologies in system design offers enormous benefits to everyone. However, the introduction of digital technologies also brings different risks than previously dealt with by SE. The case study earlier illustrates a point in time behind us, and the adversarial community continues to evolve new methods. The lesson of this case study is that the SE practitioner needs to understand the threats toward their system(s), be cognizant that attacks can and will occur, and be proactive in protecting their system(s). Robust and dynamic system security needs full engagement of SE (see Section 3.1.12). A database that the SE practitioners should be aware of is maintained by the National Institute of Standards and Technology (NIST, 2012).

6.4   CASE 4: DESIGN FOR MAINTAINABILITY—INCUBATORS

Note: This case study is excerpted from "Where Good Ideas Come From: The Natural History of Innovation" (Johnson, 2010).

Background   In the late 1870s, a Parisian obstetrician named Stephane Tarnier was visiting the Paris Zoo where they had farm animals. While there, he conceived the idea of adapting a chicken incubator to use for human newborns, and he hired "the zoo's poultry raiser to construct a device that would perform a similar function for human newborns." At the time, infant mortality was staggeringly high "even in a city as sophisticated as Paris. One in five babies died before learning to crawl, and the odds were far worse for premature babies born with low birth weights." Tarnier installed his incubator for newborns at Maternité de Paris and embarked on a quick study of 500 babies. "The results shocked the Parisian medical establishment: while 66 percent of low-weight babies died within weeks of birth, only 38 percent died if they were housed in Tarnier's incubating box.… Tarnier's statistical analysis gave newborn incubation the push that it needed: within a few years the Paris municipal board required that incubators be installed in all the city's maternity hospitals."

"Modern incubators, supplemented with high-oxygen therapy and other advances, became standard equipment in all American hospitals after the end of World War II, triggering a spectacular 75 percent decline in infant mortality rates between 1950 and 1998."… "In the developing world, however, the infant mortality story remains bleak. Whereas infant deaths are below ten per thousand births throughout Europe and the United States, over a hundred infants die per thousand (births) in countries like Liberia and Ethiopia, many of them premature babies that would have survived with access to incubators. But modern incubators are complex, expensive things. A standard incubator in an American hospital might cost more than $40,000 (about €30,000). But the expense is arguably the smaller hurdle to overcome. Complex equipment breaks, and when it breaks, you need the technical expertise to fix it. You also need replacement parts. In the year that followed the 2004 Indian Ocean tsunami, the Indonesian city of Meulaboh received eight incubators from a range of international relief organizations. By late 2008, when an MIT professor named Timothy Prestero visited the hospital, all eight were out of order, the victims of power surges and tropical humidity, along with the hospital staff's inability to read the English repair manual. The Meulaboh incubators were a representative sample: some studies suggest that as much as 95% of medical technology donated to developing countries breaks within the first 5 years of use."

Approach   "Prestero had a vested interest in those broken incubators, because the organization he founded, Design that Matters, had been working for several years on a scheme for a more reliable, and less expensive, incubator, one that recognized complex medical technology was likely to have a very different tenure in a developing world context than it would in an American or European hospital. Designing an incubator for a developing country wasn't just a matter of creating something that worked; it was also a matter of designing something that would break in a non-catastrophic way. You couldn't guarantee a steady supply of spare parts, or trained repair technicians. So instead, Prestero and his team decided to build an incubator out of parts that were already abundant in the developing world. The idea had originated with a Boston doctor named Jonathan Rosen, who had observed that even the smaller towns of the developing world seemed to be able to keep automobiles in working order. The towns might lack air conditioning and laptops and cable television, but they managed to keep their Toyota 4Runners on the road. So, Rosen approached Prestero with an idea: What if you made an incubator out of automobile parts?"

"Three years after Rosen suggested the idea, the Design that Matters team introduced a prototype device called NeoNurture. From the outside, it looked like a streamlined modern incubator, but its guts were automotive. Sealed-beam headlights supplied the crucial warmth; dashboard fans provided filtered air circulation; door chimes sounded alarms. You could power the device via an adapted cigarette lighter, or a standard-issue motorcycle battery. Building the NeoNurture out of car parts was doubly efficient, because it tapped both the local supply of parts themselves and the local knowledge of automobile repair. These were both abundant resources in the developing world context, as Rosen liked to say. You didn't have to be a trained medical technician to fix the NeoNurture; you didn't even have to read the manual. You just needed to know how to replace a broken headlight."

Conclusions   Sometime the highest technology solution is not the best. SE practitioners need to consider issues like maintainability and logistics at the project outset in the concept and development stages. It is too late to address these in later stages.

6.5   CASE 5: ARTIFICIAL INTELLIGENCE IN SYSTEMS ENGINEERING—AUTONOMOUS VEHICLES

Note: Much of the information in this case study is derived from the United States National Transportation Safety Board (NTSB) report on automation (2019a). Page numbers are indicated.

Background   On March 18, 2018, a pedestrian walking a bicycle was fatally struck by a 2017 Volvo XC90 Uber vehicle operating an Automated Driving System (ADS) then under development by Uber's Advanced Technologies Group (ATG). The Volvo's advanced driver assistance system was disabled to prevent conflicts with its radar which operated on the same frequency as the radar for Uber's ATG ADS (p. 15).

At the time of the pedestrian fatality, the ATG-ADS had used one lidar and eight radars to measure distance; several cameras for detecting vehicles, pedestrians, reading traffic lights, and classifying detected objects; and various sensors that had been recently calibrated for telemetry, positioning, monitoring of people and objects, communication, acceleration, and angular rates. It also had a human-machine interface (HMI) tablet and a Global Positioning System (GPS) used solely to assure that the car was on an approved and pre-mapped route before engaging the ADS. The ADS allowed the vehicle to operate at a maximum speed of 45 mph (p. 7), to travel only on urban and rural roads, and under all lighting and weather conditions except for snow accumulation. The ADS system was easily disengaged; until then, almost all of its data was recorded, with the exception noted below of lost data occurred whenever an alternative determination of an object was made by ADS (e.g., shifting from an "object" in the road to an oncoming "vehicle" ahead).

Approach   Designing the interactions of a human and a machine to form into a team (or system) that also acts autonomously requires significant shifts in thinking, modeling, and practice. This begins with changing the unit of analysis from individual humans or programmable machines to teams.

The ADS constructed a virtual environment from the objects that its sensors detected, tracked, classified, and then prioritized based on fusion processes (p. 8). ADS predicted and detected any perceived object's goals and paths as part of its classification system. However, if classifications were made and then changed, as happened in this case (e.g., from "object" to "vehicle" and back to "object"), the prior tracking history was discarded. A flaw since corrected. Also, pedestrians outside of a crosswalk were not assigned a predicted track. Another flaw since corrected.

When ADS detected an emergency (p. 9), it suppressed taking any action for one second to avoid false alarms. After the one second delay, the car's self-braking or evasion could begin. Another major flaw since corrected (p. 15). If a collision could not have been avoided, an auditory warning was to be given to the operator at the same time that the vehicle was to be slowed (in this case, the vehicle may have also begun to slow because an intersection was being approached).

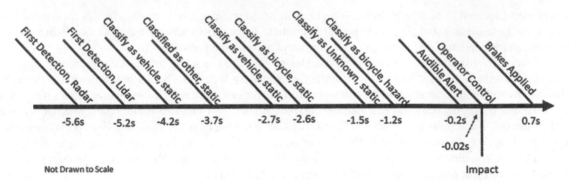

FIGURE 6.1   Timeline of vehicle impact. From NTSB (2019a). Used with permission. All other rights reserved.

As shown in Figure 6.1, using the recorded data to replay the accident, before impact: Radar first detected the pedestrian 5.6 seconds; Lidar made its first detection at 5.2 seconds, classified the object as unknown and static, changed to a static vehicle at 4.2 seconds on a path predicted to be a miss, reclassified to "other" and static but back again to vehicle between 3.7 to 2.7 seconds, each re-classification discarding its previous prediction history for that object; then a bicycle, but static and a miss at 2.6 seconds; then unknown, static and a miss at 1.5 seconds; then classified the object a bicycle and an unavoidable hazard at 1.2 seconds, the categorization of a hazard immediately initiating "action suppression;" after the 1 seconds pause, finally an auditory alert was sounded at 0.2 seconds; the operator took control at 0.02 seconds before impact; and the operator selected brakes at 0.7 seconds after impact.

The ADS failed to correctly predict the detected object's path, and only determined it to be a hazard at 1.2 seconds before impact, causing any action to be suppressed for one second but, and as a consequence of the impact anticipated in the shortened time-interval remaining before impact, exceeding the ADS design specifications for braking, and thus not enacted; after this self-imposed one second delay, an auditory alert was sounded (p. 12). For almost 20 minutes before impact, the HMI presented no requests for its human operator's input (p. 13), likely contributing to the human operator's sense of complacency.

Conclusions   The following lessons can be taken from this case study:

- The operator was distracted by their personal cell phone; the pedestrian's blood indicated that they were impaired from drugs and that they violated Arizona State's policy by jaywalking.
- The indecisiveness of the ADS was partly attributed to the pedestrian not being in a crosswalk, a feature the system was not designed to address (p. 12), since corrected.
- Uber had inadequate safety risk assessments of its procedures, ineffective oversight in real-time of its vehicle operators to determine whether they were being complacent and exhibited overall an inadequate safety culture (p. vi; see also NTSB, 2019b).
- The Uber ADS was functionally limited, unable to correctly classify the object as a pedestrian, to predict their path, or to adequately assess its risk until almost impact.
- The ADS's decision to suppress action for one second to avoid false alarms increased the risk of driving on the roads and prevented the brakes from being applied immediately to avoid a hazardous situation.
- By disconnecting the Volvo car's own safety systems, Uber increased risk by eliminating the redundant safety systems for its ADS, since corrected (p. vii).

According to NTSB's decision, although the National Highway Traffic Safety Administration (NHTSA) had published a third version of its automated vehicles policy, NHTSA provided no means to a self-driving company of evaluating its vehicle's ADS to meet national or State safety regulations, or to provide a company with the detailed guidance to design an adequate ADS to operate safely. NTSB recommended that safety assessment reports submitted to NHTSA, voluntary at the time of NTSB's final report, be made mandatory (p. viii) and uniform across all states (e.g., Arizona had taken no action by the time of NTSB's final report).

At its most basic level, this case study explores the fundamental tool of debate used for eons by autonomous humans confronting uncertainty. It concludes that machines using artificial intelligence (AI) to operate as members of a team must be able to tell its human partners whenever the machines perceive a change in the context that affects their team's performance; in turn, AI machines must be able to understand humans conversing with them, their interactions indicating the team's intelligence (Cooke, 2020). Moreover, as part of a team, once these AI governed machines learn what humans want them to learn, they will know when the human members of their team are either complacent or malicious in the human's performance of their roles (Lawless, et al., 2017), a capability not yet available in real time, but possibly over the next five years (Sofge, et al., 2019).

From a human-machine team's perspective, the Uber car was a poor team player (Lawless, 2019). Human teams are autonomous. The operator and vehicle operated independently of each other. Facing uncertain situations, the NTSB report confirmed that no single human or machine agent can determine context alone. Resolving uncertainty requires at a minimum a state of shared interdependence to build context, to adapt to rapid changes in context, and, overall, to operate safely and ethically autonomous human-machine systems. We also know from Cummings (2015) that the best science teams are fully interdependent. Cooke (2020) locates a team's intelligence in the interdependent interactions among its teammates. And to reduce uncertainty in an autonomous system necessitates that human and machine teammates are able to explain to each other, however imperfectly, their view of reality in causal terms (Pearl, 2002; Pearl and Mackenzie, 2018).

As SE moves into the future with the design and operation of autonomous human-machine teams and systems, an interdependence among the parts of a team not only makes the parts reactive to the presence of each other, but it also allows for the team to emerge as a single unit, increasing the team's performance. SE practitioners must be cognizant of the benefits and downfalls of AI and autonomy.

6.6 OTHER CASE STUDIES

Additional case studies can be found in other sources. The SEBoK maintains a set of case studies identified as "implementation examples" (https://www.sebokwiki.org/wiki/Systems_Engineering_Implementation_Examples). The SEBoK case studies span domains including defense systems, information systems, management systems, medical systems, space systems, transportation systems, and utility systems.

APPENDIX A: REFERENCES

ABET (2017). Principles of Diversity and Inclusion website, *The Accreditation Board for Engineering and Technology*. Retrieved from https://www.abet.org/about-abet/diversity-equity-and-inclusion.

Abuzied, H., Senbel, H., Awad, M., and Abbas, A. (2020). A Review of Advances in Design for Disassembly with Active Disassembly Applications, *Engineering Science and Technology, an International Journal*, 23(3), 618–624.

Adkins, H., Beyer, B., Blankinship, P., Lewandowski, P., Oprea, A., and Stubblefield, A. (2020). *Building Secure & Reliable Systems, Best Practices for Designing, Implementing and Maintaining Systems*, O'Reilly Media, Inc.

Akroyd-Wallis, K. (2018). Agile Systems Engineering Guide, *The UK Chapter of the International Council on Systems Engineering (INCOSE UK)*.

Alavi, M. and Leidner, D. E. (1999). Knowledge Management Systems: Issues, Challenges, and Benefits, *Communications of the Association for Information Systems*, 1(2), 7.

Albright, D., Brannan, P., and Walrond, C. (2010). Did Stuxnet Take Out 1,000 Centrifuges at the Natanz Enrichment Plant? *Institute for Science and International Security*. Retrieved from http://isis-online.org/isis-reports/detail/did-stuxnet-take-out-1000-centrifuges-at-the-natanz-enrichment-plant 10 March 2023.

Albright, D., Brannan, P., and Walrond, C. (2011). Stuxnet Malware and Natanz: Update of ISIS December 22, 2010 Report. *Institute for Science and International Security*. Retrieved from http://isis-online.org/uploads/isis-reports/documents/stuxnet_update_15Feb2011.pdf 10 March 2023.

Alexander, C., Ishikawa, S., Silverstein, M., Jacobson, M., Fiksdahl-King, I., and Angel, S. (1977). *A Pattern Language*, Oxford University Press.

Alwi, S., Manan, Z., Klemeš, J., and Huisingh, D. (2014). Sustainability engineering for the future, *Journal of Cleaner Production*, 71, 1–10.

ANSI/AIAA G-043B (2018). *Guide to the Preparation of Operational Concept Documents*, The American National Standards Institute and The American Institute of Aeronautics and Astronautics.

ANSI/EIA 632 (2003). *Processes for Engineering a System*, The American National Standards Institute and The Electronic Industries Alliance.

INCOSE Systems Engineering Handbook: A Guide for System Life Cycle Processes and Activities, Fifth Edition.
Edited by David D. Walden, Thomas M. Shortell, Garry J. Roedler, Bernardo A. Delicado, Odile Mornas, Yip Yew-Seng, and David Endler.
© 2023 John Wiley & Sons Ltd. Published 2023 by John Wiley & Sons Ltd.

ANSI/GEIA-STD-0009 (2008). *Reliability Program Standard for Systems Design, Development, and Manufacturing*, The American National Standards Institute and The Government Electronics and Information Technology Association.

Anx, Q. (2021). The DevSecOps Cultural Transformation, *PagerDuty Blog*. Retrieved from https://www.pagerduty.com/blog/devsecops-ops-guide/ 10 March 2023.

Arnold, S. and Lawson, H. (2003). Viewing Systems from a Business Management Perspective: The ISO/IEC 15288 Standard, *Systems Engineering*, 7(3), 229–242.

ARP 4754A (2010). *Aerospace Recommended Practice ARP4754A, Guidelines for Development of Civil Aircraft and Systems, SAE International.*

Asare, P., Broman, D., Lee, E., Torngren, M., and Sunder, S. (2012). *Cyber-Physical Systems - a concept map, Cyber-Physical Systems*. Retrieved from https://ptolemy.berkeley.edu/projects/cps 10 March 2023.

ASQ (2007). *Quality Progress, In Quality Glossary, American Society for Quality*. Retrieved from https://asq.org 10 March 2023.

Avraamidou, S., Baratsas, S., Tian, Y., and Pistikopoulos, E. (2020). Circular Economy-A challenge and an opportunity for Process Systems Engineering, *Computers & Chemical Engineering*, 133, 106629.

Axehill, J., Herzog, E., Tingström, J., and Bengtsson, M. (2021). From Brownfield to Greenfield Development - Understanding and Managing the Transition, *Proceedings of the 31st Annual International Symposium of the International Council on Systems Engineering*, The International Council on Systems Engineering (INCOSE).

Axelsson, J. (2020). Achieving System-of-Systems Interoperability Levels Using Linked Data and Ontologies, *Proceedings of the 30th Annual International Symposium of the International Council on Systems Engineering*, The International Council on Systems Engineering (INCOSE).

Baley, K. and Belcham, D. (2010). *Brownfield Application Development in .NET*, Manning Publishing.

Banach, Z. (2019). What is DevSecOps: How to integrate security into DevOps, *Invicti Web Security Blog*. Retrieved from https://www.invicti.com/blog/web-security/what-is-devsecops/ 10 March 2023.

Barnard, R. (2008). What Is Wrong with Reliability Engineering? *Proceedings of the 18th Annual International Symposium of the International Council on Systems Engineering*, The International Council on Systems Engineering (INCOSE).

Bass, L., Clements, P., and Kazman, R. (2012). *Software Architecture in Practice*, Third Edition. Addison-Wesley Professional.

Becker, O., Ben-Ashe, J., and Ackerman, I. (2000). A Method for Systems Interface Reduction Using N2 Charts, *Systems Engineering*, 3(1), 27–37.

Beery, P. and Paulo, E. (2019). Application of model-based systems engineering concepts to support mission engineering, *Systems*, 7(3), 44. Retrieved from https://www.mdpi.com/2079-8954/7/3/44 10 March 2023.

Bernus, P. (1999). GERAM – Generalized Enterprise Reference – Architecture and Methodology, *IFIP-IFAC Task Force on Architectures for Enterprise Integration*. Retrieved from http://dx.doi.org/10.13140/RG.2.2.35937.33120 10 March 2023.

Bidanda, B. (2022). *Maynard's Industrial and Systems Engineering Handbook*, Sixth Edition. McGraw Hill.

Biomimicry Institute. (2022). Biomimicry Toolbox website, *The Biomimicry Institute*. Retrieved from https://toolbox.biomimicry.org 10 March 2023.

Birman, K. (2012). *Guide to Reliable Distributed Systems: Building High-Assurance Applications and Cloud-Hosted Services*, Springer.

Blanchard, B. (1967). Cost Effectiveness, System Effectiveness, Integrated Logistics Support, and Maintainability, *IEEE Transactions on Reliability*, 163(3), 117–126

Blanchard, B. (2004). *Logistics Engineering and Management*, Sixth Edition. Pearson Prentice Hall.

Blanchard, B. and Fabrycky, W. (2011). *Systems Engineering and Analysis*, Fifth. Prentice Hall.

Bobinis, J., Haimowitz, J., Tuttle, P., Garrison, C., Mitchell, T., and Klingberg, J. (2013). Affordability Considerations: Cost Effective Capability, *Proceedings of the 23rd Annual International Symposium of the International Council on Systems Engineering*, The International Council on Systems Engineering (INCOSE).

Boehm, B. (1987). A spiral model of software development and enhancement, *ACM SIGSOFT Software Engineering Notes*, 11(4), 14–24.

Boehm, B., Lane, J., Koolmanojwong, S., and Turner, R. (2014). *The Incremental Commitment Spiral Model: Principles and Practices for Successful Systems and Software*, Addison-Wesley Professional.

Boehm, B. and Turner, R. (2004). *Balancing Agility and Discipline*, Addison-Wesley.

Boehm-Davis, D., Durso, F., and Lee, J. (2015). *APA Handbook of Human Systems Integration*, American Psychological Association.

Bogdanich, W. (2010). *Radiation Offers New Cures, and Ways to Do Harm*, The New York Times.

Bonnet, S., Voirin, J.-L., Exertier, D., and Normand, V. (2017). Modeling system modes, states, configurations with Arcadia and Capella: Method and tool perspectives, *Proceedings of the 27th Annual International Symposium of the International Council on Systems Engineering*, The International Council on Systems Engineering (INCOSE).

Boy, G. (2013). *Orchestrating Human-Centered Design*, Springer.

Boy, G. (2020). *Human Systems Integration: From Virtual to Tangible*, CRC Press – Taylor & Francis Group.

Bradley, J., Hughes, M., and Schindel, W. (2010). Optimizing Delivery of Global Pharmaceutical Packaging Solutions, Using Systems Engineering Patterns, *Proceedings of the 20th Annual International Symposium of the International Council on Systems Engineering*, The International Council on Systems Engineering (INCOSE).

Brtis, J. S. (2016). *How to Think About Resilience, MITRE Technical Report*, MITRE Corporation.

Brtis, J. S. (2020). Loss-Driven Systems Engineering (LDSE), *INSIGHT*, 23(4), The International Council on Systems Engineering (INCOSE).

Brtis, J. S. and McEvilley, M. A. (2019). Systems Engineering for Resilience, *MITRE Technical Report*, The MITRE Corporation.

Buede, D. and Miller, W. (2016). *The Engineering Design of Systems: Models and Methods*, Third Edition. John Wiley & Sons.

Cabrera, D., Cabrera, L., and Powers, E. (2015). A unifying theory of systems thinking with psychosocial applications, *Systems Research and Behavioral Science*, 32, 534–545.

Calvo-Amodio, J. and Rousseau, D. (2019). The Human Activity System: Emergence from Purpose, Boundaries, Relationships, and Context, *Procedia Computer Science*, 153, 91–99.

Cantwell. (2021). Aviation Safety Whistleblower Report, *U.S. Senate Committee on Commerce, Science, and Transportation*.

Carayon, P., Schoofs Hundt, A., Karsh, B.-T., Gurses, A., Alvarado, C., Smith, M., and Flatley Brennan, P. (2006). Work system design for patient safety: The SEIPS model, *Quality and Safety in Health Care*, 15(Suppl 1), i50–i58.

Carpenter, S., Delugach, H., Etzkorn, L., Fortune, J., Utley, D., and Virani, S. (2010). The Effect of Shared Mental Models on Team Performance, *Industrial Engineering Research Conference*, Cancun, Mexico, Institute of Industrial Engineers.

Chapman, W., Bahill, A., and Wymore, A. (1992). *Engineering Modeling and Design*, CRC Press. Retrieved from https://doi.org/10.1201/9780203757314.

Checkland, P. (1999). *Systems Thinking, Systems Practice*, John Wiley.

Choi, S., Suh, E., and Park, C. (2020). Value chain and stakeholder-driven product platform design, *Systems Engineering*, 23(3), 312–326. Retrieved from https://doi:10.1002/sys.21527.

Clausing, D. and Frey, D. (2005). Improving System Reliability by Failure Mode Avoidance including Four Concept Design Strategies, *Systems Engineering*, 8(3), 245–261.

Cloutier, R., DiMario, M., and Pozer, H. (2009). Net Centricity and Systems of Systems, in M. Jamshidi (Ed.) *Systems of Systems Engineering*, CRC Press/Taylor & Francis Group.

CMMI (2018). *Capability Maturity Model Integration (CMMI®) for Development V2.0, the CMMI Institute, a subsidiary of the Information Systems Audit and Control Association (ISACA)*. Retrieved from https://cmmiinstitute.com/cmmi/dev 10 March 2023.

CMU/SEI (2000). *ATAM – Architecture Tradeoff Analysis Method - TR-004 ESC-TR-2000-004, the Software Engineering Institute at Carnegie Mellon University*. Retrieved from https://resources.sei.cmu.edu/asset_files/TechnicalReport/2000_005_001_13706.pdf 10 March 2023.

Conrow, E. (2003). *Effective Risk Management*, Second Edition. American Institute of Aeronautics and Astronautics, Inc.

Cook, C. and Unewisse, M. (2019). *A SoS Approach for Engineering Capability Programs*. Retrieved from https://doi.org/10.1002/j.2334-5837.2017.00342.x 10 March 2023.

Cook, D. and Schindel, W. (2015). Utilizing MBSE Patterns to Accelerate System Verification, *Proceedings of the 25th Annual International Symposium of the International Council on Systems Engineering*, The International Council on Systems Engineering (INCOSE).

Cooke, N. (2020). Effective human-artificial intelligence teaming, *Proceedings of the AAAI Spring Symposium*, Stanford, CA.

Crosby, P. (1979). *Quality Is Free: The Art of Making Quality Certain*, McGraw Hill.

Cross, N. (2000). *Engineering Design Methods – Strategies for Product Design*. Chichester UK: Wiley. Third edition.

CST (2020). *A Systems Approach to Delivering Net Zero: Recommendations From The Prime Minister's Council For Science And Technology, the UK Council for Science and Technology*. Retrieved from https://assets.publishing.service.gov.uk/government/uploads/system/uploads/attachment_data/file/910446/cst-net-zero-report-30-january-2020.pdf 10 March 2023.

Cullen, T. (1990). *The Public Inquiry into the Piper Alpha Disaster*, HMSO.

Cummings, J. (2015). *Team Science successes and challenges, National Science Foundation Sponsored Workshop on Fundamentals of Team Science & Science of Team Science*, Bethesda MD.

Dahmann, J. (2014). System of Systems Pain Points. *Proceedings of the 24th Annual International Symposium of the International Council on Systems Engineering*, The International Council on Systems Engineering (INCOSE).

Dahmann, J., Rebovich, G., Lane, J., Lowry, R., and Baldwin, K. (2011). An Implementer's View of Systems Engineering for Systems of Systems, *IEEE Systems Conference*, April 4–7, 2011, Montreal, Canada, p. 212–217.

Daskin, M. (2010). *Service Science*, John Wiley & Sons, Inc.

DAU (1993). *Committed Life Cycle Cost against Time*, US Defense Acquisition University.

DAU (2022). *Glossary*, US Defense Acquisition University. Retrieved from https://www.dau.edu/glossary/Pages/Glossary.aspx 10 March 2023.

Defoe, J. (1993). An Identification of Pragmatic Principles, *Final Report, Systems Engineering Practice Working Group, Subgroup on Pragmatic Principles*, National Council on Systems Engineering (NCOSE) WMA Chapter.

Dekker, S. (2014). *Safety Differently: Human Factors for a New Era*, Second Edition. Routledge.

Delicado, B., Salado, A., and Mompó, R. (2018). Conceptualization of a T-Shaped engineering competency model in collaborative organizational settings: Problem and status in the Spanish aircraft industry, *Systems Engineering*, 21, 534–554, Retrieved from https://doi.org/10.1002/sys.21453.

Deming, W. (1986). *Out of the Crisis, Massachusetts Institute of Technology Center for Advanced Engineering Study*.

Dennis, A., Wixom, B., and Tegarden, D. (2020). *Systems Analysis and Design: An Object-Oriented Approach with UML*, Sixth Edition. John Wiley & Sons.

Desapriya, E., Kerr, J., Hewapathirane, D., Peiris, C., Mann, B., Gomes, N., Peiris, K., Scime, G., and Jones, J. (2012). Bull Bars and Vulnerable Road Users, *Traffic Injury Prevention*, 13, 86–92, Retrieved from http://dx.doi.org/10.1080/15389588.2011.624143.

DoDAF (2010). *DoD Architecture Framework, Version 2.02*, US Department of Defense. Retrieved from https://dodcio.defense.gov/Library/DoD-Architecture-Framework.

Domingue, J., Fensel, D., Davies, J., Gonzalez-Cabero, R., and Pedrinaci, C. (2009). The Service Web: A Web of Billions of Services, in G. Tselentis, et al. (Ed.) *Toward the Future Internet—A European Research Perspective*, IOS Press. Retrieved from https://doi.org/10.3233/978-1-60750-007-0-203.

Donovan, J. and Prabhu, K. (2017). *Building the Network of the Future: Getting Smarter, Faster and More Flexible with a Software Centric Approach*, CRC Press.

Dorst, K. (2015). *Frame Innovation: Create New Thinking by Design*. Cambridge MA: MIT Press.

Dove, R. and LaBarge, R. (2014). Fundamentals of Agile Systems Engineering – Part 1. *Proceedings of the 24th Annual International Symposium of the International Council on Systems Engineering*, The International Council on Systems Engineering (INCOSE).

Dove, R. and Schindel, W. (2017). Case Study: Agile SE Process for Centralized SoS Sustainment at Northrop Grumman, *Proceedings of the 27th Annual International Symposium of the International Council on Systems Engineering*, The International Council on Systems Engineering (INCOSE).

Dove, R. and Schindel, W. (2019). Agile Systems Engineering Life Cycle Model for Mixed Discipline Engineering, *Proceedings of the 29th Annual International Symposium of the International Council on Systems Engineering*, The International Council on Systems Engineering (INCOSE).

Dove, R., Schindel, W., and Garlington, K. (2018). Case Study: Agile Systems Engineering at Lockheed Martin Aeronautics Integrated Fighter Group, *Proceedings of the 28th Annual International Symposium of the International Council on Systems Engineering*, The International Council on Systems Engineering (INCOSE).

Dove, R., Schindel, W., and Hartney, R. (2017). Case Study: Agile Hardware/Firmware/Software Product Line Engineering at Rockwell Collins, *Proceedings 11th Annual IEEE International Systems Conference*. Retrieved from https://ieeexplore.ieee.org/document/7934807/authors#authors 10 March 2023.

Dove, R., Schindel, W., and Scrapper, C. (2016). Agile Systems Engineering Process Features Collective Culture, Consciousness, and Conscience at SSC Pacific Unmanned Systems Group. *Proceedings of the 26th Annual International Symposium of the International Council on Systems Engineering*, The International Council on Systems Engineering (INCOSE).

DSMC (1983). *Systems Engineering Management Guide*, US Defense Systems Management College.

Edwards, W., Miles Jr., R., and Von Winterfeldt, D. (2007). *Advances in Decision Analysis: From Foundations to Applications*, Cambridge University Press.

Elm, J. and Goldenson, D. (2012). The Business Case for Systems Engineering Study: Results of the Systems Engineering Effectiveness Study, *The Software Engineering Institute at Carnegie Mellon University*. Retrieved from http://resources.sei.cmu.edu/library/asset-view.cfm?assetID=34061.

Eppinger, S. and Browning, T. (2012). *Design Structure Matrix Methods and Applications*, MIT Press.

EPRS (2016). *The European Parliament Research Service website*. Retrieved from https://www.europarl.europa.eu/at-your-service/en/stay-informed/research-and-analysis 10 March 2023.

Estefan, J. (2008). *Survey of Model-Based Systems Engineering (MBSE) Methodologies*, Rev. B, Section 3.2, NASA Jet Propulsion Laboratory.

Evans, D. (2014). Styles of Architecting - A smarter approach to architecting the Defence Enterprise, *Niteworks White Paper*. Retrieved from Styles of architecting: a smarter approach to architecting the Defence Enterprise (white paper) - GOV.UK (www.gov.uk) 10 March 2023.

FAA (2014). *FAA Systems Engineering Manual*, US Federal Aviation Administration.

Fagen, M. (1978). *A History of engineering and science in the Bell System: National Service in War and Peace (1925–1975)*, Bell Telephone Laboratories (Author), M. Fagen (Editor).

Failliere, N. (2011). *W32, Stuxnet Dossier, Version 1.4*. Retrieved from https://archive.org/details/w32_stuxnet_dossier 10 March 2023.

Fairley, R. (2019). *Systems Engineering of Software-Enabled Systems*, John Wiley & Sons, Inc.

Firesmith, D., Capell, P., Falkenthal, D., Hammons, C., Latimer IV, T., and Merendino, T. (2008). *Method Framework for Engineering System Architectures (MFESA)*, Auerbach Publications.

Fishman, G. (1997). *Monte Carlo: Concepts, Algorithms, and Applications*, pp. 1–4. Retrieved from http://www.sei.cmu.edu/library/assets/quality-assess.pdf.

Forsberg, K. (1995). If I Could Do That, Then I Could …: Systems Engineering in a Research and Development Environment, *Proceedings of the 5th Annual International Symposium of the International Council on Systems Engineering*, The International Council on Systems Engineering (INCOSE).

Forsberg, K. and Mooz, H. (1991). The Relationship of System Engineering to the Project Cycle, *Proceedings of the 1st Annual Symposium of the National Council on Systems Engineering*, The National Council on Systems Engineering (NCOSE).

Forsberg, K., Mooz, H., and Cotterman, H. (2005). *Visualizing Project Management*, Third Edition. John Wiley & Sons, Inc.

Fossnes, T. (2005). Lessons from Mt. Everest Applicable to Project Leadership, *Proceedings of the 15th Annual International Symposium of the International Council on Systems Engineering*, The International Council on Systems Engineering (INCOSE).

Friedland, B. (2012). *Control System Design*, Dover Publications.

Fries, M. and Paal, B. (2019). *Smart contracts*. Retrieved from http://library.oapen.org/handle/20.500.12657/24858 10 March 2023.

Frost, S. (2016). *Compound Eye Sensor for Real-time Aircraft Wing Deflection Measurement*, American Institute of Aeronautics and Astronautics.

Gallup Report (2017). *State of the Global Workplace - Gallup Report*. Retrieved from https://www.gallup.com/home.aspx 10 March 2023.

Gallup Report (2020). *State of the Global Workplace - Gallup Report*. Retrieved from https://www.gallup.com/workplace/285818/state-american-workplace-report.aspx 10 March 2023.

Gamma, E., Helm, R., Johnson, R., and Vlissides, J. (1995). *Design Patterns: Elements of Reusable Object-Oriented Software*, Addison-Wesley Publishing Company.

GAO (2008). *Best Practices—Increased Focus on Requirements and Oversight Needed to Improve DOD's Acquisition Environment and Weapon System Quality*, US Government Accountability Office. Retrieved from https://www.gao.gov/new.items/d08294.pdf 10 March 2023.

Garlan, D. and Shaw, M. (1994). *Advances in Software Engineering and Knowledge Engineering*, Volume 1.

Garlan, D. and Shaw, M. (1996). *Advances in Software Engineering and Knowledge Engineering*, Volume 2.

Gartner (2022). *Gartner Glossary*. Retrieved from https://www.gartner.com/en/information-technology/glossary/ie-information-engineering 10 March 2023.

Gass, S. and Harris, C. (2001). *Encyclopedia of Operations Research and Management Science*, xli Journal of the Operational Research Society.

Gebhardt, A. and Hötte, J.-S. (2016). *Additive Manufacturing: 3D Printing for Prototyping and Manufacturing*, Hanser Publications.

GEIA HB649A (2016). *Configuration Management Standard Implementation Guide*, The Government Electronics and Information Technology Association.

Geissdoerfer, M., Pieroni, M., Pigosso, D., and Soufani, K. (2020). Circular business models: A review, *Journal of Cleaner Production*, 277, 123741.

Gilb, T. and Graham, D. (1993). *Software Inspection*, Addison-Wesley-Longman.

Goode, H. and Machol, R. (1957). *System Engineering; An Introduction to the Design of Large-scale Systems*, McGraw-Hill.

Grady, J. O. (1994). *System Integration*. Boca Raton, FL: CRC Press.

Gregg, S., Scharadin, R., and Clements, P. (2015). The More You Do, the More You Save: The Superlinear Cost Avoidance Effect of Systems and Software Product Line Engineering, *Proceedings of the Software Product Line Conference*, Nashville.

Griffin, M. (2010). How do we fix system engineering? *Proceedings of the 61st International Astronautical Congress*, Prague, CZ.

Guariniello, C., Raz, A. K., Fang, Z., and DeLaurentis, D. (2020). System-of-systems tools and techniques for the analysis of cyber-physical systems, *Systems Engineering*, 23(4), 480–491. Retrieved from https://incose.onlinelibrary.wiley.com/doi/10.1002/sys.21539.

Gupta, J. and Sharma, S. (2004). *Creating Knowledge Based Organizations*, Ida Group Publishing.

Haberfellner, R. and de Weck, O. (2005). Agile SYSTEMS ENGINEERING versus AGILE SYSTEMS Engineering, *Proceedings of the 15th Annual International Symposium of the International Council on Systems Engineering*, The International Council on Systems Engineering (INCOSE).

Hall, A. (1962). *A Methodology for Systems Engineering*, Van Nostrand.

Hallman, C. (2022). *50 Cognitive Biases to be Aware of so You Can be the Very Best Version of You.* Retrieved from https://www.titlemax.com/discovery-center/lifestyle/50-cognitive-biases-to-be-aware-of-so-you-can-be-the-very-best-version-of-you 10 March 2023.

Hammond, D. (2003). *The Science of Synthesis: Exploring the Social Implications of General Systems Theory*, University Press of Colorado.

Harding, A. and Pickard, A. (2019). Towards a More Diverse INCOSE, *INSIGHT*, 22(3), The International Council on Systems Engineering (INCOSE).

Haskins, C. (2021). Systems Engineering for Sustainable Development Goals, *Sustainability*, 13(18), 10293.

He, B., Pan, Q., and Deng, Z. (2018). Product carbon footprint for product life cycle under uncertainty, *Journal of Cleaner Production*, 187, 459–472.

Henshaw, M. (2016). Systems of Systems, Cyber-Physical Systems, the Internet-of-Things… Whatever Next?, *INSIGHT*, 19(3), The International Council on Systems Engineering (INCOSE).

Hidden, A. (1989). *Investigation into the Clapham Junction Railway Accident*, HMSO.

Hoehne, O. (2020). Case Study: Achieving System Integration through Interoperability in a large System of Systems (SoS), *Proceedings of the 30th Annual International Symposium of the International Council on Systems Engineering*, The International Council on Systems Engineering (INCOSE).

Hoeller, N., Farnsworth, M., Jacobs, S., Chirazi, J., Mead, T., Goel, A., and Salustri, F. (2016). A Systems View of Bio-inspiration: Bridging the Gaps, *INSIGHT*, 19(1), The International Council on Systems Engineering (INCOSE).

Honour, E. (2013). Systems Engineering Return on Investment, Ph.D. Thesis, *Defense and Systems Institute*, University of South Wales. Retrieved from https://incose.onlinelibrary.wiley.com/doi/10.1002/j.2334-5837.2010.tb01150.x 10 March 2023.

Hopkins, R. and Jenkins, K. (2008). *Eating the IT Elephant: Moving from Greenfield Development to Brownfield*, IBM Press/Pearson.

Howard, R. (1968). The Foundations of Decision Analysis, *IEEE Transactions on Systems Science and Cyberntics*, 4(3), 211–219. Retrieved from https://doi.org/10.1109/TSSC.1968.300115.

Hudson, P. (2001). *Safety Management and Safety Culture: The Long, Hard and Winding Road, Occupational Health & Safety Management Systems*.

ICE (2023). *Proceedings of the Institution of Civil Engineers - Engineering Sustainability*, The Institution of Civil Engineers. Retrieved from https://www.icevirtuallibrary.com/toc/jensu/current 10 March 2023.

IEC 60601 (2020). *Medical device safety*, The International Electrotechnical Commission.

IEC 62196 (2022). *Plugs, Socket-Outlets, Vehicle Connectors And Vehicle Inlets – Conductive Charging Of Electric Vehicles (Multipart Standard)*, The International Electrotechnical Commission.

IEC 62304 (2006). *Medical Device software - Software life cycle processes*, The International Electrotechnical Commission.

IEC 62366 (2015). *Application of usability engineering to medical devices*, The International Electrotechnical Commission.

IEC 62402 (2019). *Obsolescence management*, The International Electrotechnical Commission.

IEEE 1517 (2010). *IEEE Standard for Information Technology - System and Software Life Cycle Processes - Reuse Processes*, The Institute of Electrical and Electronics Engineers.

IEEE 610.12 (1990). *IEEE Standard Glossary of Software Engineering Terminology*, The Institute of Electrical and Electronics Engineers. Retrieved from https://doi.org/10.1109/IEEESTD.1990.101064.

IEEE 828 (2012). *IEEE Standard for Configuration Management in Systems and Software Engineering*, The Institute of Electrical and Electronics Engineers.

IEEE Reliability Society (2023). *IEEE Reliability Society website*, The Institute of Electrical and Electronics Engineers. Retrieved from https://rs.ieee.org 10 March 2023.

IISE (2022). *The Industrial and Systems Engineering Body of Knowledge (ISEBoK)*, The Institute of Industrial and Systems Engineers. Retrieved from https://www.iise.org/details.aspx?id=43631 10 March 2023.

INCOSE Automotive SE Vision (2020). *INCOSE Automotive SE Vision 2025*, The International Council on Systems Engineering (INCOSE).

INCOSE Code of Ethics (2023). *INCOSE Code of Ethics website*, The International Council on Systems Engineering (INCOSE). Retrieved from https://www.incose.org/about-incose/Leadership-Organization/code-of-ethics 10 March 2023.

INCOSE Complexity Primer (2021). *INCOSE Complexity Primer*, The International Council on Systems Engineering (INCOSE).

INCOSE Definitions (2019). *INCOSE Definitions*, The International Council on Systems Engineering (INCOSE).

INCOSE GtNR (2022). *INCOSE Guide to Needs and Requirements*, The International Council on Systems Engineering (INCOSE).

INCOSE GtVV (2022). *INCOSE Guide to Verification and Validation*, The International Council on Systems Engineering (INCOSE).

INCOSE GtWR (2022). *INCOSE Guide to Writing Requirements*, The International Council on Systems Engineering (INCOSE).

INCOSE HSI Primer (2023). *Human Systems Integration Primer,* The International Council on Systems Engineering (INCOSE).

INCOSE Innovation Ecosystem (2022). *Adaptive, Learning Ecosystems: Introduction to the Innovation Ecosystem S*Pattern*, Version 1.0, The International Council on Systems Engineering (INCOSE).

INCOSE MBSE Wiki (2022). *INCOSE MBSE Wiki website*, The International Council on Systems Engineering (INCOSE). Retrieved from http://www.omgwiki.org/MBSE/doku.php 10 March 2023.

INCOSE NS Primer (2023). *Natural Systems And The Systems Engineering Process: A Primer,* The International Council on Systems Engineering (INCOSE).

INCOSE Measurement Primer (2010). *Systems Engineering Measurement Primer, A Basic Introduction to Measurement Concepts and Use for Systems Engineering*, The International Council on Systems Engineering (INCOSE).

INCOSE NRM (2022). *- INCOSE Needs and Requirements Manual*, The International Council on Systems Engineering (INCOSE).

INCOSE OOSEM (2022). *INCOSE OOSEM Working Group website,* The International Council on Systems Engineering (INCOSE). Retrieved from https://www.incose.org/incose-member-resources/working-groups/transformational/object-oriented-se-method 10 March 2023.

INCOSE PLE Primer (2019). *INCOSE PLE Primer*, The International Council on Systems Engineering (INCOSE).

INCOSE PMGtSEMfPS (2015). *Project Manager's Guide to Systems Engineering Measurement for Project Success, A Basic Introduction to Systems Engineering Measures for Use by Project Managers*, The International Council on Systems Engineering (INCOSE).

INCOSE S*MBSE (2022) *Minimal S*Models: A Primer*. The International Council on Systems Engineering (INCOSE).

INCOSE S*MCP (2019). *The Model Characterization Pattern: A Universal Characterization & Labeling S*Pattern for All Models, V1.9.3*, The International Council on Systems Engineering (INCOSE).

INCOSE S*Patterns Primer (2022). *An MBSE S*Patterns Primer, Version 1.0*, The International Council on Systems Engineering (INCOSE).

INCOSE SE Principles (2022). *INCOSE SE Principles*, The International Council on Systems Engineering (INCOSE).

INCOSE SECAG (2023). *INCOSE Systems Engineering Competency Assessment Guide*, The International Council on Systems Engineering (INCOSE), Wiley.

INCOSE SECF (2018). INCOSE SE Competency Framework, The International Council on Systems Engineering (INCOSE).

INCOSE SETDB (2021). *INCOSE Systems Engineering Tools Database*, The International Council on Systems Engineering (INCOSE).

INCOSE Value Report (2021). *INCOSE Value Strategic Initiative Report*, The International Council on Systems Engineering (INCOSE).

INCOSE Vision 2020 (2007). *Systems Engineering Vision 2020*, The International Council on Systems Engineering (INCOSE).

INCOSE Vision 2035 (2022). *Systems Engineering Vision 2035*, The International Council on Systems Engineering (INCOSE).

INCOSE, PSM, NDIA, SERC, OSD R&E, AIA, and Aerospace (2022). *Practical Software and Systems Measurement (PSM) Digital Engineering Measurement Framework,* The International Council on Systems Engineering, Practical Software & Systems Measurement Support Center, The National Defense Industrial Association, Systems Engineering Research Center, US Office of the Secretary of Defense Research & Engineering, The Aerospace Industries Association, and The Aerospace Corporation.

INFORMS (2022). *Institute for Operations Research and the Management Sciences website*. Retrieved from https://www.informs.org 10 March 2023.

IOGP (2021). *International Association of Oil & Gas Producers website*. Retrieved from https://www.iogp.org 10 March 2023.

ISO 10007 (2017). *Quality management - Guidelines for configuration management*, The International Organization for Standardization.

ISO 10303-233 (2012). *Industrial automation systems and integration — Product data representation and exchange - Part 233: Application protocol: Systems engineering*, The International Organization for Standardization.

ISO 13008 (2022). *Information and documentation — Digital records conversion and migration process*, The International Organization for Standardization.

ISO 13485 (2016). *Medical devices — Quality management systems — Requirements for regulatory purposes*, The International Organization for Standardization.

ISO 14000 (2015). *ISO 14001 and related standards Environmental management*, The International Organization for Standardization.

ISO 14001 (2015). *Environmental management systems — Requirements with guidance for use*, The International Organization for Standardization.

ISO 14971 (2019). *Medical devices — Application of risk management to medical devices*, The International Organization for Standardization.

ISO 15704 (2019). *Enterprise modelling and architecture — Requirements for enterprise-referencing architectures and methodologies*, The International Organization for Standardization.

ISO 17757 (2019). *Earth-moving machinery and mining — Autonomous and semi-autonomous machine system safety*, The International Organization for Standardization.

ISO 21347 (2005). *Space systems — Fracture and damage control*, The International Organization for Standardization.

ISO 22367 (2020). *Medical laboratories — Application of risk management to medical laboratories*, The International Organization for Standardization.

ISO 26262 (2018). *Road vehicles — Functional safety (Multi-part Standard)*, The International Organization for Standardization.

ISO 27000 (2018). *Series of standards for information security matters*, The International Organization for Standardization.

ISO 8930 (2021). *General principles on reliability for structures — Vocabulary*, The International Organization for Standardization.

ISO 9000 (2015). *Quality management*, The International Organization for Standardization.

ISO 9001 (2015). *Quality management — Requirements*, The International Organization for Standardization.

ISO Guide 73 (2009). *Risk management — Vocabulary*, The International Organization for Standardization.

ISO/IEC 17799 (2005). *Information technology — Security techniques — Code of practice for information security management*, The International Organization for Standardization and The International Electrotechnical Commission.

ISO/IEC 19759 (2015). *Software Engineering — Guide to the software engineering body of knowledge (SWEBOK)*, The International Organization for Standardization and The International Electrotechnical Commission.

ISO/IEC 19970-1 (2017). *Information Technology - IT Asset Management - Part 1: IT Asset Management Systems - Requirements*, The International Organization for Standardization and The International Electrotechnical Commission.

ISO/IEC 2382 (2015). *Information technology — Vocabulary*, The International Organization for Standardization and The International Electrotechnical Commission.

ISO/IEC 29110 (2016). *Systems and Software Engineering Standards and Guides for Very Small Entities (VSEs) (Multi-part set)*, The International Organization for Standardization and The International Electrotechnical Commission.

ISO/IEC 31000 (2018). *Risk management*, The International Organization for Standardization and The International Electrotechnical Commission.

ISO/IEC 31010 (2019). *Risk management – Risk assessment techniques*, The International Organization for Standardization and The International Electrotechnical Commission.

ISO/IEC 33060 (2020). *Process Assessment - Process Assessment Model for System Life Cycle Processes*, The International Organization for Standardization and The International Electrotechnical Commission.

ISO/IEC Guide 51 (2014). *Safety Aspects — Guidelines for their Inclusion in Standards*, The International Organization for Standardization and The International Electrotechnical Commission.

ISO/IEC/IEEE 12207 (2017). *Systems and Software Engineering - Software Life Cycle Processes*, The International Organization for Standardization, The International Electrotechnical Commission, and The Institute of Electrical and Electronics Engineers.

ISO/IEC/IEEE 15026 (2019). *Systems and Software Engineering - Systems and Software Assurance (Multi-Part Standard)*, The International Organization for Standardization, The International Electrotechnical Commission, and The Institute of Electrical and Electronics Engineers.

ISO/IEC/IEEE 15288 (2023). *Systems and Software Engineering - System Life Cycle Processes*, The International Organization for Standardization, The International Electrotechnical Commission, and The Institute of Electrical and Electronics Engineers.

ISO/IEC/IEEE 15289 (2019). *Systems and Software Engineering - Content of Life Cycle Information Items (Documentation)*, The International Organization for Standardization, The International Electrotechnical Commission, and The Institute of Electrical and Electronics Engineers.

ISO/IEC/IEEE 15939 (2017). *Systems and Software Engineering - Measurement Process*, The International Organization for Standardization, The International Electrotechnical Commission, and The Institute of Electrical and Electronics Engineers.

ISO/IEC/IEEE 16085 (2021). *Systems and Software Engineering - Life Cycle Processes – Risk Management*, The International Organization for Standardization, The International Electrotechnical Commission, and The Institute of Electrical and Electronics Engineers.

ISO/IEC/IEEE 16326 (2019). *Systems and Software Engineering - Life Cycle Processes – Project Management*, The International Organization for Standardization, The International Electrotechnical Commission, and The Institute of Electrical and Electronics Engineers.

ISO/IEC/IEEE 21839 (2019). *Systems and software engineering - System of systems (SoS) considerations in life cycle stages of a system*, The International Organization for Standardization, The International Electrotechnical Commission, and The Institute of Electrical and Electronics Engineers.

ISO/IEC/IEEE 21840 (2019). *Systems and software engineering - Guidelines for the utilization of ISO/IEC/IEEE 15288 in the context of system of systems (SoS)*, The International Organization for Standardization, The International Electrotechnical Commission, and The Institute of Electrical and Electronics Engineers.

ISO/IEC/IEEE 21841 (2019). *Systems and software engineering — Taxonomy of systems of systems*, The International Organization for Standardization, The International Electrotechnical Commission, and The Institute of Electrical and Electronics Engineers.

ISO/IEC/IEEE 24641 (2022) *Systems and Software Engineering — Methods and Tools for Model-Based Systems and Software Engineering, The International Organization for Standardization*, The International Electrotechnical Commission, and The Institute of Electrical and Electronics Engineers.

ISO/IEC/IEEE 24748-1 (2018). *Systems and Software Engineering - Life Cycle Management - Part 1: Guidelines for Life Cycle Management*, The International Organization for Standardization, The International Electrotechnical Commission, and The Institute of Electrical and Electronics Engineers.

ISO/IEC/IEEE 24748-2 (2018). *Systems and Software Engineering - Life Cycle Management - Part 2: Guidelines for the Application of ISO/IEC/IEEE 15288 (System Life Cycle Processes)*, The International Organization for Standardization, The International Electrotechnical Commission, and The Institute of Electrical and Electronics Engineers.

ISO/IEC/IEEE 24748-4 (2016). *Systems and Software Engineering - Life Cycle Management - Part 4: Systems Engineering Planning*, The International Organization for Standardization, The International Electrotechnical Commission, and The Institute of Electrical and Electronics Engineers.

ISO/IEC/IEEE 24748-5 (2017). *Systems and Software Engineering - Life cycle Management - Part 5: Software Development Planning*, The International Organization for Standardization, The International Electrotechnical Commission, and The Institute of Electrical and Electronics Engineers.

ISO/IEC/IEEE 24748-6 (2016). *Systems and software engineering - Life cycle management - Part 6: System integration engineering*, The International Organization for Standardization, The International Electrotechnical Commission, and The Institute of Electrical and Electronics Engineers.

ISO/IEC/IEEE 24748-7 (2019). *Systems and software engineering - Life cycle management - Part 7: Application of systems engineering on defense programs*, The International Organization for Standardization, The International Electrotechnical Commission, and The Institute of Electrical and Electronics Engineers.

ISO/IEC/IEEE 24748-8/IEEE15288.2 (2014). *Systems and Software Engineering - Life Cycle Management - Part 8: Technical Reviews and Audits on Defense Programs*, The International Organization for Standardization, The International Electrotechnical Commission, and The Institute of Electrical and Electronics Engineers.

ISO/IEC/IEEE 24765 (2017). *Systems and Software Engineering - Vocabulary*, The International Organization for Standardization, The International Electrotechnical Commission, and The Institute of Electrical and Electronics Engineers.

ISO/IEC/IEEE 26550 (2015). *Software and Systems Engineering - Reference Model for Product Line Engineering and Management*, The International Organization for Standardization, The International Electrotechnical Commission, and The Institute of Electrical and Electronics Engineers.

ISO/IEC/IEEE 26580 (2021). *Software and Systems Engineering - Methods and Tools for the Feature-Based Approach to Software and Systems Product Line Engineering*, The International Organization for Standardization, The International Electrotechnical Commission, and The Institute of Electrical and Electronics Engineers.

ISO/IEC/IEEE 29148 (2018). *Systems and Software Engineering - Life Cycle Processes - Requirements Engineering*, The International Organization for Standardization, The International Electrotechnical Commission, and The Institute of Electrical and Electronics Engineers.

ISO/IEC/IEEE 42010 (2022). *Systems and Software Engineering - Architecture Description*, The International Organization for Standardization, The International Electrotechnical Commission, and The Institute of Electrical and Electronics Engineers.

ISO/IEC/IEEE 42020 (2019). *Software, Systems and Enterprise - Architecture Processes*, The International Organization for Standardization, The International Electrotechnical Commission, and The Institute of Electrical and Electronics Engineers.

ISO/IEC/IEEE 42030 (2019). *Software, Systems and Enterprise - Architecture Evaluation Framework*, The International Organization for Standardization, The International Electrotechnical Commission, and The Institute of Electrical and Electronics Engineers.

ISO/PAS 19450 (2015). *Automation systems and integration — Object-Process Methodology (OPM)*, The International Organization for Standardization.

ISO/SAE 21434 (2021). *Road vehicles — Cybersecurity engineering*, The International Organization for Standardization and SAE International.

ISO/TS 16949 (2009). *Quality management systems — Particular requirements for the application of ISO 9001:2008 for automotive production and relevant service part organizations*, The International Organization for Standardization.

Jackson, M. (2003). *Systems Thinking: Creative Holism for Managers*, Wiley.

Jackson, M. (2019). *Critical Systems Thinking and the Management of Complexity*, Wiley.

Jackson, M. and Keys, P. (1984). Towards a system of systems methodologies, *Journal of the Operational Research Society*, 35, 473–486.

Jackson, S. and Ferris, T. (2013). Resilience Principles for Engineered Systems, *Systems Engineering*, 19(2), 152–164.

Jackson, S. and Ferris, T. (2016). *Proactive and Reactive Resilience: A Comparison of Perspectives*. Retrieved from https://www.academia.edu/34079700/Proactive_and_Reactive_Resilience_A_Comparison_of_Perspectives 10 March 2023.

Jacky, J. (1989). Programmed for Disaster, *The Sciences*, 29, 22–27.

Jensen, J. (2014). *The Øresund Bridge—Linking Two Nations*. Retrieved from https://www.cowi.dk 10 March 2023.

Johnson, S. (2010). *Where Good Ideas Come From: The Natural History of Innovation*, New York, NY: Riverhead Books.

Journal of Cleaner Production (2023). *Journal of Cleaner Production website*. Retrieved from www.journals.elsevier.com/journal-of-cleaner-production 10 March 2023.

Journal of Environmental Management (2023). *Journal of Environmental Management website*. Retrieved from www.sciencedirect.com/journal/journal-of-environmental-management 10 March 2023.

Journal of Organizational Behavior (2004). *Journal of Organizational Behavior*, 25, 439–459, Wiley.

Juran, J. (1974). *Quality Control Handbook*, Third Edition. McGraw-Hill.

Kagermann, H., Wahlster, W., and Helbig, J. (2013). *Securing the Future of German Manufacturing Industry: Recommendations for Implementing the Strategic Initiative INDUSTRIE 4.0, Final Report of the INDUSTRIE 4.0 Working Group, Vol. 4.*

Kaposi, A. and Myers, M. (2001). *Systems for All*, Imperial College Press.

Keeney, R. (2002). Common Mistakes in Making Value Trade-Offs, *Operations Research*, 50(6), 935–945.

Keeney, R. and Gregory, R. (2005). Selecting Attributes to Measure the Achievement of Objectives, *Operations Research*, 15(1), 1–11.

Kemp, D. (2010). So what is "in service systems engineering"? *Proceedings of the 20th Annual International Symposium of the International Council on Systems Engineering*, The International Council on Systems Engineering (INCOSE).

Kemp, D. and O'Neil, M. (2018). Breaking Casandra's Curse: Understanding Unsafe Mental Models to Build a Safe Systems Engineering Culture, *Proceedings of the 28th Annual International Symposium of the International Council on Systems Engineering*, The International Council on Systems Engineering (INCOSE).

Kennedy, L. (2005). *Keeping the promise: A work ethic for doing things right, QMI Books.* Retrieved from https://qualitymanagementinstitute.com/lk/biolk2.aspx.

Krueger, C. (2022). From Systems Engineering to System Family Engineering, *Proceedings of the 32nd Annual International Symposium of the International Council on Systems Engineering*, The International Council on Systems Engineering (INCOSE).

Kumar, A. (2020). Delivering System Value: A Systematic Approach, in G. Metcalf, et al. (Eds.), *Handbook of Systems Sciences*, Springer. Retrieved from https://doi.org/10.1007/978-981-13-0370-8_21-1.

Kurtz, C. and Snowden, D. (2003). The new dynamics of strategy: Sense-making in a complex and complicated world, *IBM Systems Journal*, 42, 462–483.

LAI (2013). *Lean Enterprise Value Phase 1, The Massachusetts Institute of Technology Lean Advancement Initiative.* Retrieved from https://dspace.mit.edu/handle/1721.1/1785 10 March 2023.

Langley, M., Robitaille, S., and Thomas, J. (2011). Toward a New Mindset: Bridging the Gap Between Program Management and Systems Engineering, *PM Network*, 25(9),24–26. Retrieved from http://www.pmi.org 10 March 2023.

Langner, R. (2012). Stuxnet Deep Dive, *SCADA Security Scientific Symposium (S4).* Retrieved from. https://www.youtube.com/watch?v=zBjmm48zwQU 10 March 2023.

Lano, R. (1977). *The N² Chart*, TRW, Inc.

Larman, C. and Basili, V. (2003). Iterative and Incremental Development: A Brief History, *IEEE Software*, 36(6), 47–56.

Lawless, W. (2019). The Interdependence of Autonomous Human-Machine Teams: The Entropy of Teams, But Not Individuals, *Advances Science, Entropy*, 21(12), 1195.

Lawless, W., Mittu, R., Sofge, D., and Russell, S. (2017). *Autonomy and Artificial Intelligence: A Threat or Savior?* Springer.

Lawson, B. (1997). How Designers Think – The Design Process Demystified. Oxford: Architectural Press. Third edition.

Lee, E. (2015). The Past, Present and Future of Cyber-Physical Systems: A Focus on Models, *Sensors*, 15(3), 4837–4869. Retrieved from http://www.mdpi.com/1424-8220/15/3/4837 10 Mar.

LEfMEP (2012). *The Guide to Lean Enablers for Managing Engineering Programs, Version 1.0. Joint MIT-PMI-INCOSE Community of Practice on Lean in Program Management.* Retrieved from http://hdl.handle.net/1721.1/70495.

Leveson, N. (1995). *Safeware: System Safety and Computers*, Addison Wesley.

Leveson, N. (2011). *Engineering a Safer World*, MIT Press.

Leveson, N. and Turner, C. (1993). An Investigation of the Therac-25 Accidents, *IEEE Computer*, 26(7), 18–41.

Lewis, T. (2019). *Communication, Chapter 5 in Critical Infrastructure Protection in Homeland Security: Defending a Networked Nation*, Third Edition. Wiley, 102–122.

Li, M. and Vitanyi, P. (2009). *An Introduction to Kolmogorov Complexity and Its Applications*, Third edition. Springer-Verlag, 102.

Long, J. (2000). COTS: What You Get (In Addition to the Potential Development Savings), *Proceedings of the 10th Annual International Symposium of the International Council on Systems Engineering*, The International Council on Systems Engineering (INCOSE).

Lu, J. (2019). *A Framework for Cyber-physical System Tool-chain Development: A Service-oriented and Model-based Systems Engineering Approach*, KTH Royal Institute of Technology. Retrieved from https://kth.diva-portal.org/smash/get/diva2:1316044/FULLTEXT01.pdf.

Lustig, I., Dietrich, B., Johnson, C., and Dziekan, C. (2010). *The Analytics Journey*, Oxford University Press.

Maier, M. (1998). Architecting Principles for Systems of Systems, *Systems Engineering*, 1(4), 267–284.

Maier, M. W. and Rechtin, E. (2009). *The Art of Systems Architecting*, Third Edition. CRC Press.

Martin, J. (1996). *Systems Engineering Guidebook: A Process for Developing Systems and Products*, CRC Press.

Martin, J. (2003). On the Use of Knowledge Modeling Tools and Techniques to Characterize the NOAA Observing System Architecture, *Proceedings of the 13th Annual International Symposium of the International Council on Systems Engineering*, The International Council on Systems Engineering (INCOSE).

Martin, J. (2005). Using an Enterprise Architecture to Assess the Societal Benefits of Earth Science Research, *Proceedings of the 15th Annual International Symposium of the International Council on Systems Engineering*, The International Council on Systems Engineering (INCOSE).

Martin, J., Conklin, J., Evans, J., Robinson, C., Doggrell, L., and Diehl, J. (2004). The Capability Integration Framework: A New Way of doing Enterprise Architecture, *Proceedings of the 14th Annual International Symposium of the International Council on Systems Engineering*, The International Council on Systems Engineering (INCOSE).

McDermott, T., Folds, D., and Hallo, L. (2020). Addressing Cognitive Bias in Systems Engineering Teams, *Proceedings of the 30th Annual International Symposium of the International Council on Systems Engineering*, The International Council on Systems Engineering (INCOSE).

McDonough, W. (2013). McDonough Innovations: Design for the Ecological Century. Retrieved from http://www.mcdonough.com. 10 March 2023.

McGarry et al. (2001). Practical Software Measurement: Objective Information for Decision Makers

McManus, H. (2005). *Product Development Transition to Lean (PDTTL) Roadmap, LAI Release Beta, Massachusetts Institute of Technology Lean Advancement Initiative.*

McNicholas, R. (2021). Application of PLE to US Army Live Training, *Momentum2021 Symposium*. Retrieved from https://vimeo.com/548839826 10 March 2023.

MDPI (2023). *Sustainability Journal website, Multidisciplinary Digital Publishing Institute (MDPI)*. Retrieved from https://www.mdpi.com/journal/sustainability 10 March 2023.

MIL-HDBK-61 (2020). *Configuration Management Guidance*, US Department of Defense.

Miller, G. (1956). The Magical Number Seven, Plus or Minus Two: Some Limits on our Capacity for Processing Information, *Psychological Review*, 63(2), 81.

Minsky, M. (1965). Models, Minds, Machines, *Proceedings of the IFIP Congress*, pp. 45–49.

NAFEMS (2021). *NAFEMS Resource Centre website, The International Association for Engineering Modelling, Analysis and Simulation (NAFEMS)*. Retrieved from https://www.nafems.org/publications/resource_center 10 March 2023.

NAFEMS and INCOSE (2019). *What is Systems Modeling and Simulation, The International Association for Engineering Modelling, Analysis and Simulation (NAFEMS) WT10 and The International Council on Systems Engineering Systems (INCOSE) Modeling and Simulation Working Group (SMSWG)*. Retrieved from https://www.nafems.org/publications/resource_center/wt10 10 March 2023.

NAO (2008). *Chinook Mk3 Helicopters, NAO Report HC 512 2007-08*, UK Ministry of Defence, National Audit Office.

NASA (2003). *Columbia Accident Investigation Report*, US National Aeronautic and Space Administration.

NASA (2007a). *NASA Pilot Benchmarking Initiative: Exploring Design Excellence Leading to Improved Safety and Reliability, Final Report*, US National Aeronautic and Space Administration.

NASA (2007b). *NASA Systems Engineering Handbook*, US National Aeronautic and Space Administration.

NASA JPL (2013). *ON-OFF Adhesive Grippers for Earth Orbit*, US National Aeronautic and Space Administration, Jet Propulsion Laboratory. Retrieved from https://www.nasa.gov/sites/default/files/files/A_Parness-Gecko_Like_Adhesives_for_InSpace_Inspection.pdf 10 March 2023.

NASA JPL (2014). *Gecko Grippers Get a Microgravity Test Flight*, US National Aeronautic and Space Administration, Jet Propulsion Laboratory. Retrieved from https://www.nasa.gov/jpl/tech/gecko-grippers-microgravity-flight 10 March 2023.

NASA JPL (2015). *Gecko Grippers Moving On Up*, US National Aeronautic and Space Administration, Jet Propulsion Laboratory. Retrieved from https://www.nasa.gov/jpl/gecko-grippers-moving-on-up 10 March 2023.

NATO (2018). *NATO Architecture Framework (NAF), Version 4, the North Atlantic Treaty Organization.*

NDIA, INCOSE, and PSM (2011). *System Development Performance Measurement Report, National Defense Industrial Association*, The International Council on Systems Engineering, and Practical Software & Systems Measurement Support Center.

Nejib, P., Yakabovicz, E., and Beyer, D. (2017). System Security Engineering: What Every System Engineer Needs to Know, *Proceedings of the 27th Annual International Symposium of the International Council on Systems Engineering*, The International Council on Systems Engineering (INCOSE).

Nissen, J. (2006). The Øresund Link, *The Arup Journal*, 31(2), 37–41.

NIST NVB (2012). National Vulnerability Database, Version 2.2, *US National Institute of Standards and Technology, Computer Science Division.*

NIST SP 500-230 (1996). *Application Portability Profile (APP): The U.S. Government's Open System Environment Profile, Version 3.0*, US National Institute of Standards and Technology Special Publication.

NIST SP 800-53 (2020). Security and Privacy Controls for Information Systems and Organizations, US National Institute of Standards and Technology Special

NIST SP 800-128 (2019). *Guide for Security-Focused Configuration Management of Information Systems*, US National Institute of Standards and Technology Special Publication.

NIST SP 800-160 Vol. 1 (2022). *Systems Security Engineering: Considerations for a Multidisciplinary Approach in the Engineering of Trustworthy Secure Systems, Rev 1*, US National Institute of Standards and Technology Special Publication.

NIST SP 800-160 Vol. 2 (2021). *Developing Cyber-Resilient Systems: A Systems Security Engineering Approach*, US National Institute of Standards and Technology Special Publication.

Noorani, R. (2008). *Rapid Prototyping: Principles and Applications*, Wiley.

NTSB (2019a). *Vehicle Automation Report*, US National Transportation Safety Board.

NTSB (2019b). *Inadequate Safety Culture Contributed to Uber Automated Test Vehicle Crash - NTSB Calls for Federal Review Process for Automated Vehicle Testing on Public Roads*, US National Transportation Safety Board.

NTSB (2019c). *Accident Report, Collision Between Vehicle Controlled by Developmental Automated Driving System and Pedestrian*, Tempe, AZ: US National Transportation Safety Board. March 18, 2018, NTSB/HAR-19/03, PB2019-101402.

O'Connor, P. and Kleyner, A. (2012). *Practical Reliability Engineering*, Fifth Edition. John Wiley & Sons, Inc.

OASIS (2012). *Reference Architecture Foundation for Service Oriented Architecture (SOA-RAF), Version 1.0, OASIS Open TC 01.*

Oberndorf, T., Brownsword, L., and Sledge, C. (2000). *An Activity Framework for COTS-Based Systems, CMU/SEI-2000-TR-010*, The Software Engineering Institute (SEI) at Carnegie Mellon University.

OMG MBSE Wiki (2023). *OMG MBSE Wiki, MBSE Events and Related Meetings website*. Retrieved from http://www.omgwiki.org/MBSE/doku.php?id=mbse:methodology 10 March 2023.

OMG SysML™ (2021). *Systems Modeling Language (SysML)*. Retrieved from https://omgsysml.org 10 March 2023.

OMG TOGAF (2023). *TOGAF - The Open Group Architecture Framework*. Retrieved from http://www.opengroup.org/togaf 10 March 2023.

OMG UAF (2023). *Unified Architecture Framework*. Retrieved from http://www.omg.org/uaf 10 March 2023.

Oppenheim, B. (2004). Lean Product Development Flow, *Systems Engineering*, 7(4), 352–376.

Oppenheim, B. (2011). *Lean for Systems Engineering with Lean Enablers for Systems Engineering*, Wiley.

Oppenheim, B. (2021). *Lean Healthcare Systems Engineering Process for Clinical Environments, A Step-by-step Process for Managing Workflow and Care Improvement Projects*, Routledge, Taylor & Francis Group, Productivity Press.

Oxford (2020). *Oxford English Dictionary*, Oxford University Press.

Parnell, G. (2016). *Trade-off Analytics: Creating and Evaluating the Tradespace*, Wiley & Sons.

Parnell, G., Bresnick, T., Tani, S., and Johnson, E. (2013). *Handbook of Decision Analysis*, John Wiley & Sons, Inc.

Parnell, G., Kenley, C., Specking, E., and Pohl, E. (2022). Systems Engineering and Industrial Engineering, *Proceedings of the 32nd Annual International Symposium of the International Council on Systems Engineering*, The International Council on Systems Engineering (INCOSE).

Paté-Cornell, M. (1990). Organizational Aspects of Engineering System Safety: The Case of Offshore Platforms, *Science*, 250.

Pax Water Technologies (2022). *Biomimicry website, Pax Water Technologies*. Retrieved from https://www.paxwater.com/biomimicry 10 March 2023.

Pearce, O., Murry, N., and Broyd, T. (2012). Halstar: Systems engineering for sustainable development, Proceedings of the Institution of Civil Engineers, *Engineering Sustainability*, 165(2), 129–140, Thomas Telford Ltd.

Pearl, J. (2002). Reasoning with Cause and Effect, *AI Magazine*, 23(1), 95–95. Retrieved from https://doi.org/10.1609/aimag.v23i1.1612.

Pearl, J. and Mackenzie, D. (2018). AI can't reason why, *Wall Street Journal*. Retrieved from https://www.wsj.com/articles/ai-cant-reason-why-1526657442.

Peterson, T. (2019). Systems Engineering: Transforming Digital Transformation, *Proceedings of the 29th Annual International Symposium of the International Council on Systems Engineering*, The International Council on Systems Engineering (INCOSE).

Petersen, T. and Sutcliffe, P. (1992). Systems engineering as applied to the Boeing 777, AIAA 92-1010, *AIAA Aerospace Design Conference*, Irvine, CA.

Pineda, R. (2010). *Understanding Complex Systems of Systems Engineering, Fourth General Assembly Cartagena Network of Engineering*, Metz, France.

PMI (2013). *The Standard for Program Management*, Third Edition, Project Management Institute.

PMI (2016). *Requirements Management: A Practice Guide*, Project Management Institute.

PMI (2017). *PMI Standard for Portfolio Management*, Fourth Edition, Project Management Institute.

PMI (2021). *Project Management Body of Knowledge (PMBOK®)*, Project Management Institute.

PMI (2022). *Situation Context Framework (SCF)*, Project Management Institute.

Pragmatic 365 (2023). *Pragmatic Enterprise Architecture Framework (PEAF) website*. Retrieved from https://www.pragmatic365.org/peaf-intro.asp 10 March 2023.

PSM (2003). *Practical Software and Systems Measurement (PSM) Guide, Version 4.0c, Practical Software and System Measurement Support Center*. Retrieved from https://www.psmsc.com/psmguide.asp 10 March 2023.

PSM, NDIA, and INCOSE (2021). *Continuous Iterative Development Measurement Framework, Practical Software & Systems Measurement Support Center, National Defense Industrial Association, and The International Council on Systems Engineering*.

Quade, E. and Boucher, W. (1968). *Systems Analysis and Policy Planning: Applications in Defense, RAND R-439-PR (Abridged)*.

Rasoulkahni, K. (2018). Resilience as an emergent property of human infrastructure dynamics; A multi-agent simulation model for characterizing regime shifts and tipping point behaviours in infrastructure systems, *Plos One*, 13.

Raworth, K. (2017). *Doughnut Economics: Seven Ways to Think like a 21st-Century Economist*, Random House.

Raz, A., Wood, P., Mockus, L., and DeLaurentis, D. (2020). System of systems uncertainty quantification using machine learning techniques with smart grid application, *Systems Engineering*, 23(6), 770–782.

Rebentisch, E. (2017). *Integrating Program Management and Systems Engineering: Methods, Tools, and Organizational Systems for Improving Performance*, Wiley.

Rebovich, G. and White, B. (2011). *Enterprise Systems Engineering: Advances in the Theory and Practice*, CRC Press.

Roedler, G. (2010). Knowledge Management Position, *Proceedings of the 20th Annual International Symposium of the International Council on Systems Engineering*, The International Council on Systems Engineering (INCOSE).

Roedler, G. and Jones, C. (2005). *INCOSE Technical Measurement Guide*, The International Council on Systems Engineering.

Roedler, G., Rhodes, D., Schimmoler, H., and Jones, C. (2010). *Systems Engineering Leading Indicators Guide, v2.0, Massachusetts Institute of Technology*, The International Council on Systems Engineering, and Practical Software & Systems Measurement Support Center.

Rogers, E. and Mitchell, S. (2021). MBSE delivers significant return on investment in evolutionary development of complex SoS, *Systems Engineering*, 24(6), 385–408.

Rosen, M. (2012). Engineering sustainability: A technical approach to sustainability, *Sustainability*, 4(9), 2270–2292.

Rouse, W. (2009). Engineering the Enterprise as a System, in A. Sage & W. Rouse (Eds.), *Handbook of Systems Engineering and Management*, Second Edition. John Wiley & Sons, Inc.

Rousseau, D. (2018a). A framework for understanding systems principles and methods, *INSIGHT*, 21(3), The International Council on Systems Engineering (INCOSE).

Rousseau, D. (2018b). Three general systems principles and their derivation: Insights from the philosophy of science applied to systems concepts, in Madni, et al. (Eds.) *Disciplinary convergence in systems engineering research* (pp. 665–681). Springer International Publishing. Retrieved from https://link.springer.com/book/10.1007/978-3-319-62217-0 10 March 2023.

Rousseau, D., Billingham, J., and Calvo-Amodio, J. (2018). Systemic Semantics: A Systems Approach to Building Ontologies and Concept Maps, *Systems*, 6(3).

Rousseau, D., Billingham, J., and Calvo-Amodio, J. (2019). Systemic virtues as a foundation for a general theory of design elegance, *Systems Research and Behavioural Science*, 36(5), 656–667.

Rousseau, D., Pennotti, M., and Brook, P. (2022). Systems Engineering's Evolving Guidelines. *Report of the INCOSE Bridge Team presented on January 31, 2022 to the INCOSE Systems Science Working Group meeting*, The International Council on Systems Engineering, published on the SSWG's Google Drive. Retrieved from https://drive.google.com/file/d/1JibL44sUh0ztefZQ5Rfy4kGiodXIy63n/view 10 March 2023.

Royce, W. (1970). Managing the Development of Large Software Systems, *Proceedings of IEEE WESCON*, pp. 1–9.

SAE 1001 (2018). *Integrated Project Processes for Engineering a System*, SAE International. (Note: Replaced ANSI/EIA 632.)

SAE-EIA 649C (2019). *Configuration Management Standard*, SAE International and The Electronic Industries Alliance.

Salado, A. and Kannan, H. (2018). A mathematical model of verification strategies, *Systems Engineering*, 21(6), 593–608.

Salter, K. (2003). Presentation Given at the Jet Propulsion Laboratory, *US National Aeronautic and Space Administration, Jet Propulsion Laboratory*.

Schindel, W. (2005). Requirement Statements Are Transfer Functions: An Insight from Model-Based Systems Engineering), *Proceedings of the 15th Annual International Symposium of the International Council on Systems Engineering*, The International Council on Systems Engineering (INCOSE).

Schindel, W. (2010). Failure Analysis: Insights from Model-Based Systems Engineering, *Proceedings of the 20th Annual International Symposium of the International Council on Systems Engineering*, The International Council on Systems Engineering (INCOSE).

Schindel, W. (2011). What Is the Smallest Model of a System? *Proceedings of the 21st Annual International Symposium of the International Council on Systems Engineering*, The International Council on Systems Engineering (INCOSE).

Schindel, W. (2012). Integrating Materials, Process & Product Portfolios: Lessons from Pattern-Based Systems Engineering, *Proceedings of the Society for the Advancement of Material and Process Engineering*, Society for Advancement of Material and Process Engineering.

Schindel, W. (2013). System Interactions: Making the Heart of Systems More Visible, *Proceedings of the INCOSE 2013 Great Lakes Regional Conference on Systems Engineering*, The International Council on Systems Engineering (INCOSE).

Schindel, W. (2016). Got Phenomena? Science-Based Disciplines for Emerging Systems Challenges, *Proceedings of the 26th Annual International Symposium of the International Council on Systems Engineering*, The International Council on Systems Engineering (INCOSE).

Schindel, W. (2020). System Patterns in Engineering and Science, in G. Metcalf, et al. (Eds.), *Handbook of System Sciences*, Springer Nature.

Schindel, W. (2022a). Pattern-Based Methods and MBSE, in A. Madni, et al. (Eds.), *Handbook of Model-Based Systems Engineering*, Springer. Retrieved from https://doi.org/10.1007/978-3-030-27486-3_73-1.

Schindel, W. (2022b). Realizing the Value Promise of Digital Engineering: Planning, Implementing, and Evolving the Ecosystem, *INSIGHT*, 25(1), The International Council on Systems Engineering (INCOSE).

Schindel, W. and Dove, R. (2016). Introduction to the INCOSE Agile Systems Engineering Life Cycle Management (ASELCM) Pattern, *Proceedings of the 26th Annual International Symposium of the International Council on Systems Engineering*, The International Council on Systems Engineering (INCOSE).

Schindel, W. and Peterson, T. (2013). Introduction to Pattern-Based Systems Engineering (PBSE): Leveraging MBSE Techniques, *Proceedings of the 23rd Annual International Symposium of the International Council on Systems Engineering*, The International Council on Systems Engineering (INCOSE).

Schindel, W. and Smith, V. (2002). Results of Applying a Families-of-Systems Approach to Systems Engineering of Product Line Families, *Technical Report 2002-01-3086*, SAE International.

Schlager, K. (1956). Systems engineering – Key to modern development, *IRE Transactions of Professional Group Engineering Management*, 3, 64–66.

Seacord, R., Plakosh, D., and Lewis, G. (2003). *Modernizing Legacy Systems*, Pearson.

SEBoK (2023). Guide to the Systems Engineering Body of Knowledge (SEBoK) website, *The International Council on Systems Engineering (INCOSE), the IEEE Systems Council, and Stevens Institute of Technology*. Retrieved from https://sebokwiki.org/wiki/Guide_to_the_Systems_Engineering_Body_of_Knowledge_(SEBoK) 10 March.

Senge, P. (1990). *The Fifth Discipline: The Art & Practice of the Learning Organization*, Crown Business.

SEVOCAB (2023). *SEVOCAB: Software and Systems Engineering Vocabulary website, The Institute of Electrical and Electronics Engineers Computer Society*, The International Organization for Standardization, and The International Electrotechnical Commission.

Shafaat, A. and Kenley, C. (2020). Model-based design of project systems, *Modes, and States, Systems Engineering*, 23(2), 165–176.

Sharma, R., Jabbour, C., and Lopes de Sousa Jabbour, A. (2020). Sustainable manufacturing and industry 4.0: What we know and what we don't, *Journal of Enterprise Information Management*, Retrieved from, https://doi.org/10.1108/JEIM-01-2020-0024.

Siegel, N. (2019). *Engineering Project Management*, John Wiley and Sons, Inc.

Sillitto, H. (2010). Design Principles for Ultra-Large Scale (ULS) Systems, *Proceedings of the 20th Annual International Symposium of the International Council on Systems Engineering*, The International Council on Systems Engineering (INCOSE).

Sillitto, H. and Dori, D. (2017). Defining 'System': A Comprehensive Approach, *Proceedings of the 27th Annual International Symposium of the International Council on Systems Engineering*, The International Council on Systems Engineering (INCOSE).

Skanska (2013). *Øresund Bridge: Improving Daily Life for Commuters, Travelers, and Frogs*, The Skanska Group.

Slack, R. (1998). *Application of Lean Principles to the Military Aerospace Product Development Process, Master of Science—Engineering and Management Thesis*, Massachusetts Institute of Technology.

Snowden, D. and Boone, M. (2007). A Leader's Framework for Decision Making, *Harvard Business Review*, 85(11), 68–76.

Sofge, D., Mittu, R., and Lawless, W. (2019). AI Bookie Bet: How likely is it that an AI-based system will self-authorize taking control from a human operator? *AI Magazine*, 40(3), 79–84.

Sowa, J. and Zachman, J. (1992). Extending and formalizing the framework for information systems architecture, *IBM Systems Journal*, 31(3), 590–616.

Specking, E., Parnell, G., Pohl, E., and Buchanan, R. (2018). Early Design Space Exploration with Model-Based System Engineering and Set-Based Design, *Systems*, 6(4), 45.

Spohrer, J. (2011). Service Science: Progress & Direction, *International Joint Conference on Service Science*, Taipei, Taiwan.

Suh, N. (2001). Axiomatic Design: Advances and Applications, Oxford University Press

Studor, G. (2016). What is NASA's Interest in Natural Systems? *INSIGHT*, 19(1), The International Council on Systems Engineering (INCOSE).

Taleb, N. (2018). *Skin in the Game: Hidden Asymmetries in Daily Life*, Random House.

Tapscott, D. and Tapscott, A. (2018). *Blockchain Revolution: How the Technology Behind Bitcoin and Other Cryptocurrencies Is Changing the World*, Portfolio.

Thaler, R. and Sunstein, C. (2008). *Nudge: Improving Decisions About Health, Wealth, and Happiness*, Penguin Books.

Tortorella, M. (2015). *Reliability, Maintainability, and Supportability: Best Practices for Systems Engineers*, John Wiley & Sons.

Toyota (2009). *Toyota Production System: Just-in-Time—Productivity Improvement*, Toyota. Retrieved from https://global.toyota/en/company/vision-and-philosophy/production-system 10 March 2023.

Tuttle, P. and Bobinis, J. (2013). Specifying Affordability, *Proceedings of the 23rd Annual International Symposium of the International Council on Systems Engineering*, The International Council on Systems Engineering (INCOSE).

Tversky, A. and Kahneman, D. (1974). Judgment under Uncertainty: Heuristics and Biases, *Science*, 185(4157), 1124–1131.

Tyson, B., Albert, C., and Brownsword, L. (2003). *Interpreting Capability Maturity Model Integration (CMMI) for COTS-Based Systems, Technical Report CMU/SEI-2003-TR-022*, The Software Engineering Institute at Carnegie Mellon University.

UAM (2022). *The Unified Architecture Method (UAM) website*. Retrieved from https://www.unified-am.com 10 March 2023.

Urwick, L. (1956). The Manager's Span of Control, *Harvard Business Review*.

US Army (1997). *Army Technical Architecture, Version 4.9.5X, Draft*, US Department of the Army.

US DoD (2021). *DoD Dictionary of Military and Related Terms*, US Department of Defense.

Van De Ven, M., Talik, J., and Hulse, J. (2012). An Introduction to Applying Systems Engineering to In-Service Systems, *Proceedings of the 22nd Annual International Symposium of the International Council on Systems Engineering*, The International Council on Systems Engineering (INCOSE).

Velcro (2023). *A Mind-Blowing Biomimicry Example, Velcro Brand.* Retrieved from https://www.velcro.com/news-and-blog/2020/07/a-mind-blowing-biomimicry-examples 10 March 2023.

von Bertalanffy, L. (1950). The Theory of Open Systems in Physics and Biology, *Science*, 111(2872), 23–29.

von Bertalanffy, L. (1968). *General System Theory: Foundations, Development, Applications,* Braziller.

von Bertalanffy, L. (1969). The theory of open systems in physics and biology, in F. Emery (Ed.), *Systems Thinking*, 70–85, Penguin.

von Bertalanffy, L. (1971). *General System Theory*, Penguin.

VV&A (2021). *Verification, Validation, & Accreditation (VV&A), US DoD Modeling and Simulation Enterprise.* Retrieved from https://vva.msco.mil 10 March 2023.

Walden, D. (2007). YADSES: Yet Another Darn Systems Engineering Standard, *Proceedings of the 17th Annual International Symposium of the International Council on Systems Engineering*, The International Council on Systems Engineering (INCOSE).

Walden, D. (2019). Brownfield Systems Development: Moving from the Vee Model to the N Model for Legacy Systems, *Proceedings of the 29th Annual International Symposium of the International Council on Systems Engineering*, The International Council on Systems Engineering (INCOSE).

Warfield, J. (2006). *An Introduction to Systems Science*, World Scientific Publishing Company.

Wasson, C. (2016). *System Engineering Analysis, Design, and Development: Concepts, Principles, and Practices*, Second Edition, John Wiley & Sons, Inc.

Watson, I. O. T. (2017). *Descriptive, Predictive, Prescriptive: Transforming Asset and Facilities Management with Analytics*, Software Group, IBM Corporation.

Watson, M. (2020). *Engineering Elegant Systems: Theory of Systems Engineering, NASA_TP_20205003644, NASA*, Washington, D.C., August 2020.

Watson, M., Mesmer, B., Roedler, G., Rousseau, D., Gold, R., Calvo-Amodio, J., Jones, C., Miller, W., Long, D., Lucero, S., Russell, R., Sedmak, A., and Verma, D. (2019). Systems Engineering Principles and Hypotheses, *INSIGHT*, 21(1), The International Council on Systems Engineering (INCOSE).

White, R. and Tantsura, J. (2016). *Navigating Network Complexity*, Addison Wesley.

Womack, J. and Jones, D. (1996). *Lean Thinking*, Simon & Schuster.

Wood, R., Zhu, J., and Liu, G. (2023). *Journal of Industrial Ecology*. Retrieved from https://onlinelibrary.wiley.com/journal/15309290 10 March 2023.

Wymore, A. (1993). *Model-Based Systems Engineering*, CRC Press.

Xu, T. and Zhou, Y. (2015). Systems Approaches to Tackling Configuration Errors: A Survey, *ACM Computing Survey*, 47(4), 41.

APPENDIX B: ACRONYMS

Note: Other acronyms may be defined as used under their respective appendices. Abbreviations used for references are described in Appendix A. Abbreviations used for the system life cycle processes are described in Appendix D.

A_a	Achieved availability
A_i	Inherent availability
A_o	Operational availability
ADS	Automated Driving Systems
AECL	Atomic Energy Commission Limited [Canada]
AF	Architecture Framework
AI	Artificial Intelligence
AIAA	American Institute of Aeronautics and Astronautics [United States]
ALARP	As low as reasonably practicable
ALT	Accelerated Life Testing
ANSI	American National Standards Institute [United States]
API	American Petroleum Institute
API	Application programming interface
ARAP	As Resilient as Practicable
ARP	Aerospace Recommended Practice
ASEP	Associate Systems Engineering Professional [INCOSE]
ASOT	Authoritative Source of Truth
ASQ	American Society for Quality
ATAM	Architecture tradeoff analysis method
AWG	Automotive Working Group [INCOSE]
BIT	Built-In Test
CAD	Computer-aided design

INCOSE Systems Engineering Handbook: A Guide for System Life Cycle Processes and Activities, Fifth Edition.
Edited by David D. Walden, Thomas M. Shortell, Garry J. Roedler, Bernardo A. Delicado, Odile Mornas, Yip Yew-Seng, and David Endler.
© 2023 John Wiley & Sons Ltd. Published 2023 by John Wiley & Sons Ltd.

CAIV	Cost as an independent variable
CBA	Cost–benefit analysis
CBS	Cost Breakdown Structure
CCB	Configuration Control Board
CE	Conformité Européenne [EU]
CE	Cost Effectiveness
CEA	Cost-effectiveness analysis
CFD	Computational Fluid Dynamics
CI	Configuration Item
CI/CD	Continuous Integration/Continuous Delivery
CM	Configuration Management
CMMI®	Capability Maturity Model® Integration [CMMI Institute]
COCOMO	Constructive Cost Model
ConOps	Concept of operations
COSYSMO	Constructive Systems Engineering Cost Model
COTS	Commercial off-the-shelf
CPS	Cyber-physical system
CRB	Configuration Review Board
CSEP	Certified Systems Engineering Professional [INCOSE]
DANSE	Designing for Adaptability and evolutioN in System of Systems Engineering
DAU	Defense Acquisition University [United States]
DD	Data Dictionary
DE	Digital Engineering
DevOps	Development, Operations
DevSecOps	Development, Security, Operations
DFD	Data flow diagrams
DFM	Design For Manufacturing
DFT	Design For Testability
DFX	Design For X
DMSMS	Diminishing manufacturing sources and material shortages
DoD	Department of Defense [United States]
DoDAF	Department of Defense Architecture Framework [United States]
DSM	Design Structure Matrix
DT	Design Thinking
DTC	Design to cost
EIA	Electronic Industries Alliance
EPD	Environmental Product Declaration
ESEP	Expert Systems Engineering Professional [INCOSE]
EU	European Union
FAA	Federal Aviation Administration [United States]
FBS	Functional Breakdown Structure
FCA	Functional configuration audit
FD/FI	Failure detection/Failure isolation or Fault Detection/Fault Isolation
FEA	Finite Element Analysis
FEP	Fuel enrichment plant
FFBD	Functional flow block diagram
FMEA	Failure Mode and Effects Analysis
FMECA	Failure modes, effects, and criticality analysis

FTA	Fault tree analysis
FuSE	Future of Systems Engineering [INCOSE]
G&A	General and administrative
GAO	Government Accountability Office [United States]
GEIA	Government Electronics & Information Technology Association
GERAM	Generalized Enterprise Reference Architecture and Methodology
GNP	Gross national product
GPS	Global Positioning System
GRIP	Governance for Railway Investment Projects
HALT	Highly accelerated life testing
HCD	Human-Centered Design
HFE	Human factors engineering
HITL	Human-in-the-loop
HMI	Human Machine Interface
HPC	High Performance Computing
HSI	Human systems integration
IAD	Interface Agreement Document
IC	Initial cost
ICD	Interface Control Document
ICS	Industrial control system
ICSM	Incremental Commitment Spiral Model
ICWG	Interface Control Working Group
IDD	Interface Definition Document
IE	Industrial engineering
IEC	International Electrotechnical Commission
IEEE	The IEEE [formerly the Institute of Electrical and Electronics Engineers]
IISE	Institute of Industrial and Systems Engineers
ILS	Integrated logistics support
INCOSE	International Council on Systems Engineering
INFORMS	Institute for Operations Research and the Management Sciences
IoT	Internet of Things
IPAL	INCOSE Product Asset Library [INCOSE]
IPDT	Integrated Product Development Team
IPO	Input–process–output
IPT	Integrated Product Team
ISEBoK	Industrial and SE Body of Knowledge
ISO	International Organization for Standardization
IT	Information technology
ITS	Intelligent Transportation System
JAXA	Japan Aerospace Exploration Agency [Japan]
JERG	JAXA Engineering Requirement, Guideline [Japan]
KM	Knowledge management
KPP	Key Performance Parameter
LAI	Lean Advancement Initiative
LCA	Life cycle assessment
LCC	Life cycle cost
LCIA	Life cycle impact assessment

LCM	Life cycle management
LCO	Life cycle optimization
LDSE	Loss-Driven Systems Engineering
LEfMEP	Lean Enablers for Managing Engineering Programs
LINAC	Linear accelerator
LORA	Level of Repair Analysis
MA&S	Modeling, analysis, and simulation
MaaS	Mobility as a Service
MBSE	Model-based systems engineering
MFESA	Method Framework for Engineering System Architectures
MIT	Massachusetts Institute of Technology [USA]
ML	Machine Learning
MODA	Multiple objective decision approach
MOE	Measure of effectiveness
MOP	Measure of performance
MTBF	Mean time between failure
MTTR	Mean time to repair
MVP	Minimum viable product
N^2	N-squared diagram
NAF	NATO Architecture Framework
NAFEMS	The International Association for the Engineering Modelling, Analysis and Simulation Community
NASA	National Aeronautics and Space Administration [United States]
NCOSE	National Council on Systems Engineering (INCOSE, pre-1995)
NCS	Network-Centric Systems
NDI	Non-developmental item
NDIA	National Defense Industrial Association [United States]
NIH	Not Invented Here
NIST	National Institute of Standards and Technology [United States]
O&G	Oil and Gas
OBS	Organizational Breakdown Structure
OEM	Original Equipment Manufacturer
OMG	Object Management Group
OOSEM	Object-Oriented Systems Engineering Method
OpsCon	Operational concept
OR	Operations Research
PBS	Product Breakdown Structure
PCA	Physical configuration audit
PEAF	Pragmatic Enterprise Architecture Framework
PESTEL	Political, Economic, Social, Technological, Environmental, and Legal
PHS&T	Packaging, handling, storage, and transportation
PLC	Programmable logic controller
PLE	Product line engineering
PLM	Product line management
PMBoK	Project Management Body of Knowledge [PMI]
PMI	Project Management Institute
PMP	Project Management Plan
PPP	Public–Private Partnership
PSM	Practical Software and Systems Measurement

QA	Quality assurance
QC	Quality characteristics
QC	Quality control
QM	Quality management
R&D	Research and development
RAM	Reliability, availability, and maintainability
RBD	Reliability block diagram
RCM	Reliability-centered maintenance
RFP	Request for proposal
RFQ	Request for quotation
RMP	Risk management plan
ROI	Return on investment
SAE	SAE International [formerly the Society of Automotive Engineers]
SAFe	Scaled agile framework
SaMDs	Software as Medical Devices
SBD	Set-Based Design
SC	Sustainment cost
SCF	Situation Context Framework
SCM	Supply chain management
SE	System effectiveness
SE	Systems engineering
SEBoK	Guide to the Systems Engineering Body of Knowledge
SECAG	Systems Engineering Competency Assessment Guide [INCOSE]
SECF	Systems Engineering Competency Framework [INCOSE]
SEH	Systems Engineering Handbook [INCOSE]
SEIPS	Systems Engineering Intervention for Patient Safety
SEIT	Systems Engineering and Integration Team
SEMP	Systems Engineering Management Plan
SEMS	Systems Engineering Master Schedule
SEP	Systems Engineering Plan
SEQM	System Engineering Quality Management
SLA	Service-level agreement
SMSWG	Systems Modeling and Simulation Working Group [NAFEMS and INCOSE]
SoI	System of interest
SoS	System of systems
SOW	Statement of work
SPC	Statistical Process Control
SROI	Social return on investment
SSE	System Security Engineering
STEM	Science, technology, engineering, and mathematics
STPA	System-theoretic process analysis
SWaP	Size, weight, and power
SWE	Software engineering
SWOT	Strengths, Weaknesses, Opportunities, Threats
SysML™	Systems Modeling Language [OMG]
TADSS	Training Aids, Devices, Simulators, and Simulations
TCO	Total cost of ownership
TOC	Total ownership cost

TOGAF	The Open Group Architecture Framework [The Open Group]
TOP	Technology, Organization, People
TOWS	Threats, Opportunities, Weaknesses, and Strengths
TP	Transaction-processing
TPM	Technical performance measure
TLI	Technical Leadership Institute [INCOSE]
TR	Technical report
TRL	Technology readiness level
UAF	Unified Architecture Framework
UAM	Unified Architecture Method
UI	User Interface
UIC	International Union of Railways
UK	United Kingdom
UL	Underwriters Laboratory [United States and Canada]
US/USA	United States/ United States of America
USB	Universal Serial Bus
USD	US dollars [United States]
UX	User Experience or User eXperience
V&V	Verification and Validation or Verify and Validate
VSE	Very Small Entities or Very Small Enterprises
VV&A	Verification, validation, and accreditation
WBS	Work Breakdown Structure
WG	Working group
WLC	Whole Life Cost
WP	Work package
WPA	Work Process Analysis
XP	Extreme Programming
ZD	Zero Defect
ZDA	Zero Defect Attitude

APPENDIX C: TERMS AND DEFINITIONS

Note: Terms that carry meanings consistent with general dictionary definitions are not included in this glossary. Sources of definitions are as indicated. Other related terms can be found in ISO/IEC/IEEE 24765 (2017) *and* SEVOCAB (2023). *Definitions of the typical inputs and outputs on the IPO diagrams can be found in Appendix E.*

Term	Definition
Ability	A term used in human resource management denoting an acquired or natural capacity or talent that enables an individual to perform a particular task successfully. (INCOSE SECF)
Acquirer	The stakeholder that acquires or procures a product or service from a supplier. (ISO/IEC/IEEE 15288, 2023)
Activity	A set of cohesive tasks of a process. (ISO/IEC/IEEE 15288, 2023)
Agile systems-engineering	An SE process using agile approach.
Agile-systems engineering	An engineering process producing agile systems.
Agreement	The mutual acknowledgment of terms and conditions under which a working relationship is conducted. (ISO/IEC/IEEE 15288, 2023)
Architect	See System architect.
Architecture	See System architecture.
Artifact	Work product that is produced and used during a project to capture and convey information. (ISO/IEC/IEEE 15288, 2023)
Attribute	An attribute of a system (or system element) is an observable characteristic or property of the system (or system element).

(Continued)

INCOSE Systems Engineering Handbook: A Guide for System Life Cycle Processes and Activities, Fifth Edition.
Edited by David D. Walden, Thomas M. Shortell, Garry J. Roedler, Bernardo A. Delicado, Odile Mornas, Yip Yew-Seng, and David Endler.
© 2023 John Wiley & Sons Ltd. Published 2023 by John Wiley & Sons Ltd.

(Continued)

Term	Definition
Baseline	An agreed-to description of the attributes of a product at a point in time, which serves as a basis for defining change. (EIA-649C, 2019)
Behavior	The way in which one acts or conducts oneself, especially towards others. (INCOSE SECF)
Black box	Black box represents an external view of the system (attributes). Also referred to as opaque box.
Brownfield SE	Development of "to-be" system or system elements in the presence of existing or legacy "as-is" system or system elements. Note: A brownfield approach is usually used to extend, improve, or replace a system that is in use or to reuse system elements that will not be impacted by the desired changes. The new system architecture must take into account the existing system elements and functions, which impose constraints on the overall system definition.
Capability	An expression of a system, product, function, or process ability to achieve a specific objective under stated conditions.
Commonality	(Of a product line) refers to functional and non-functional characteristics that can be shared with all member products within a product line. (ISO/IEC/IEEE 26550, 2015)
Competence	The measure of specified ability to do something well. (INCOSE SECF)
Competency	An observable, measurable set of skills, knowledge, abilities, behaviors, and other characteristics an individual needs to successfully perform work roles or occupational functions. Competencies are typically required at different levels of proficiency depending on the specific work role or occupational function. Competencies can help ensure individual and team performance aligns with the organization's mission and strategic direction. (INCOSE SECF)
Configuration item (CI)	A system, system element, or artifact designated for configuration management.
Customer	See Acquirer.
Decision gate	A decision gate is an approval event (may be associated with a review). Entry and exit criteria are established for each decision gate; continuation beyond the decision gate is contingent on the agreement of decision makers.
Design constraints	The boundary conditions, externally or internally imposed, for the SoI within which the organization must remain when executing the processes during the concept and development stages.
Engineered System	A system designed or adapted to interact with an anticipated operational environment to achieve one or more intended purposes while complying with applicable constraints. (INCOSE Definitions, 2019)
Enterprise	A purposeful combination of interdependent resources that interact with each other to achieve business and operational goals. (Rebovich and White, 2011)
Environment	The surroundings (natural or man-made) in which the SoI is utilized and supported or in which the system is being developed, produced, and retired.
Facility	The physical means or equipment for facilitating the performance of an action, for example, buildings, instruments, and tools.
Failure	The event in which any part of a system or system element does not perform as required by its specification. Note: The failure may occur at a value in excess of the minimum required in the specification, that is, past design limits or beyond the margin of safety.

(Continued)

(Continued)

Term	Definition
Functional configuration audit (FCA)	An evaluation to ensure that the product meets baseline functional and performance capabilities. (Adapted from ISO/IEC/IEEE 15288, 2023)
Greenfield SE	Development of a system for a new environment and set of user scenarios and requirements.Note: A greenfield approach typically has no significant legacy constraints or dependencies within the system boundary. However, it is rare that there are no constraints or dependencies from external interfaces or enabling systems.
Human factors	The systematic application of relevant information about human abilities, characteristics, behavior, motivation, and performance.
Interface	A shared boundary between two systems or system elements, defined by functional characteristics, common physical interconnection characteristics, signal characteristics, or other characteristics, as appropriate. (Adapted from ISO/IEC 2382, 2015)
IPO diagram	Figures in this handbook that provide a high-level view of the process of interest. The diagram summarizes the process activities and their typical inputs and typical outputs from/to other processes or external actors.
Knowledge	A body of information applied directly to the performance of a function. (INCOSE SECF)
Life cycle cost (LCC)	The total cost of a system over its entire life. Note: It includes all costs associated with the system and its use in the concept, development, production, utilization, support, and retirement stages.
Life cycle model	A framework of processes and activities concerned with the life cycle, which also acts as a common reference for communication and understanding. (ISO/IEC/IEEE 15288, 2023)
Measure	Variable to which a value is assigned as the result of measurement. (ISO/IEC/IEEE 15939, 2017)
Measurement	Set of operations having the object of determining a value of a measure. (ISO/IEC/IEEE 15939, 2017)
Measures of effectiveness (MOEs)	Measures that define the acquirer's key indicators of achieving the mission needs for performance, suitability, and affordability across the life cycle.
Measures of performance (MOPs)	Measures to assess whether the system meets design or performance requirements and has the capability to achieve operational objectives.
N^2 diagrams	Graphical representation used to define the internal operational relationships or external interfaces of the SoI.
Need statement	The result of a formal transformation of one or more life cycle concepts into an agreed-to expectation for an entity to perform some function or possess some quality. (INCOSE GtWR, 2022)
Operator	See User.
Organization	Person, or a group of people, and facilities with an arrangement of responsibilities, authorities, and relationships. (Adapted from ISO 9001, 2015)
Performance	A quantitative measure characterizing a physical or functional attribute relating to the execution of a process, function, activity, or task; performance attributes include quantity (how many or how much), quality (how well), timeliness (how responsive, how frequent), and readiness (when, under which circumstances).

(Continued)

(Continued)

Term	Definition
Physical configuration audit (PCA)	An evaluation to ensure that the operational system conforms to the operational and configuration documentation. (Adapted from ISO/IEC/IEEE 15288, 2023)
Process	A set of interrelated or interacting activities that transforms inputs into outputs. (Adapted from ISO 9001, 2015)
Product line	Group of products or services sharing a common, managed set of features that satisfy specific needs of a selected market or mission. (ISO/IEC/IEEE 24765, 2017)
Project	An endeavor with defined start and finish criteria undertaken to create a product or service in accordance with specified resources and requirements. (ISO/IEC/IEEE 15288, 2023)
Proof of concept	A realization of an idea or technology to demonstrate its feasibility.
Quality Characteristics	Inherent characteristic of a product, process, or system related to a requirement. (ISO/IEC/IEEE 15288, 2023)
Requirement statement	The result of a formal transformation of one or more needs or parent requirements into an agreed-to obligation for an entity to perform some function or possess some quality. (INCOSE GtWR, 2022)
Resource	An asset that is utilized or consumed during the execution of a process. (ISO/IEC/IEEE 15288, 2023)
Return on investment	Ratio of revenue from output (product or service) to development and production costs, which determines whether an organization benefits from performing an action to produce something. (ISO/IEC/IEEE 24765, 2017)
Reuse	The use of an asset in the solution of different problems. (IEEE 1517, 2010)
Skills	An observable competence to perform a learned psychomotor act.
Stage	A period within the life cycle of an entity that relates to the state of its description or realization. Note: Typical life cycle stages include concept, development, production, utilization, support, and retirement.
Stakeholder	A party having a right, share, or claim in a system or in its possession of characteristics that meet that party's needs and expectations.
Supplier	An organization or an individual that enters into an agreement with an acquirer for the supply of a product or service. (ISO/IEC/IEEE 15288, 2023)
System	An arrangement of parts or elements that together exhibit behavior or meaning that the individual constituents do not. (INCOSE Definitions, 2019)
System architect	The person, team, or organization responsible for a system's architecture, for coordinating engineering effort towards devising solutions to complex problems, and overseeing their implementations.
System architecture	The fundamental concepts or properties of an entity in its environment and governing principles for the realization and evolution of this entity and its related life cycle processes. (ISO/IEC/IEEE 42020, 2019)
System element	Member of a set of elements that constitutes a system. (ISO/IEC/IEEE 15288, 2023)
System life cycle	The evolution with time of a SoI from conception to retirement.
System of interest (SoI)	The system whose life cycle is under consideration. (ISO/IEC/IEEE 15288, 2023)

(Continued)

(Continued)

Term	Definition
System of systems	A SoI whose system elements are themselves systems; typically, these entail large-scale interdisciplinary problems with multiple, heterogeneous, distributed systems.
Systems engineering	A transdisciplinary and integrative approach to enable the successful realization, use, and retirement of engineered systems, using systems principles and concepts, and scientific, technological, and management methods. (INCOSE Definitions, 2019)
Tailoring	The manner in which any selected issue is addressed in a particular project. Tailoring may be applied to various aspects of the project, including project documentation, processes, and activities performed in each life cycle stage, the time and scope of reviews, analysis, and decision making consistent with all applicable statutory requirements.
Technical performance measures (TPMs)	Measures to assess design progress, compliance to performance requirements, or technical risks and provide visibility into the status of important project technical parameters to enable effective management, thus enhancing the likelihood of achieving the technical objectives of the project.
Trade-off	Decision-making actions that selects from various alternatives on the basis of net benefit to the stakeholders.
User	An individual who, or an organization that, contributes to the functionality of a system and draws on knowledge, skills, and procedures to contribute to the function. Individual who or group that benefits from a system during its utilization.
Validation	Confirmation, through the provision of objective evidence, that the requirements for a specific intended use or application have been fulfilled. (ISO/IEC/IEEE 15288, 2023)
Value	A measure of worth (e.g., benefit divided by cost) of a specific product or service by a customer, and potentially other stakeholders. (McManus, 2005)
Variability	(Of a product line) refers to characteristics that may differ among members of the product line. (ISO/IEC/IEEE 26550, 2015)
Verification	Confirmation, through the provision of objective evidence, that specified requirements have been fulfilled. (ISO/IEC/IEEE 15288, 2023)
Waste	Work that adds no value to the product or service in the eyes of the customer. (Womack and Jones, 1996)
White box	White box represents an internal view of the system (attributes and structure of the elements). Also referred to as transparent box.

APPENDIX D: N² DIAGRAM OF SYSTEMS ENGINEERING PROCESSES

Note: Figure D.1 in this appendix provides an N² diagram (see Section 3.2.4) of the typical inputs and outputs that appear in the IPO diagrams in this handbook. The off-diagonal squares represent the typical inputs/outputs shared by the processes that intersect at a given square. Outputs flow horizontally, inputs flow vertically, and the diagram can be read in a clockwise fashion. These typical inputs and outputs represent "a" way that the SE processes can be performed, but not necessarily "the" way that they must be performed. The absence of a relationship between any two processes does not preclude tailoring to create a relationship. Definitions of the typical inputs and outputs on the IPO diagrams can be found in Appendix E.

The system life cycle processes are placed on the diagonal, abbreviated as follows:

Abbreviation	Life Cycle Process	Handbook Section
ACQ	Acquisition	2.3.2.1
SUP	Supply	2.3.2.2
LCMM	Life Cycle Model Management	2.3.3.1
INFRAM	Infrastructure Management	2.3.3.2
PM	Portfolio Management	2.3.3.3
HRM	Human Resource Management	2.3.3.4
QM	Quality Management	2.3.3.5
KM	Knowledge Management	2.3.3.6
PP	Project Planning	2.3.4.1
PAC	Project Assessment and Control	2.3.4.2
DM	Decision Management	2.3.4.3
RM	Risk Management	2.3.4.4
CM	Configuration Management	2.3.4.5
INFOM	Information Management	2.3.4.6

(Continued)

INCOSE Systems Engineering Handbook: A Guide for System Life Cycle Processes and Activities, Fifth Edition.
Edited by David D. Walden, Thomas M. Shortell, Garry J. Roedler, Bernardo A. Delicado, Odile Mornas, Yip Yew-Seng, and David Endler.
© 2023 John Wiley & Sons Ltd. Published 2023 by John Wiley & Sons Ltd.

(Continued)

Abbreviation	Life Cycle Process	Handbook Section
MEAS	Measurement	2.3.4.7
QA	Quality Assurance	2.3.4.8
BMA	Business or Mission Analysis	2.3.5.1
SNRD	Stakeholder Needs and Requirements Definition	2.3.5.2
SRD	System Requirements Definition	2.3.5.3
SAD	System Architecture Definition	2.3.5.4
DD	Design Definition	2.3.5.5
SA	System Analysis	2.3.5.6
IMPL	Implementation	2.3.5.7
INT	Integration	2.3.5.8
VER	Verification	2.3.5.9
TRAN	Transition	2.3.5.10
VAL	Validation	2.3.5.11
OPER	Operation	2.3.5.12
MAINT	Maintenance	2.3.5.13
DISP	Disposal	2.3.5.14
TLR	Tailoring	4.1

In addition to the individual system life cycle processes, the following are also placed on the diagonal, abbreviated as follows:

Abbreviation	Name	Description
EXT	External	External represents those typical inputs and outputs that come from, or go to, beyond the set of system life cycle processes (i.e., they do not come from, or go to, another system life cycle process). Note that these can be either internal (e.g., Organization strategic plan) or external (e.g., Applicable laws and regulations) to the organization.
CTL	Controls	Controls represent those typical inputs and outputs that control, or limit, the execution of the system life cycle processes. They either come in as an external (EXT) typical input or from one or more life cycle processes. They go into every system life cycle process and are shown in Figure 2.11.
ENAB	Enablers	Enablers represent those typical inputs and outputs that enable, or assist in, the execution of the system life cycle processes. They either come in as an external (EXT) typical input or from one or more life cycle processes. They go into every system life cycle process and are shown in Figure 2.11.
SIT	Situational	Situational represents those typical inputs and outputs that are situational with respect to the execution of the system life cycle processes (i.e., they are invoked when needed). They can come from any life cycle process. They go into a select number of system life cycle processes, specifically: Decision Management, Risk Management, Configuration Management, Information Management, and System Analysis.

EXT	X	X		X	X	X	X	X	X	X	X	X	X			X	X			X	X						X							X
	CTL			X	X	X	X	X	X	X	X	X	X	X	X	X	X	X	X	X	X	X	X	X	X	X	X	X	X	X	X	X	X	X
		ENAB		X	X	X	X	X	X	X	X	X	X	X	X	X	X	X	X	X	X	X	X	X	X	X	X	X	X	X	X	X	X	X
			SIT											X	X	X	X								X									
X	X		X	ACQ	X						X	X	X			X	X										X							
X	X		X	X	SUP			X			X	X	X			X	X																	
X	X	X	X			LCMM					X	X						X																X
X		X	X				INFRAM				X																							
X	X		X			X	X	PM	X		X	X																						
X			X						HRM		X	X																						
X		X	X			X				QM	X								X															
X		X	X								KM																X							
	X		X	X			X	X	X		X	PP	X																					
X	X		X	X	X			X			X	X	PAC			X	X	X																
		X	X								X	X	X	DM											X									
		X	X								X	X	X		RM																			
X		X	X	X	X			X			X	X	X			CM	X																	
X		X	X	X	X						X	X	X			X	INFOM																	
		X	X								X	X	X					MEAS																
		X	X							X	X	X	X						QA															
X		X	X								X	X	X					X		BMA	X	X	X	X	X	X	X	X	X	X	X	X	X	
X		X	X								X	X	X					X		X	SNRD	X	X	X	X	X	X	X	X	X	X	X	X	
X		X	X								X	X	X					X				SRD	X	X		X		X						
X		X	X								X	X	X					X				X	SAD	X		X	X	X	X	X				
X		X	X								X	X	X					X				X	X	DD		X	X	X	X	X				
X		X	X								X	X	X	X						X	X	X	X	X	SA									
X		X	X								X	X	X	X						X	X	X	X	X	X	IMPL	X		X					
X		X	X								X	X	X	X						X	X	X	X	X	X	X	INT	X	X	X				
X		X	X								X	X	X	X						X	X	X	X	X	X			VER	X	X				
X		X	X		X						X	X	X	X						X	X	X	X	X	X				TRAN	X	X	X	X	
X		X	X		X						X	X	X	X						X	X	X	X	X	X				X	VAL				
X		X	X								X	X	X	X						X	X	X	X	X	X						OPER	X	X	
X		X	X								X	X	X	X						X	X	X	X	X	X						X	MAINT	X	
X		X	X		X						X	X	X	X						X	X	X	X	X	X								DISP	
			X			X					X	X																						TLR

FIGURE D.1 Input/output relationships between the various SE processes. INCOSE SEH original figure created by Shortell, Walden, and Yip. Usage per the INCOSE Notices pages.

APPENDIX E: INPUT/OUTPUT DESCRIPTIONS

Note: This appendix is a alphabetical list of all the typical inputs and outputs that appear in the IPO diagrams in this handbook. Sources of descriptions are as indicated. These typical inputs and outputs represent "a" way that the SE processes can be performed, but not necessarily "the" way that they must be performed. Other related terms and definitions can be found in Appendix B.

Typical Input/Output	Description
Accepted system or system element	System or system element (product or service) accepted by an acquirer from a supplier consistent with the delivery conditions of the supply agreement.
Acquired system or system element	System or system element (product or service) delivered to the acquirer from a supplier consistent with the delivery conditions of the acquisition agreement.
Acquisition agreement	Mutual acknowledgment of terms and conditions under which a working relationship is conducted between an acquirer and a supplier. (Adapted from ISO/IEC/IEEE 15288, 2023).
Acquisition need	Identified need that cannot be met within the organization encountering the need or a need that can be met in a more economical way by a supplier.
Acquisition payment	Payments or other compensations for an acquired system.
Acquisition records/ artifacts	Permanent, readable form of data, information, or knowledge related to acquisition.
Acquisition report	An account prepared for interested parties in order to communicate the status, results, and outcomes of the acquisition activities.
Acquisition strategy/ approach	Approaches, schedules, resources, and specific considerations required to perform acquisition.

(Continued)

INCOSE Systems Engineering Handbook: A Guide for System Life Cycle Processes and Activities, Fifth Edition.
Edited by David D. Walden, Thomas M. Shortell, Garry J. Roedler, Bernardo A. Delicado, Odile Mornas, Yip Yew-Seng, and David Endler.
© 2023 John Wiley & Sons Ltd. Published 2023 by John Wiley & Sons Ltd.

(Continued)

Typical Input/Output	Description
Agreements	Agreements from all applicable life cycle processes, including: acquisition agreement and supply agreement.
Alternative solution classes	Identifies and describes the classes of solutions that may address the problem or opportunity.
Analysis situations	Analyses that arise from any stakeholder. Can originate from any life cycle process.
Applicable laws and regulations	International, national, or local laws or regulations.
Breakdown structures	Hierarchical representations of project aspects into smaller components providing the necessary frameworks to accomplish the project objectives and create the required deliverables.
Business or mission analysis records/ artifacts	Permanent, readable form of data, information, or knowledge related to business or mission analysis.
Business or mission analysis report	An account prepared for interested parties in order to communicate the status, results, and outcomes of the business or mission analysis activities.
Business or mission analysis strategy/ approach	Approaches, schedules, resources, and specific considerations required to perform business or mission analysis.
Business plan	The overall organization business plan, including the business objectives.
Candidate items for configuration management	Items for configuration control. Can originate from any life cycle process.
Candidate items for information management	Items for information control. Can originate from any life cycle process.
Candidate risks and opportunities	Risks and opportunities that arise from any stakeholder. Can originate from any life cycle process.
Change request	Identified anomaly, required, or recommended enhancement to a project, from the time an idea is recorded until the disposition by a designated change authority. (Adapted from ISO/IEC/ IEEE 24765, 2017).
Concept of operations (ConOps)	At the organization level, addresses the leadership's intended way of operating the organization (ISO/IEC/IEEE 29148, 2018).
Configuration baseline	Configuration information formally designated at a specific time during the life of a product, product component, service, or service component. (Adapted from ISO/IEC/IEEE 24765, 2017).
Configuration management records/artifacts	Permanent, readable form of data, information, or knowledge related to configuration management.
Configuration management report	An account prepared for interested parties in order to communicate the status, results, and outcomes of the configuration management activities.

(Continued)

(Continued)

Typical Input/Output	Description
Configuration management strategy/approach	Approaches, schedules, resources, and specific considerations required to perform configuration management.
Configuration management system	System used to support and enable configuration management.
Configuration verification and audit report	Provides results of configuration management verifications and audits to ensure adequate traceability, control, and visibility. It includes evaluation criteria.
Constraints on solution	Externally imposed limitation on the system, its design, or implementation or on the process used to develop or modify a system. (ISO/IEC/IEEE 29148, 2018).
Critical performance measurement data	Data provided for the identified system-of-interest measurement needs.
Critical performance measurement needs	Identified information needs of the decision makers with respect to system-of-interest expectations.
Customer satisfaction inputs	Responses to customer satisfaction surveys or other instruments.
Decision management records/artifacts	Permanent, readable form of data, information, or knowledge related to decision management.
Decision management report	An account prepared for interested parties in order to communicate the status, results, and outcomes of the decision management activities.
Decision management strategy/approach	Approaches, schedules, resources, and specific considerations required to perform decision management.
Decision register	A repository that supports the availability for use and communication of all relevant decision information in a timely, complete, valid, and, if required, confidential manner.
Decision situations	Decisions that arise from any stakeholder. Can originate from any life cycle process.
Design definition records/artifacts	Permanent, readable form of data, information, or knowledge related to design definition.
Design definition report	An account prepared for interested parties in order to communicate the status, results, and outcomes of the design definition activities.
Design definition strategy/approach	Approaches, schedules, resources, and specific considerations required to perform design definition.
Disposal procedure	Presents an ordered series of steps to perform disposal.
Disposal records/ artifacts	Permanent, readable form of data, information, or knowledge related to disposal.
Disposal report	An account prepared for interested parties in order to communicate the status, results, and outcomes of the disposal activities.
Disposal strategy/ approach	Approaches, schedules, resources, and specific considerations required to perform disposal.

(Continued)

(Continued)

Typical Input/Output	Description
Disposed system	System (product or service) that has been deactivated, disassembled, and removed from operations and been properly disposed.
Enabling systems	External systems that facilitate the life cycle activities of the SoI but are not a direct element of the operational environment.
Human resource management records/artifacts	Permanent, readable form of data, information, or knowledge related to human resource management.
Human resource management report	An account prepared for interested parties in order to communicate the status, results, and outcomes of the human resource management activities.
Human resource management strategy/approach	Approaches, schedules, resources, and specific considerations required to perform human resource management.
Implementation records/artifacts	Permanent, readable form of data, information, or knowledge related to implementation.
Implementation report	An account prepared for interested parties in order to communicate the status, results, and outcomes of the implementation activities.
Implementation strategy/approach	Approaches, schedules, resources, and specific considerations required to perform implementation.
Information management records/artifacts	Permanent, readable form of data, information, or knowledge related to information management.
Information management report	An account prepared for interested parties in order to communicate the status, results, and outcomes of the information management activities.
Information management strategy/approach	Approaches, schedules, resources, and specific considerations required to perform information management.
Information register	A repository that supports the availability for use and communication of all relevant project information artifacts in a timely, complete, valid, and, if required, restricted manner.
Infrastructure management records/artifacts	Permanent, readable form of data, information, or knowledge related to infrastructure management.
Infrastructure management report	An account prepared for interested parties in order to communicate the status, results, and outcomes of the infrastructure management activities.
Infrastructure management strategy/approach	Approaches, schedules, resources, and specific considerations required to perform infrastructure management.
Installation procedure	Presents an ordered series of steps to perform transition.
Installed system	System (product or service) that has been installed in its operational environment.

(Continued)

(Continued)

Typical Input/Output	Description
Integrated system or system element	System or system element (product or service) that has been aggregated from system elements.
Integration procedure	Presents an ordered series of steps to perform integration.
Integration records/ artifacts	Permanent, readable form of data, information, or knowledge related to integration.
Integration report	An account prepared for interested parties in order to communicate the status, results, and outcomes of the integration activities.
Integration strategy/ approach	Approaches, schedules, resources, and specific considerations required to perform integration.
Knowledge management records/artifacts	Permanent, readable form of data, information, or knowledge related to knowledge management.
Knowledge management report	An account prepared for interested parties in order to communicate the status, results, and outcomes of the knowledge management activities.
Knowledge management strategy/approach	Approaches, schedules, resources, and specific considerations required to perform knowledge management.
Knowledge management system	System used to support and enable knowledge management.
Life cycle concepts	Articulation and refinement of the various life cycle concepts consistent with the stakeholder needs. Typical concepts include: acquisition concept; deployment concept; operational concept (OpsCon); support concept; retirement concept.
Life cycle model management records/artifacts	Permanent, readable form of data, information, or knowledge related to life cycle model management.
Life cycle model management report	An account prepared for interested parties in order to communicate the status, results, and outcomes of the life cycle model management activities.
Life cycle model management strategy/approach	Approaches, schedules, resources, and specific considerations required to perform life cycle model management.
Life cycle models	Framework of processes and activities concerned with the life cycle that can be organized into stages, acting as a common reference for communication and understanding. (ISO/IEC/ IEEE 15288, 2023)
Maintained and sustained system	System (product or service) that has been maintained for use in its operational environment.
Maintenance and logistics procedure	Presents an ordered series of steps to perform maintenance.
Maintenance and logistics records/ artifacts	Permanent, readable form of data, information, or knowledge related to maintenance.

(Continued)

(Continued)

Typical Input/Output	Description
Maintenance and logistics report	An account prepared for interested parties in order to communicate the status, results, and outcomes of the maintenance activities.
Maintenance and logistics strategy/ approach	Approaches, schedules, resources, and specific considerations required to perform maintenance.
Measurement data	Measurement data from all applicable life cycle processes, including: critical performance measurement data, organizational measurement data, and project measurement data.
Measurement needs	Measurement needs from all applicable life cycle processes, including: critical performance measurement needs, organizational measurement needs, and project measurement needs.
Measurement records/artifacts	Permanent, readable form of data, information, or knowledge related to measurement.
Measurement register	A repository that supports the availability for use and communication of all relevant measures in a timely, complete, valid, and, if required, confidential manner.
Measurement report	An account prepared for interested parties in order to communicate the status, results, and outcomes of the measurement activities.
Measurement strategy/approach	Approaches, schedules, resources, and specific considerations required to perform measurement.
Operation procedure	Presents an ordered series of steps to perform operation.
Operation records/ artifacts	Permanent, readable form of data, information, or knowledge related to operation.
Operation report	An account prepared for interested parties in order to communicate the status, results, and outcomes of the operation activities.
Operation strategy/ approach	Approaches, schedules, resources, and specific considerations required to perform operation.
Operational system	System (product or service) being used in its operational environment.
Organization infrastructure	Resources, facilities, personnel, and/or services that support the organization.
Organization infrastructure needs	Identified organizational infrastructure needs.
Organization lessons learned	Organizational-related lessons learned. Results from an evaluation or observation of an implemented corrective action that contributed to improved performance or increased capability. A lesson learned also results from an evaluation or observation of a positive finding that did not necessarily require corrective action other than sustainment.
Organization policies	High-level direction at the organizational level consistent with the organization's strategies. (Adapted from ISO/IEC/IEEE 15289, 2019)
Organization portfolio direction and constraints	Organization direction and constraints related to the project portfolio.
Organization procedures	Presents an ordered series of steps to perform a process, activity, or task for an organization. (Adapted from ISO/IEC/IEEE 15289, 2019)
Organization processes	Set of interrelated or interacting activities that transform inputs into outputs for an organization. (Adapted from ISO/IEC/IEEE 15288, 2023)

(Continued)

Typical Input/Output	Description
Organization reports	Reports from all applicable organization life cycle processes, including: life cycle model management report, infrastructure management report, portfolio management report, human resource management report, quality management report, and knowledge management report.
Organization strategic plan	The overall organization strategy, including the business mission or vision and strategic goals and objectives.
Organization strategies/ approaches	Strategies/approaches for all applicable organization life processes, including: life cycle model management strategy/approach, infrastructure management strategy/approach, portfolio management strategy/approach, human resource management strategy/approach, quality management strategy/approach, and knowledge management strategy/approach.
Organization tailoring strategy/ approach	Organization's specific strategy and approach to tailoring required to incorporate new or updated external standards.
Organizational measurement data	Data provided for the identified organizational measurement needs.
Organizational measurement needs	Identified information needs of the decision makers with respect to organizational expectations.
Other validated artifacts	Artifacts that are validated
Other verified artifacts	Artifacts that are verified
Portfolio management records/artifacts	Permanent, readable form of data, information, or knowledge related to portfolio management.
Portfolio management report	An account prepared for interested parties in order to communicate the status, results, and outcomes of the portfolio management activities.
Portfolio management strategy/approach	Approaches, schedules, resources, and specific considerations required to perform portfolio management.
Problem or opportunity statement	Description of the problem or opportunity. Should be derived from the organization strategy and provide enough detail to understand the gap or new capability that is being considered
Project assessment and control records/artifacts	Permanent, readable form of data, information, or knowledge related to project assessment and control.
Project assessment and control strategy/approach	Approaches, schedules, resources, and specific considerations required to perform project assessment and control.
Project authorization	Authorization from the organization to proceed per the agreed-to project plan.
Project authorization request	Request from the project to the organization to authorize the project.

(Continued)

(Continued)

Typical Input/Output	Description
Project budget	Estimate of the costs associated with a particular project. Includes labor, infrastructure, acquisition, and enabling system costs along with reserves for risk management.
Project constraints	Externally imposed limitation on the project developing or modifying a system. (ISO/IEC/IEEE 29148, 2018)
Project control request	Project directives based on action required due to deviations from the project plan. If assessments are associated with a decision gate, a decision to proceed or not to proceed, is taken.
Project decision gate/review result	Decision gate and review artifacts that are expected through conduct of the decision gate or technical review and that can be considered elements of exit criteria. (Adapted from ISO/IEC/IEEE 24748–8, 2019)
Project direction	Organizational direction to the project. Includes sustainment of projects meeting objectives and redirection or termination of projects not meeting objectives.
Project human resource needs	Identified human resource needs of the project.
Project infrastructure	Resources, facilities, personnel, and/or services that support the project.
Project infrastructure needs	Identified infrastructure needs of the project.
Project lessons learned	Project-related lessons learned. Results from an evaluation or observation of an implemented corrective action that contributed to improved performance or increased capability. A lesson learned also results from an evaluation or observation of a positive finding that did not necessarily require corrective action other than sustainment.
Project measurement data	Data provided for the identified project measurement needs.
Project measurement needs	Identified information needs of the decision makers with respect to project expectations.
Project objectives	The objectives or goals for the project.
Project planning records/artifacts	Permanent, readable form of data, information, or knowledge related to project planning.
Project portfolio	Collection of projects that addresses the strategic objectives of the organization. (ISO/IEC/IEEE 12207, 2017)
Project procedures	Procedures from all applicable life cycle processes, including: integration procedure, verification procedure, installation procedure, validation procedure, operation procedure, maintenance and logistics procedure, and disposal procedure.
Project reports	Reports from all applicable project life cycle processes, including: acquisition report, supply report, decision management report, risk management report, configuration management report, configuration verification and audit report, information management report, measurement report, quality assurance report, quality assurance evaluation report, business or mission analysis report, stakeholder needs and requirements definition report, system requirements definition report, system architecture definition report, system architecture assessment report, design definition report, system design assessment report, system analysis report, implementation report, integration report, verification report, transition report, validation report, operation report, maintenance and logistics report, and disposal report.

(Continued)

(Continued)

Typical Input/Output	Description
Project schedule	A linked list of a project's milestones, activities, and deliverables with intended start and finish dates May include a top-level milestone schedule and multiple levels (also called tiers) of schedules of increasing detail and task descriptions with completion criteria and work authorizations.
Project status report/ dashboard	Provides results of monitoring the execution of the defined plan or processes for internal or external distribution. It includes a summary of decisions, monitoring results, action items, process or performance data, and recorded process improvements. It assesses the degree of adherence to the plans. (Adapted from ISO/IEC/IEEE 15289, 2019)
Project strategies/ approaches	Strategies/approaches for all applicable project life processes, including: acquisition strategy/ approach, supply strategy & approach, project assessment and control strategy/approach, decision management strategy/approach, risk management strategy/approach, configuration management strategy/approach, information management strategy/approach, measurement strategy/approach, quality assurance strategy/approach, business or mission analysis strategy/approach, stakeholder needs and requirements definition strategy/approach, system requirements definition strategy/approach, system architecture definition strategy/approach, design definition strategy/approach, system analysis strategy/approach, implementation strategy/approach, integration strategy/approach, verification strategy/approach, transition strategy/approach, validation strategy/approach, operation strategy/approach, maintenance and logistics strategy/approach, and disposal strategy/approach.
Project tailoring strategy/approach	Project's specific strategy and approach to tailoring required to incorporate new or updated life cycle models.
Qualified personnel	Individuals equipped to perform duties on behalf of the organization, including officers, employees, and contractors. (Adapted from ISO/IEC/IEEE 24765, 2017)
Quality assurance corrective action	Action to eliminate the cause or reduce the likelihood of recurrence of a detected project nonconformity or other undesirable situation. (Adapted from ISO/IEC 19770–1, 2017)
Quality assurance evaluation report	Provides results of quality assurance evaluations. It includes evaluation criteria. (Adapted from ISO/IEC/IEEE 15289, 2019)
Quality assurance records/artifacts	Permanent, readable form of data, information, or knowledge related to quality assurance.
Quality assurance report	An account prepared for interested parties in order to communicate the status, results, and outcomes of the quality assurance activities.
Quality assurance strategy/approach	Approaches, schedules, resources, and specific considerations required to perform quality assurance.
Quality assurance system	System used to support and enable quality assurance.
Quality management corrective action	Action to eliminate the cause or reduce the likelihood of recurrence of a detected organizational nonconformity or other undesirable situation. (Adapted from ISO/IEC 19770–1, 2017)
Quality management criteria and methods	Rules on which a judgment or decision can be based, or by which an organization can be evaluated. (Adapted from ISO/IEC/IEEE 15289, 2019)
Quality management evaluation report	Provides results of quality management evaluations. It includes evaluation criteria. (Adapted from ISO/IEC/IEEE 15289, 2019)

(Continued)

(Continued)

Typical Input/Output	Description
Quality management records/artifacts	Permanent, readable form of data, information, or knowledge related to quality management.
Quality management report	An account prepared for interested parties in order to communicate the status, results, and outcomes of the quality management activities.
Quality management strategy/approach	Approaches, schedules, resources, and specific considerations required to perform quality management.
Quality management system	System used to support and enable quality management.
Records/artifacts	Records from all applicable life cycle processes, including: acquisition records/artifacts, supply records/artifacts, life cycle model management records/artifacts, infrastructure management records/artifacts, portfolio management records/artifacts, human resource management records/artifacts, quality management records/artifacts, knowledge management records/artifacts, project planning records/artifacts, project assessment and control records/artifacts, decision management records/artifacts, risk management records/artifacts, configuration management records/artifacts, information management records/artifacts, measurement records/artifacts, quality assurance records/artifacts, business or mission analysis records/artifacts, stakeholder needs and requirements definition records/artifacts, system requirements definition records/artifacts, system architecture definition records/artifacts, design definition records/artifacts, system analysis records/artifacts, implementation records/artifacts, integration records/artifacts, verification records/artifacts, transition records/artifacts, validation records/artifacts, operation records/artifacts, maintenance and logistics records/artifacts, disposal records/artifacts, tailoring records/artifacts.
Request for supply	Acquirer's request for information and commitments needed from the supplier that are required to be included in the potential supplier's proposal. It announces the acquirer's intention to potential bidders to acquire a specified system or system element (product or service). (Adapted from ISO/IEC/IEEE 15289, 2019)
Requirements imposed on enabling systems	Identified requirements for enabling systems of the system-of-interest.
Reused system or system element	System or system element (product or service) reused by an organization consistent with its system element requirements.
Risk management records/artifacts	Permanent, readable form of data, information, or knowledge related to risk management.
Risk management report	An account prepared for interested parties in order to communicate the status, results, and outcomes of the risk management activities.
Risk management strategy/approach	Approaches, schedules, resources, and specific considerations required to perform risk management.
Risk register	A repository that supports the availability for use and communication of all relevant risk information in a timely, complete, valid, and, if required, confidential manner.
Source documents	External documents relevant to the particular stage of the system of interest.
Stakeholder identification	List of individuals or organizations having a right, share, claim, or interest in a system or in its possession of characteristics that meet their needs and expectations. (Adapted from ISO/IEC/IEEE 15288, 2023)

(Continued)

(Continued)

Typical Input/Output	Description
Stakeholder needs and requirements	Structured collection of the requirements [characteristics, context, concepts, constraints and priorities] of the stakeholder and the relationship to the external environment. (ISO/IEC/IEEE 29148, 2018)
Stakeholder needs and requirements definition records/ artifacts	Permanent, readable form of data, information, or knowledge related to stakeholder needs and requirements definition.
Stakeholder needs and requirements definition report	An account prepared for interested parties in order to communicate the status, results, and outcomes of the stakeholder needs and requirements definition activities.
Stakeholder needs and requirements definition strategy/ approach	Approaches, schedules, resources, and specific considerations required to perform stakeholder needs and requirements definition.
Standards	This handbook and relevant industry, country, military, acquirer, and other specifications and standards. Includes new knowledge from industry sponsored knowledge networks.
Supplied system or system element	System or system element (product or service) delivered from a supplier to an acquirer consistent with the delivery conditions of the supply agreement.
Supply agreement	Mutual acknowledgment of terms and conditions under which a working relationship is conducted between a supplier and an acquirer. (Adapted from ISO/IEC/IEEE 15288, 2023)
Supply payment	Payments or other compensations for the supplied system.
Supply records/ artifacts	Permanent, readable form of data, information, or knowledge related to supply.
Supply report	An account prepared for interested parties in order to communicate the status, results, and outcomes of the supply activities.
Supply response	Prepared by a potential supplier to support the offer of a contract bid, including cost, schedule, risk statements, methodology to satisfy the request for supply, experiences and capabilities, any recommendations to tailor the request for supply or contract, and the signature of the supplier's approving authority. Informally, may be prepared within an organization. (Adapted from ISO/IEC/IEEE 15289, 2019)
Supply strategy/ approach	Approaches, schedules, resources, and specific considerations required to perform supply.
System analysis records/artifacts	Permanent, readable form of data, information, or knowledge related to system analysis.
System analysis report	An account prepared for interested parties in order to communicate the status, results, and outcomes of the system analysis activities.
System analysis request	A request to conduct a system analysis.

(Continued)

(Continued)

Typical Input/Output	Description
System analysis strategy/approach	Approaches, schedules, resources, and specific considerations required to perform system analysis.
System architecture assessment report	Provides results of system architecture assessments. It includes evaluation criteria. (Adapted from ISO/IEC/IEEE 15289, 2019)
System architecture definition records/ artifacts	Permanent, readable form of data, information, or knowledge related to system architecture definition.
System architecture definition report	An account prepared for interested parties in order to communicate the status, results, and outcomes of the system architecture definition activities.
System architecture definition strategy/ approach	Approaches, schedules, resources, and specific considerations required to perform system architecture definition.
System architecture description	The fundamental conception of a system-of-interest in terms of its purpose, system qualities (such as feasibility, performance, safety, and interoperability), constraints, and design decisions and rationale. Identification of the architecture's stakeholders and the stakeholders' architecture-related concerns. (Adapted from ISO/IEC/IEEE 15289, 2019)
System architecture rationale	Rationale for architecture selection, technological/technical system element selection, and allocation between system requirements and architectural entities.
System design assessment report	Provides results of system design assessments. It includes evaluation criteria. (Adapted from ISO/IEC/IEEE 15289, 2019)
System design characteristics	Design attributes or distinguishing features that pertain to a measurable description of a product or service. (ISO/IEC/IEEE 15288, 2023)
System design description	Describes the design of a system or element. (Adapted from ISO/IEC/IEEE 24765, 2017)
System design rationale	Rationale for design selection, system element selection, and allocation between system requirements and system elements. Includes rationale of major selected implementation options and enablers.
System element	System element (product or service) implemented consistent with its system element requirements.
System element description	Applies the system architecture description to the low-level system configuration items and elements. It is at a level of detail to permit design, implementation, and test. (Adapted from ISO/IEC/IEEE 15289, 2019)
System interface definition	Description of the interfaces between systems and system elements. (Adapted from ISO/IEC/IEEE 24765, 2017)
System requirements	Structured collection of the requirements [functions, performance, design constraints, and other attributes] for the system and its operational environments and external interfaces. (ISO/IEC/IEEE 29148, 2018)
System requirements definition records/ artifacts	Permanent, readable form of data, information, or knowledge related to system requirements definition.
System requirements definition report	An account prepared for interested parties in order to communicate the status, results, and outcomes of the system requirements definition activities.

(Continued)

(Continued)

Typical Input/Output	Description
System requirements definition strategy/approach	Approaches, schedules, resources, and specific considerations required to perform system requirements definition.
System viewpoints, views, and models	Definitions of viewpoints to document the procedures for creating, interpreting, analyzing, and evaluating architectural data. One or more views of the system. Each architectural view is a representation of the complete system from the perspective of one or more system concerns, for its stakeholders. (Adapted from ISO/IEC/IEEE 15289, 2019)
Systems engineering management plan (SEMP)	Presents how the project processes and activities are executed to assure the project's successful completion, and the quality of the deliverable product or service. (Adapted from ISO/IEC/IEEE 15289, 2019)
Tailoring records/artifacts	Permanent, readable form of data, information, or knowledge related to tailoring.
Traceability mapping	Records the relationship between two or more artifacts of the development process (e.g., requirements, functions, system elements, verifications, and validations, tasks). (Adapted from ISO/IEC/IEEE 24765, 2017)
Trained personnel	Trained individuals or organizations that perform the operation, maintenance, or other functions of or for a system.
Training materials	Materials for the provision of formal and informal learning activities. (Adapted from ISO/IEC/IEEE 24765, 2017)
Transition records/artifacts	Permanent, readable form of data, information, or knowledge related to transition.
Transition report	An account prepared for interested parties in order to communicate the status, results, and outcomes of the transition activities.
Transition strategy/approach	Approaches, schedules, resources, and specific considerations required to perform transition.
Validated stakeholder needs and requirements	Set of stakeholder needs and requirements that have been validated.
Validated system	System (product or service) that has been validated.
Validated system architecture and design	System architecture and design that has been validated.
Validation criteria	The validation criteria (the measures to be assessed), who will perform validation activities, and the validation environments of the system-of-interest.
Validation procedure	Presents an ordered series of steps to perform validation.
Validation records/artifacts	Permanent, readable form of data, information, or knowledge related to validation.
Validation report	An account prepared for interested parties in order to communicate the status, results, and outcomes of the validation activities.
Validation strategy/approach	Approaches, schedules, resources, and specific considerations required to perform validation.

(Continued)

(Continued)

Typical Input/Output	Description
Variance/deviation/ waiver request	Request, temporary or permanent, to accept a configuration item or other designated item which, during production or after having been submitted for inspection, is found to depart from specified requirements, but is nevertheless considered suitable for use as is or after rework by an approved method. (Adapted from ISO/IEC/IEEE 24765, 2017)
Verification criteria	The verification criteria (the measures to be assessed), who will perform verification activities, and the verification environments of the system-of-interest.
Verification procedure	Presents an ordered series of steps to perform verification.
Verification records/ artifacts	Permanent, readable form of data, information, or knowledge related to verification.
Verification report	An account prepared for interested parties in order to communicate the status, results, and outcomes of the verification activities.
Verification strategy/ approach	Approaches, schedules, resources, and specific considerations required to perform verification.
Verified system	System (product or service) that has been verified.
Verified system architecture and design	System architecture and design that has been verified.
Verified system requirements	Set of system requirements that have been verified.

APPENDIX F: ACKNOWLEDGMENTS

The INCOSE Systems Engineering Handbook Fifth Edition editorial team owes a debt of gratitude to all the contributors to prior editions (versions 1, 2, 2A, 3.n, and 4). Tim Robertson led the effort to create Version 1 of the handbook. Version 2 was led by: James Whalen, ESEP and Richard Wray, ESEP. Version 3 was led at various times by: Kevin Forsberg, ESEP; Terje Fossnes, ESEP; Douglas Hamelin; Cecilia Haskins, ESEP; Michael Krueger, ESEP; and David Walden, ESEP. The Fourth Edition was led by David Walden, ESEP. The framework they provided gave a solid basis for moving ahead with this edition. This revision reflects changes to the previous version based on three primary objectives: first, to reflect the updated ISO/IEC/IEEE 15288:2023 standard; second, to reflect the state-of-the-good-practice based on inputs from the relevant INCOSE Working Groups (WGs); and third, reflect changes suggested by the INCOSE community.

A great deal of effort and enthusiasm was provided by the key authors, many of whom also serve as INCOSE WG Chairs or SEBoK authors. We acknowledge them in alphabetical order: Juan Amenabar, ESEP; Randy Anway; James Armstrong, ESEP; Albertyn Barnard; William Bearden, CSEP; Peter Bernus; Dawn Beyer; Mike Boardman; Guy-André Boy; Barclay Brown, ESEP; Dale Brown; Jeffrey Brown; Christopher Browne, CSEP; John Brtis, CSEP; Javier Calvo-Amodio; Yann Chazal; Cindy Chen; John Clark, CSEP; Daniel Cobb, CSEP; Peter Coleman; Iain Cunningham; Kenneth Cureton; Cihan Dagli; Judith Dahmann; Alain Dauron; Hans Peter de Koning; William Donaldson; Rick Dove; Rod Dreisbach; Adrianna D'Souza, CSEP; Daniel Eisenberg; Richard Fairley; Paul Frenz, ESEP; Sanford Friedenthal; Jean-Luc Garnier; Donald Gelosh, ESEP; Peter Graham, ASEP; Alan Harding; Cecilia Haskins, ESEP; Porter Haskins, CSEP; Michael Henshaw; David Hetherington; Oliver Hoehne, CSEP; Adam Hulse, CSEP; Mike Jackson; Scott Jackson; Chamara Johnson, CSEP; John Juhasz; Alexander Karl; David Kaslow; Tami Katz, ESEP; Duncan Kemp; Bob Kenley, ESEP; Grace Kennedy, CSEP; Larry Kennedy; Ron Kenett; Bill Klimack, CSEP; Alain Kouassi, CSEP; Charles Krueger; Eric Krueger; Anand Kumar; William Lawless; Alejandro Levi, CSEP; Ivan Mactaggart; Ray Madachy; Robert Malins; Thomas Manley, CSEP; James Martin; Sean McCoy, CSEP; Dorothy McKinney; Curt McNamara; William Miller; Ricardo Moraes; Perri Nejib, ESEP; Meaghan O'Neil; Bohdan Oppenheim; Gregory Parnell, CSEP; Bob Parro; Tasha Penner, CSEP; Michael Pennotti; Troy Peterson, CSEP; Andrew Pickard; Edward Pohl; Stephen Powley; Tim Rabbets; Susan Ronning, ASEP; Larri Rosser, ASEP; David Rousseau;

INCOSE Systems Engineering Handbook: A Guide for System Life Cycle Processes and Activities, Fifth Edition.
Edited by David D. Walden, Thomas M. Shortell, Garry J. Roedler, Bernardo A. Delicado, Odile Mornas, Yip Yew-Seng, and David Endler.
© 2023 John Wiley & Sons Ltd. Published 2023 by John Wiley & Sons Ltd.

Jean-Claude Roussel, ESEP; Gary Rushton; Michael Ryan; Frank Salvatore, ESEP; Bill Scheible, ESEP; William Schindel, CSEP; Christopher Schreiber; Zane Scott, ASEP; Dr. Alice F. Squires, ESEP-ACQ; Dr. Tina P. Srivastava; Kim Stansfield; Jack Stein; Drew Stovall; Bob Swarz; Corrie Taljaard; Maurice Theobald; Sergey Tozik; Hubertus Tummescheit; Laura Uden, CSEP; Christopher Unger, ESEP; Ricardo Valerdi; Marcel van de Ven, CSEP; Harry van der Velde, CSEP; Andreas van Zyl; Michael Vinarcik, ESEP; Charles Wasson, ESEP; Michael Watson; Louis Wheatcraft; Clifford Whitcomb; Raymond Wolfgang, CSEP; Hazel Woodcock, ESEP; Edward Yakabovicz; Michael Yokell, ESEP; Lori Zipes, ESEP; and Avigdor Zonnenshain. We also acknowledge the INCOSE UK Energy Systems Interest Group, the INCOSE-PMI Alliance, ISO, and NAFEMS for their support.

The INCOSE Technical Operations reviews were led by TJ Ferrell and facilitated by Molly Kovaka. The reviews generated excellent comments that significantly improved the handbook. Other individual and group reviewers also generated useful review comments. Space prevents us from acknowledging them individually. We also thank the INCOSE Corporate Advisory Board (CAB), the INCOSE Certification Advisory Group (CAG), and the specific and anonymous reviewers who provided comments on the Fourth Edition. Their inputs were much appreciated.

The editors thank Vitech, A Zuken Company, for the use of their GENESYS™ tool, which was used to create an underlying process model that helped ensure consistency in the handbook IPO diagrams. We also thank Jama for the use of their Connect® tool to help manage the significant number of handbook requirements. The editors also thank Taylor Riethle for her graphical support with key handbook figures and the Wiley team for their editorial support.

Any errors introduced as part of the editorial process rest with the editors, not the contributors.

We apologize if we unintentionally omitted anyone from these lists.

Gratefully,
David D. Walden, ESEP
Thomas M. Shortell, CSEP
Garry J. Roedler, ESEP
Bernardo A. Delicado, ESEP
Odile Mornas, ESEP
Yip Yew-Seng, CSEP
David Endler, ESEP

APPENDIX G: COMMENT FORM

Comments and suggestions from users of the handbook are welcome. Please make sure your inputs are actionable by following the suggested format below.

Reviewed document:	INCOSE SE Handbook Fifth Edition
Name of submitter:	Given FAMILY (given name and family name)
Date submitted:	DD-MMM-YYYY
Contact info:	john.doe@anywhere.com (email address)
Type of submission:	Individual or Group
Group name and number of contributors:	INCOSE XYZ WG (if applicable)
Comments:	Detailed comments with reference to document section, paragraph, etc. Please include detailed recommendations, as shown in the table below

Send comments to: info@incose.org

Comment ID (if multiple comments, sequential for your set of comments)	**Category** (TH, TL, E, G—see below)	**Section number** (e.g., 3.4.2.1)	**Specific reference** (e.g., paragraph, line, figure, table)	**Issue, comment, and rationale** (rationale must make comment clearly evident and supportable)	**Proposed change/ new text** (mandatory entry, must be substantial to increase the odds of acceptance)

TH, technical high; TL, technical low; E, editorial; G, general

---

INCOSE Systems Engineering Handbook: A Guide for System Life Cycle Processes and Activities, Fifth Edition.
Edited by David D. Walden, Thomas M. Shortell, Garry J. Roedler, Bernardo A. Delicado, Odile Mornas, Yip Yew-Seng, and David Endler.
© 2023 John Wiley & Sons Ltd. Published 2023 by John Wiley & Sons Ltd.

INDEX

Note: Index only shows primary entries for commonly used terms.

accessibility, *see* reliability, availability, maintainability (RAM)

acquirer, *see also* supplier, 2, 3, 5, 33, 44–50, 65, 76, 89, 97, 106, 114–115, 128, 140, 145–152, 168, 176–179, 203, 252, 253, 311, 313, 321, 330–331

acquisition process, 45–48

adaptability, *see* agility

aerospace, *see* commercial aerospace systems

affordability, 97, 130, 160–165, 313

aggregate, 135–137, 181, 205–206, 238, 325

agile, 21, 29, 38–39, 72, 90, 137, 165–168, 188–189, 221–224, 274, 309, 311

agility, 161, 165–168, 181, 223, 274

agreement(s), 44–51, 60, 65, 70–71, 74, 76, 86, 107, 145, 148, 179, 202–203, 216, 218, 232, 237, 311, 321–322, 331

allocate/allocation, 43–44, 48, 56–63, 70, 74, 83, 99, 112–117, 121–128, 144, 174, 198, 203, 221, 232, 233, 252–253, 257, 267–268, 332

analysis, *see also* inspection, demonstration, and test, 131, 136, 139, 142, 147

architect, 20, 191, 311, 314

architecture, 2, 4, 8, 9, 42–44, 74, 78, 101, 115, 118–129, 137, 142, 150–151, 166–167, 178, 180–182, 198, 201, 203–205, 206–208, 212, 230, 232, 237, 243–244, 252, 254, 268, 311, 314, 332–334

architecture definition, *see* system architecture definition

architecture framework, 120–123, 206–208, 305

artificial intelligence (AI), 5, 170, 249, 275, 283–285, 305

assessment, 18, 20, 52–54, 59, 61–62, 64, 75–78, 81, 86, 98, 114–115, 120, 123, 126, 129, 132, 185, 189–190, 195, 204, 208, 217, 243–244, 261, 280, 332

associate systems engineering professional (ASEP), xxii, 305

attribute, 9, 14, 66, 109, 114–117, 127, 162, 266, 311, 313

audits, *see also* reviews, 31–33, 54, 72–73, 76–78, 89–90, 132–134, 248–249, 252, 313–314, 323

automotive systems, 245–247, 282–285

availability, *see* reliability, availability, maintainability (RAM)

baseline, 31–32, 35–36, 74, 89–90, 105, 110, 114, 117, 131, 133, 136, 140, 148, 200, 204, 218, 312, 322

behavioral architecture, *see* functional architecture

benchmark, 53–54

bias, 17–18, 110–111, 112, 230

big data-driven systems, *see* internet of things (IoT)

biomedical and healthcare systems, 248–249

biomimicry, 213

INCOSE Systems Engineering Handbook: A Guide for System Life Cycle Processes and Activities, Fifth Edition.
Edited by David D. Walden, Thomas M. Shortell, Garry J. Roedler, Bernardo A. Delicado, Odile Mornas, Yip Yew-Seng, and David Endler.
© 2023 John Wiley & Sons Ltd. Published 2023 by John Wiley & Sons Ltd.

black box, *see also* white box, 9, 13, 200, 312
blank sheet, *see* greenfield
boundary, 8, 10, 21, 113, 120–121, 166, 203–204, 245, 281
brainstorming, 86, 125
breakdown structures, 13, 71–75, 174, 269, 322
brownfield, *see also* greenfield, 145, 229, 230, 259, 278, 312
business or mission analysis process, 103–107
business requirements, 103–107, 252

case studies, 277–285
certified systems engineering professional (CSEP), xxii, 306
change control, 35, 56, 91, 117
changeability, *see* agility
clean sheet, *see* greenfield
cognitive bias *see* bias
commercial off-the-shelf (COTS), 69–70, 122, 134, 231–232,
    238, 251, 258
commercial aerospace systems, 8, 246, 249–250
compatibility, *see* interoperability
competence, 171, 242, 161–262, 312, 314
competency, 21, 62, 226, 242, 261–263, 312
complexity, xix, 2, 5–6, 9–10, 15, 20–24, 33–35, 53, 170, 195,
    203, 210, 216, 218, 220, 226, 229, 234, 238, 243,
    245–247, 249, 250–251, 253, 255, 257–258, 262,
    281–282
concept of operations (ConOps), *see also* operational concept
    (OpsCon) and life cycle concepts, 104, 106, 108, 152,
    154, 182–183, 223, 244, 322
concept stage, 26–28, 222, 278
concurrency, *see also* iteration and recursion, xxiii, 25, 30, 35,
    39, 42–44, 101–103, 110, 112, 115, 145, 152, 154, 176,
    192–193, 203, 215, 222, 269
configuration/change control board (CCB), 89, 117
configuration item (CI), 88–89, 155, 312, 322
configuration management process, 87–90
connectivity, *see* interoperability
consensus, 45, 110,197
constraint, 3–4, 9, 14, 19, 43, 45, 71–73, 89, 101, 103–105,
    107–111, 113–116, 118–123, 125–127, 131, 132–133,
    134–137, 139–140, 147–149, 152, 155, 156–157,
    160–165, 175, 178, 183–184, 195, 222, 228, 229, 232,
    237, 239–240, 244, 262, 272, 279, 312, 323, 326, 328
context, 8–15, 17–20, 21–24, 45, 73–75, 80, 82–83, 86, 88,
    101–102, 104–105, 108, 110–111, 117, 119–120,
    144–145, 159, 161–163, 168, 192, 204, 207, 212, 220,
    224, 237–238, 252, 253
contract(s)/subcontract(s), *see also* agreement, 11, 44–45,
    46–48, 50, 91, 99, 115, 139, 145, 147–148, 157, 203,
    208, 226, 228, 251, 252–253, 279–280
cost as an independent variable (CAIV), 162
cost breakdown structure (CBS), *see* breakdown structures
cost effectiveness, *see* affordability
cost estimating, 16, 160–165, 253
coupling, 19, 208

coupling matrix, *see also* N^2 diagram, 136, 205–206
customer, *see* acquirer
cyber-physical systems (CPS), 180, 228, 233–235,
    266, 275
cybersecurity, *see* security

decision gates, *see also* reviews, 25–27, 29–31, 47, 51, 72–73,
    75, 252, 312
decision management process, 78–81
decisions, *see also* trade study/trade-off study, 14, 15–16,
    17–18, 19, 25–26, 29–31, 42–44, 45–47, 49, 56–57, 59,
    61, 63, 76–77, 78–81, 90, 93–98, 105, 114, 119–123,
    126, 129–131, 163–164, 167, 170, 175, 178, 184–185,
    187, 189, 191, 192, 195–196, 217–218, 220, 223,
    231–232, 234, 238–239, 241, 249, 263, 268, 269, 270,
    271–272, 275, 278, 284, 323
defense systems, 4, 8, 27, 31, 91, 97, 173, 206–207, 246,
    250–251, 256, 285
demonstration, *see also* inspection, analysis, and test, 131,
    136, 139, 142, 147
derivation/derived, 9, 94–95, 97–98, 114–115, 123, 132, 149,
    154, 178–179, 188, 228, 267
design definition, 124–129
design for X (DFX), *see also* quality characteristics and
    approaches, 127, 159
design structure matrix (DSM), *see also* N^2 diagram, 205
design thinking, 127, 170, 212
design to cost (DTC), 165
desirability, *see* human systems integration (HSI)
development models, *see* life cycle model approaches
development stage, 26–28, 35, 222
DevOps, 38, 90, 221
DevSecOps, 38, 90, 221
digital engineering, 5, 81, 93, 95–96, 170, 226, 273–274, 275
digital twin, 11, 175, 193, 202, 228, 234, 273
disposability, *see* sustainability
disposal process, 156–158
diversity, equity, and inclusion (DEI), 265–266
domains/industries/sectors, 244–259

effectiveness, 51–54, 67, 71–72, 75, 83, 87, 89, 96–97, 99,
    101, 111, 118, 121–122, 131–132, 161–164, 173, 178,
    189, 209
element, *see* system element
emergence and emergent properties/behaviors, 5, 9–10, 15,
    21–23, 70, 118, 123, 129, 151, 169–170, 186, 210, 212,
    213, 230, 235, 237–238, 240, 245
enabling system, *see also* interfacing system and
    interoperating system, 2, 8, 10–11, 16, 19, 26, 28, 33,
    56–57, 63, 72, 99, 103–105, 107–111, 113–114,
    118–119. 122, 125–126, 130, 132, 134, 135–137,
    139–140, 143, 145, 147–148, 152, 154, 155–156,
    157–158, 159, 170, 175, 191, 203, 243, 254–255, 267,
    268, 324

engineered system, 3, 11–12
enterprise, *see also* organization/organizational, 8, 11–12, 44, 60, 82, 88, 96, 106, 118, 122–123, 124, 206–208, 219, 223, 226, 241–244, 273, 275, 312
environment, 2–3, 8, 9, 10, 11–12, 14, 15, 16, 17, 19, 21, 22, 28, 33, 48, 69, 84, 88, 101, 103, 104–106, 107– 111, 115, 119, 122, 129, 132, 136–137, 142, 143, 144, 146–147, 149–152, 154–155, 156–158, 159, 161, 165–166, 168–169, 171, 173–174, 176–179, 180–182, 184–185, 186–189, 192–193, 202, 204, 206–207, 213–214–215, 221–222, 230, 231, 233–234, 238, 240, 241–242, 244, 248, 250, 272, 273, 276, 279–280, 312
environmental engineering/impact, *see* sustainability
ergonomics, *see* human systems integration (HSI)
estimating, 28, 50, 53, 71–73, 83–84, 131, 164–165, 168, 177, 179, 239, 261, 272
ethics, 234, 262, 263, 264–265
evaluation criteria, 162, 208
evolutionary, *see also* incremental and sequential, 21, 29, 33–39, 77, 96, 137, 222, 235, 249, 256, 275
evolvability, *see* agility
expert systems engineering professional (ESEP), 306
extensibility, *see* agility

failure, 17, 20, 29, 30, 45, 50, 65, 75, 100, 101, 151, 155–156, 161, 171, 174–175, 176–180, 181, 186–187, 208–211, 266, 269, 312
failure modes, and effects, [and criticality] analysis (FMEA/FMECA), 86, 174, 178, 180, 204
family of systems (FoS)/system family, *see* product line engineering (PLE)/product lines
flexibility, *see* agility
flow-down/flow-up, 115, 193, 221
functional analysis, 8, 88–89, 105–106, 110, 113, 129–131, 133, 134, 159, 166–167, 174, 186–187, 193, 202, 205, 231–232, 233, 238, 239, 266
functional architecture, *see also* physical architecture, 8, 14, 74, 120–121, 124, 126–129, 136–137, 192, 212, 253
functional flow block diagram (FFBD), 204–205,
functional tree/functional breakdown structure (FBS), *see* breakdown structures
future of SE (FuSE), 275–276

gates *see* decision gate
greenfield, *see also* brownfield, 145, 229–230, 313

habitability, see human systems integration (HSI)
hardware engineering (HWE), 16, 38–39, 267–268, 278
hazard, 124, 169, 186–189, 284
healthcare, *see* biomedical and healthcare systems
heuristics, *see* systems engineering heuristics
hierarchy, 12–13, 21, 32, 35, 42–44, 115, 117, 129, 137, 211, 221

horizontal integration, *see also* vertical integration, 10, 14, 44, 90, 110, 124, 137, 203, 218
horizontal traceability, *see also* vertical traceability, 201
human-centered design, *see* human systems integration (HSI)
human-computer interaction (HCI), *see* human systems integration (HSI)
human factors, *see* human systems integration (HSI)
human-machine interface (HMI), *see* human systems integration (HSI)
human resource management process, 60–63
human systems integration (HSI), 131, 134, 168–171, 201, 212, 249, 255

–ilities, *see* quality characteristics and approaches
implementation, 132–134
incremental, *see also* evolutionary and sequential, 28, 31, 33–39, 48, 65, 77, 96, 118, 137, 144–145, 156, 170, 193, 198, 222, 229, 245, 248, 249, 256, 267
incremental commit spiral model (ICSM), 31, 37
industrial engineering (IE), 270–271
information assurance (IA), *see* security
information management process, 91–93
infrastructure, 54–57, 72–73, 75, 166–167, 174, 175, 234
infrastructure management process, 54–57
infrastructure systems, 251–253
innovation ecosystem, 11–12, 211–212
input-process-output (IPO) diagram, 40, 42, 204, 313
inspection, *see also* analysis, demonstration, and test, 131, 136, 139, 142, 147
integrated logistics support, *see* logistics
integration, 134–137
interchangeability, *see* reliability, availability, maintainability (RAM)
interface, 8, 10–11, 89–90, 107–108, 114–115, 119–121, 123–124, 125–127, 130–131, 134–137, 166–167, 172, 197, 204–206, 231, 232–233, 238, 250, 253, 266–268, 313
interfacing system, *see also* enabling system and interoperating system, 8, 10–11, 108, 136, 148, 152
international council on systems engineering (INCOSE), ix, xix
international organization for standardization (ISO), ix, xxi, xxiii, 3–4
internet of things (IoT), 172, 234–235, 238–239
interoperability, 124, 161, 171–172, 197–198, 209, 240, 248, 249, 251, 254, 256
interoperating system, *see also* enabling system and interfacing system, 8, 10–11, 19, 148
ISO/IEC/IEEE 15288, ix, xxi, xxiii, 3, 41
iteration, *see also* concurrency and recursion, xxiii, 15, 32, 35, 37, 39, 42–44, 72, 95–96, 101–103, 106, 110, 112, 115, 118, 128, 132, 170, 192–194, 203, 211, 215, 221–222, 238, 266

key performance parameters (KPPs), 80, 97
knowledge management process, 67–70

leadership, 2, 66, 229, 237, 263–264, 265
leading indicators, 95–97
lean, 224–226
legacy, *see* brownfield
lessons learned, 20, 47, 50, 54, 67–70, 136, 225, 239, 252, 326, 328
life cycle concepts, xxiii, 25–33, 175, 202–203, 325
life cycle cost (LCC), *see also* affordability, 7, 38, 132, 158, 160–165, 169, 173, 180, 231–232, 262, 313
life cycle model approaches, 33–36, 77, 110, 221
life cycle model management process, 51–54
life cycle processes, 39–158
life cycle stages, 25–29, 51, 164, 215, 222–223, 314
logical architecture, *see* functional architecture
logistics, 106, 154–156, 171, 172–175, 178–180, 230, 251, 283
loss-driven systems engineering, 191

maintainability, *see* reliability, availability, maintainability (RAM)
maintenance process, 154–156
manufacturability/producibility, 175
margin, 20, 48, 98, 151, 246
measurement process, 93–98
measures of effectiveness (MOEs), 80, 95, 97, 131–132, 313
measures of performance (MOPs), 80, 95, 97, 131–132, 313
medical/medical devices, *see* biomedical and healthcare systems
minimum viable product (MVP), 38
mission analysis *see* business or mission analysis
mode, *see also* state, 8–9, 14, 117, 120, 148, 182–184, 236
model, *see also* simulation, 11–12, 19, 21, 24, 33–39, 40, 51–53, 118–123, 124–128, 129–131, 137, 141, 150, 175, 192–201, 208–211, 238–239, 273, 274–275, 313
model-based systems engineering (MBSE), 5, 90, 96, 143, 151, 202, 209–211, 219–221, 228, 273, 274–275
modularity, *see* agility

N^2 diagram, 205–206, 252, 313, 317–319
natural systems, *see* biomimicry
non-developmental item (NDI), *see also* commercial off-the-shelf (COTS), 69

object-oriented systems engineering method (OOSEM), 219–221
oil and gas systems, 187, 253–254
opaque box, *see* black box
operation, 152–154
operational concept (OpsCon), *see also* concept of operations (ConOps), 28, 106, 108, 131, 148, 152–154, 170, 182–183, 325

operations research (OR), 270, 271–272
opportunity, *see also* risk, 27, 35, 44, 59, 78, 81–87, 103–107, 132, 242–244, 265, 327
opportunity management process, *see* risk management process
organization/organizational, *see also* enterprise, 16, 19, 44–50, 51–70, 168–171, 215–219, 241–243, 274–275, 313
organizational breakdown structure (OBS), *see* breakdown structures

patterns, 11–12, 21–22, 67–69, 83, 116, 119–120, 123, 141–142, 166, 184, 195, 198–199, 206, 208–212, 222–223, 264
physical architecture, *see also* functional architecture, 4, 8–9, 74, 78, 101–102, 118–124, 127, 132–133, 134–137, 166–167, 182, 198, 201, 203–205, 206–208, 220–221, 226–228, 230, 232, 233, 237, 252, 254, 267–268
physical model, 192, 196, 201
portfolio management process, 57–60
power and energy systems, 254–255
process, 51–54, 70–75, 215–219
producibility, *see* manufacturability/producibility
product line engineering (PLE)/product lines, 37, 58–59, 67–69, 117, 123, 164, 180, 209–211, 224, 226–229, 312, 314–315
product tree/product breakdown structure (PBS), *see* breakdown structures
production stage, 26–28, 222
professional competencies, *see also* soft skills, 40, 261–263
professional development, 62
project/program, 60, 226, 241–242, 268–270
project assessment and control process, 75–78
project dashboard, *see* status report/dashboard
project management (PM), 11–12, 44, 56, 58–60, 66, 70, 72–73, 85, 95, 111, 118, 223–224, 226, 268–271
project planning process, 71–75
prototyping, 5, 21, 28, 59, 69, 121, 128, 134, 136–137, 170, 200–201, 252, 275

qualification, 120, 130, 151
quality assurance process, 98–101
quality characteristics and approaches, 159–192
quality management process, 63–66

reconfigurability, *see* agility
recursion, *see also* iteration, concurrency, and recursion, xxiii, 35, 39, 42–44, 102, 110, 112, 115, 117, 118, 132, 137, 142, 192, 207, 215, 221, 314
reliability, availability, maintainability (RAM), 176–180
repairability, *see* reliability, availability, maintainability (RAM)
requirements, 19, 33–39, 101–117, 138–143, 146–152, 201–206
resilience, 180–184
retirement stage, 26–27, 29, 158, 222–223

return on investment (ROI), 5–6, 163
reuse, 13, 36, 44, 67–70, 117, 123, 134, 156–157, 167, 184–185, 201, 206, 218, 220, 226, 233, 245–246, 273, 314
reviews, *see also* audits and decision gates, 29, 31–33, 48, 52, 54, 58–59, 72, 76–78, 117, 126, 133, 252
risk, *see also* opportunity (and safety in biomedical and healthcare), xix, 2, 16, 18, 19, 25–26, 28, 29–30, 32–33, 35–36, 37, 38, 47–48, 50–51, 71–74, 75–78, 81–87, 94–97, 106, 112, 116–117, 130–132, 135–137, 140,149, 169, 171, 185–189, 190, 195, 201, 215–219, 220, 231–232, 248, 253, 265
risk management process, 81–87
robustness, *see* resilience

S*, 11–12, 208–212
safety, *see also* risk, 185–189
scalability, *see* agility
scenario, 107–109, 148–149, 170, 182–184, 192–193
security, 190–191
sensitivity analysis, 69, 81, 129, 131, 272
sequential, *see also* evolutionary and incremental, 8, 25, 30, 31, 33–39, 42, 221–222, 249, 254, 256
service systems, 239–240
similarity, 142
simulation, *see also* model, 141–142, 150, 192–201, 272
soft skills, *see also* professional competencies, 262–263
software engineering (SWE), 9, 16, 38–39, 90, 266–267, 278
software intensive systems, 198, 203, 232–233, 257, 266–267
space/aerospace systems, 8, 27, 31, 38, 69, 97, 151, 174, 224, 246, 249–250, 255–257, 285
specialty engineering, *see* quality characteristics and approaches
spiral, 31, 37, 229
stakeholder needs and requirements definition process, 107–112
standards, 3–4, 172, 197, 215–219, 247, 248, 249, 251, 256–257, 263
state, *see also* mode, 4, 8–9, 14, 179, 181–184, 210, 243, 278
status report/dashboard, 49, 58, 75–78, 329
supplier, *see also* acquirer, 3, 5, 33, 44–49, 65, 76–77, 89, 106, 114–115, 142, 148–151, 153, 168, 175, 178, 195, 203–204, 231, 245, 247, 252, 253, 311, 314, 321, 330–331
supply process, 48–50
support stage, 26–27, 29, 222, 250, 266
supportability *see* logistics
survivability, *see* resilience
sustainability, 184–185
SysML *see* Systems Modeling Language (SysML)
system (definition), 2–3
system analysis, 79, 109, 120, 129–132, 193, 204, 267, 331–332
system analysis process, 129–132
system architecture definition process, 118–124

system element, 2–3, 8–9, 12–14, 19, 26, 35–36, 43, 44–48, 68–69, 78–81, 101, 112, 115–117, 118–124, 125–128, 132–137, 142, 151, 156–158, 201–206, 223–224, 230, 231–232, 314
system(s) engineer, *see* systems engineering practitioner
system of interest (SoI), 8, 10–12, 13
system of systems (SoS), 234, 235–238
system requirements definition process, 112–117
system science/systems thinking, xxii, 1, 21–24, 66, 127, 170, 238, 253
systems engineering (definition), 1–2
systems engineering and integration team (SEIT), 137
systems engineering body of knowledge (SEBoK), guide to, xxi, xxiii
systems engineering heuristics, 20–21
systems engineering management plan (SEMP), 31, 71–73, 77, 88, 97, 118, 139, 147, 176, 333
systems engineering practitioner, xix, xxi-xxii, 261–266
systems engineering principles, 17–20
Systems Modeling Language (SysML), 4–5, 220–221, 228

tailoring, 215–219
taxonomy, 68, 130, 165, 181–182, 196, 236
team, 1, 63, 87, 230, 241, 262–265
technical performance measures (TPMs), 72, 80, 95, 97–98, 131–132, 137, 315
telecommunications systems, 257–258
test, *see also* inspection, analysis, and demonstration, 131, 136, 139, 142, 147
testability, *see* reliability, availability, maintainability (RAM)
testing, *see* verification
tools, xxii, 5, 23, 51, 197–198, 274
traceability, 67, 81, 87–90, 105, 107–108, 110–112, 113–114, 117, 121, 126, 131, 133,136–137, 140, 145, 148, 153, 155, 188–189, 191, 195–196, 198, 201–202, 333
trade study/trade-off study, *see also* decision management, 27–28, 43, 44, 48, 79–81, 87, 90, 94, 104, 115, 118, 120, 123, 126, 128, 130, 132, 161–162, 164–165, 170–171, 172–173, 178, 193, 195, 204, 208, 210, 231, 233, 265, 315
training, 52, 54, 61–63, 106, 133–134, 144, 152,155, 171, 173–174, 187, 193, 262
transdisciplinary, 1, 21–23, 168, 274
transformation, 96, 101–103, 109, 111, 115, 125, 138, 185, 192, 194, 197, 228, 273,274, 275–276
transition, 143–145
transparent box, *see* white box
transportation systems, 258–259
tree(s), *see* breakdown structures
trustworthiness, *see* security

uncertainty, xix, 2, 15–16, 17–18, 19, 48, 51, 59, 79, 81, 82, 84, 106, 112, 120, 129–131, 165, 192, 198, 201, 221–222, 239, 244, 272, 275, 281, 285

usability, *see* human systems integration (HSI)
user, *see* operator
user eXperience (UX), *see* human systems integration (HSI)
user interface (UI), *see* human systems integration (HSI)
utilization stage, 26–28, 145, 222

validation, 146–152
value, xxii, 5, 7–8, 9, 14, 15–16, 17–19, 22, 36, 53, 66, 72, 77,
        80–82, 85–87, 95, 97–98, 104–105, 109–110, 120,
        123–124, 126, 128, 129–131, 160–164, 170, 173–175, 181,
        183–184, 185, 191, 192–195, 208–210, 212, 224–225,
        226–229, 239, 240, 241–243, 265, 272, 275, 313, 315
value robustness, *see* affordability
variable, 9, 14, 20, 129, 131, 210, 239, 313
variability, 58, 226–229, 315

Vee model, 35–36, 222, 256
verification, 138–143
vertical integration, *see also* horizontal integration, 10, 14, 44,
        90, 124, 137, 203, 218
vertical traceability, *see also* horizontal traceability, 201
very small enterprise (VSE), 4, 219
views and viewpoints, 8–9, 14, 16, 17, 19, 23–24, 26–27, 70,
        77–78, 84, 97, 110–112, 118–129, 137, 168, 176, 180,
        182, 189, 190, 195–198, 201, 205, 206–208, 238, 265,
        333
vision, 1, 3, 4–5, 22, 192, 219, 245, 264, 275–276

waste, 97, 156–158, 184–185, 224–226, 229, 230, 255, 315
white box, *see also* black box, 9, 200, 315
work breakdown structure (WBS), *see* breakdown structures